**Machine Learning in Chemical
Safety and Health**

Machine Learning in Chemical Safety and Health

Fundamentals with Applications

Edited by

Qingsheng Wang
Department of Chemical Engineering
Texas A&M University
College Station, TX, USA

Changjie Cai
Department of Occupational and Environmental Health
Hudson College of Public Health, The University of Oklahoma
Oklahoma City, OK, USA

Registered Offices
John Wiley & Sons, Ltd., The Atrium, Southern Gate, Chichester, West Sussex, PO19 8SQ, UK

Editorial Office
The Atrium, Southern Gate, Chichester, West Sussex, PO19 8SQ, UK

For details of our global editorial offices, customer services, and more information about Wiley products visit us at www.wiley.com.

Wiley also publishes its books in a variety of electronic formats and by print-on-demand. Some content that appears in standard print versions of this book may not be available in other formats.

Library of Congress Cataloging-in-Publication Data Applied for:
HB ISBN: 9781119817482

Cover Image: © kessudap/Shutterstock
Cover Design by Wiley

Set in 9.5/12.5pt STIXTwoText by Straive, Pondicherry, India
Printed and bound by CPI Group (UK) Ltd, Croydon, CR0 4YY

C9781119817482_201022

Contents

List of Contributors

Rustam Abubarkirov
Department of Civil, Chemical,
Environmental, and Materials Engineering
University of Bologna
Bologna
Italy

Salim Ahmed
Centre for Risk, Integrity, and Safety
Engineering (C-RISE)
Faculty of Engineering and Applied Science
Memorial University of Newfoundland
St. John's, Nagaland
Canada

Rajeevan Arunthavanathan
Centre for Risk, Integrity, and Safety
Engineering (C-RISE)
Faculty of Engineering and Applied Science
Memorial University of Newfoundland
St. John's, Nagaland
Canada

Changjie Cai
Department of Occupational and
Environmental Health
Hudson College of Public Health
The University of Oklahoma
Oklahoma City, OK
USA

Yoram Cohen
Department of Chemical and Biomolecular
Engineering
University of California
Los Angeles, CA
USA
and
Institute of the Environment and
Sustainability
University of California
Los Angeles, CA
USA

Yu Feng
School of Chemical Engineering
Oklahoma State University
Stillwater, OK
USA

Lan Gao
School of Meteorology
The University of Oklahoma
Norman, OK
USA

Pingfan Hu
Artie McFerrin Department of Chemical
Engineering
Texas A&M University
College Station, TX
USA

Xiao-Ming Hu
School of Meteorology
The University of Oklahoma
Norman, OK
USA
and
Center for Analysis and Prediction of
Storms
The University of Oklahoma
Norman, OK
USA

Syed Imtiaz
Centre for Risk, Integrity, and Safety
Engineering (C-RISE)
Faculty of Engineering and Applied Science
Memorial University of Newfoundland
St. John's, Nagaland
Canada

Juncheng Jiang
College of Safety Science and Engineering
Nanjing Tech University
Nanjing
China

Zeren Jiao
Artie McFerrin Department of Chemical
Engineering
Texas A&M University
College Station, TX
USA

Faisal Khan
Mary Kay O'Connor Process Safety Center
Artie McFerrin Department of Chemical
Engineering
Texas A&M University
College Station, TX
USA

Bilal M. Khan
Department of Computer Science and
Engineering
California State University San Bernardino
San Bernardino, CA
USA
and
Institute of the Environment and
Sustainability
University of California
Los Angeles, CA
USA

Denglong Ma
School of Mechanical Engineering
Xi'an Jiaotong University
Xi'an
Shannxi, China

Yong Pan
College of Safety Science and Engineering
Nanjing Tech University
Nanjing
China

Hao Sun
Safety and Security Science Section
Department of Values, Technology, and
Innovation
Faculty of Technology, Policy, and
Management
Delft University of Technology
The Netherlands
and
College of Mechanical and Electronic
Engineering
China University of Petroleum (East China)
Qingdao
China

Qingsheng Wang
Artie McFerrin Department of Chemical
Engineering
Texas A&M University
College Station, TX
USA

Gregory L. Watson
Department of Biostatistics
University of California at Los Angeles
Los Angeles, CA
USA

Yan Yan
School of Electrical Engineering and
Computer Science
Washington State University
Pullman, WA
USA

Ming Yang
Safety and Security Science Section
Department of Values, Technology, and
Innovation
Faculty of Technology, Policy, and
Management
Delft University of Technology
The Netherlands
and
Australia Maritime College
University of Tasmania
Launceston, Tasmania
Australia

Preface

We are very pleased to offer the first edition of our book on machine learning related to chemical safety and health. The importance of chemical safety and health in professional education and industry remains critical, since professionals have a legal and ethical responsibility to prevent incidents and protect the public. Machine learning, a core subset of artificial intelligence, has developed tremendously and has been implemented in various fields of scientific research and professional practices. The need for professionals to understand the fundamentals and implementations of machine learning in chemical safety and health is essential if they are to apply them in their own work. The present book is rooted in an invited review paper published in *ACS Chemical Health and Safety* from our research group at Texas A&M University. It is also based on experience gained from numerous courses, references, and research publications. Thus, this book can provide guidance for professionals, including students, engineers, and scientists, who are interested in studying and applying machine learning methods in their studies, work, and research.

The objective of the book is to enable professionals to gain a broad overview of machine learning and its applications in chemical safety and health. Moreover, it guides readers to understand machine learning and identify the associated useful resources, such as the commonly used machine learning toolkits and public databases. One significant trend in the field of chemical safety and health is the continuously increased implementation of both shallow and deep learning. Many researchers have achieved significant improvements in chemical safety and health by adopting machine learning methods in their professional activities. The ever-changing nature of this field resulted in extensive efforts being taken in the development of this book so as to match current technology and industrial practices. It is also intended to develop a common and understandable language between specialists and non-specialists.

In addition, interdisciplinary study has gained increased interest among professionals from different fields. For example, several decades ago, chemical process safety, fire safety, industrial hygiene, occupational safety and health, environmental science, chemical emission, and other similar areas of practice were often isolated from each other. Now, many of these topics have been combined to form a more comprehensive field in both scientific research and professional practice. Therefore, this book serves to cover the machine learning applications involved with a wide variety of chemical safety and health issues while combining related topics into a broader one. This approach may serve as a basis for the further application of machine learning in various industries.

However, technologies, standards, regulations, and laws are continuously changed and updated. Machine learning and chemical safety differ from traditional engineering in that they remain dynamically evolving fields. Thus, information and guidance involving these topics is likely to become out of date relatively quickly, particularly in certain areas. The readers should recognize these types of changes and consult relevant professionals and organizations to ensure compliance with current requirements. It is our intention to update this book periodically.

We hope that this book will act as a catalyst for the development of deeper synergies among research areas, which include machine learning and chemical safety and health. We also hope that this book will serve as an important reference for solving current challenges involving chemical safety and health and contribute to a much safer and healthier future.

- Qingsheng Wang
College Station, TX, USA

- Changjie Cai
Oklahoma City, OK, USA

1

Introduction

Pingfan Hu and Qingsheng Wang

Artie McFerrin Department of Chemical Engineering, Texas A&M University, College Station, TX, USA

Machine learning (ML) is a method spanning a broad array of disciplines, involving probability theory, statistics, approximation theory, convex analysis, algorithm complexity theory, and others. Furthermore, it is the core subset of artificial intelligence (AI). The term "machine learning" was first proposed in 1959 by Arthur Samuel (Samuel 1959). Machine learning algorithms can build mathematical models based on training data to make predictions or decisions without being explicitly programmed to do so. Bayesian and Laplace's derivations of least squares and Markov chains, which date back to the seventeenth century, have previously constituted the tools and foundations widely used in ML (Andrieu et al. 2003). Since then, the ML algorithms have developed tremendously and have been widely applied in various aspects of scientific research and everyday life. These include data mining (Mitchell 1999), computer vision (Voulodimos et al. 2018), natural language processing (Cambria and White 2014), biometric recognition (Chaki et al. 2019), medical diagnosis (Bakator and Radosav 2008), detection of credit card fraud (Modi and Dayma 2017), stock market analysis (Chong et al. 2017), speech and handwriting recognition (Nassif et al. 2019), strategy games (Robertson and Watson 2015), and robotics (Pierson and Gashler 2017).

Deep learning (DL) is a relatively new branch within the field of ML. It is an algorithm that uses artificial neural networks (ANNs) as the architecture to characterize and learn data. The concept of DL originates from the research of ANNs, and a multilayer perceptron with multiple hidden layers is a DL structure (Lecun et al. 2015). DL forms a more abstract high-level representation attribute category or feature by combining low-level features to discover distributed feature representations of data. Several DL frameworks have been utilized, including deep neural network (DNN), convolutional neural network (CNN), and recurrent neural network (RNN).

The applications of ML algorithms in chemical safety and health studies date back to the mid-1990s (Lee et al. 1995). Some research used basic ML algorithms in toxicity classification and prediction studies. For other fields such as hazardous property prediction and consequence analysis, the implementation of ML/DL algorithms did not emerge until the late 2000s (Pan et al. 2008; Pan et al. 2009). Chemical safety and health, although an important field, has rarely been investigated using interdisciplinary research with applied ML. This is because at the early development stage of ML/DL, the algorithm was relatively primitive, and its excellent predictive capabilities and accuracy were not widely verified and proven. Second, due to the lack of relatively simple and easy-to-use toolkits and the high skill requirements for algorithms and programming, the applications of ML/DL algorithms in chemical safety and health research have been limited. As a result, studies implementing ML have been relatively rare in the field of chemical safety and health in the late twentieth century and first decade of the twenty-first century.

However, with the rapid advancement of AI and computer science in the past 10 years, the importance of ML/DL and their unparalleled advantages over traditional statistical methods and labor-intensive work have drawn increasing attention and hence have developed significantly. There is also growing interest in expanding the application of ML/DL in the research field of chemical safety and health in academia.

In this book, ML fundamentals as well as popular ML/DL tools for the implementation of ML/DL in chemical safety and health research are introduced (Jiao et al. 2020a). For the applications of ML/DL, the book describes flammability characteristics predictions using quantitative structure–property relationship modeling (Chapter 3), consequence prediction using quantitative property–consequence relationship modeling (Chapter 4), ML involving process safety and asset integrity management (Chapter 5), and ML for process fault detection and diagnosis (Chapter 6). Furthermore, the book describes intelligent methods for chemical emission source identification (Chapter 7), ML and DL applications in medical image analysis (Chapter 8), predictive nanotoxicology: nanoinformatics approach for toxicity analysis of nanomaterials (Chapter 9), ML in environmental exposure assessment (Chapter 10), and air quality prediction using ML (Chapter 11). This book provides useful guidance for researchers and practitioners who are interested in implementing ML/DL related to chemical safety and health. This book is an excellent reference for readers to find more information about novel ML/DL tools and algorithms.

1.1 Background

Author Tom Mitchell provides a modern definition of ML as follows: "A computer program is said to learn from experience E with respect to some task T and some performance measure P, if its performance on T, as measured by P improves with experience E" (Jordan and Mitchell 2015). In general, there are three types of ML: supervised learning, unsupervised learning, and reinforcement learning.

Supervised learning learns a function from a given training data set. When new data (validation/test data) comes, it can predict the results based on the function. The training set requirements for supervised learning include inputs (features) and outputs (targets). The targets in the training set are already labeled (with specific experimental/simulation values). Common supervised learning algorithms include regression and classification algorithms. While some algorithms are only capable of classification analysis (e.g. linear discrimination analysis, naive Bayes classification), most of them (e.g. k-nearest neighbor, random forest) are able to conduct both classification analysis and regression analysis (James et al. 2017; Witten et al. 2017).

The difference between supervised learning and unsupervised learning is whether or not the target of the training set is labeled. Compared with supervised learning, the training set of unsupervised learning has no artificially labeled results. Common unsupervised learning algorithms can be used for clustering (James et al. 2017; Witten et al. 2017). There is also semi-supervised learning, which combines elements of supervised learning and unsupervised learning. The algorithm for semi-supervised learning gradually adjusts its behavior as the environment changes.

For DL, the original work on neural networks was published by Warren McCulloch and Walter Pitts in 1943 (McCulloch and Pitts 1943). They introduced the McCulloch–Pitts neural model, also known as the "linear threshold gate." As the first computational model of a neuron, the McCulloch–Pitts neural model is very simplistic, generating only a binary output. The weights and threshold require hand-tuning. In the 1950s, the perceptron became the first model with the capability to autonomously learn the optimal weight coefficients, allowing the training of a single neuron (Rosenblatt 1958). With the help of the backpropagation algorithm, neural networks began to be trained with one or two hidden layers (Rumelhart et al. 1986).

A single hidden layer neural network consists of three layers: input layer, hidden layer, and output layer. In the neural network that is trained with supervised learning, the training set contains values for the inputs x and target outputs y. The hidden layer refers to the fact that in a training set, the true values for these nodes are not observed. As shown in Figure 1.1, a notation for the values of the input features is $a^{[0]}$, where the term "a" stands for activation. It refers to the values that different layers of the neural network pass on to the subsequent layers. After the input layer passes on the values x to the hidden layer, the hidden layer in turn generates some sets of activations, $a^{[1]}$. Finally, the output layer generates some value $a^{[2]}$, which is a real number that equals the value of $\hat{y}$. The hidden layer and

Figure 1.1 Structure of a single hidden layer neural network.

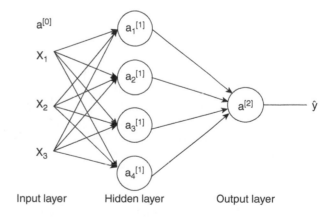

output layer are associated with the parameters w and b. In order to compute the outputs (a) of the neural network, which is a sigmoid function of z $(\sigma(z))$, it is similar to operating repeated logistic regression. The calculations are shown in Eqs. 1.1 through 1.4. Besides the sigmoid function, other activation functions can be used to compute the hidden layer values. In modern neural networks, the default recommendation is to use hyperbolic tangent (tanh) or the rectified linear unit (ReLU).

$$z_1^{[1]} = W_1^{[1]T}x + b_1^{[1]}, a_1^{[1]} = \sigma\left(z_1^{[1]}\right) \tag{1.1}$$

$$z_2^{[1]} = W_2^{[1]T}x + b_2^{[1]}, a_2^{[1]} = \sigma\left(z_2^{[1]}\right) \tag{1.2}$$

$$z_3^{[1]} = W_3^{[1]T}x + b_3^{[1]}, a_3^{[1]} = \sigma\left(z_3^{[1]}\right) \tag{1.3}$$

$$z_4^{[1]} = W_4^{[1]T}x + b_4^{[1]}, a_4^{[1]} = \sigma\left(z_3^{[1]}\right) \tag{1.4}$$

In recent years, the ML community has determined that some cases can only be learned using DNNs rather than the single hidden layer neural networks (Hinton et al. 2006). DNNs with multiple hidden layers can use earlier layers to learn about low-level simpler features and then use the later deeper layers to detect more complex features. Compared with the shallower neural networks, DNNs require significantly less hidden units to compute. Although for any given problem it can be hard to predict in advance exactly how deep a neural network should be, the number of hidden layers can be treated as a hyperparameter and be evaluated by holding out cross-validation data. The DNN relies on both forward propagation and backpropagation. The forward propagation allows the input to provide the initial information and to propagate up to the subsequent layers, while the backpropagation allows the information to flow backward from the cost to compute the gradient more efficiently. Figure 1.2 summarizes the calculation of a DNN with four hidden layers using both forward and backward propagation. Figure 1.3 shows the structures of some typical neural networks that are useful in the chemical engineering field.

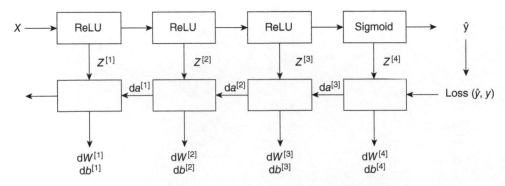

Figure 1.2 Calculation of a deep neural network with four hidden layers.

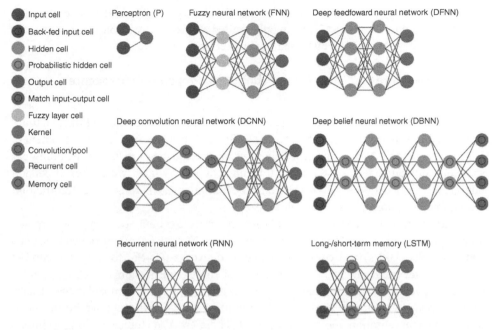

Figure 1.3 Structures of some typical neural networks.

1.2 Current State

1.2.1 Flammability Characteristics Prediction Using Quantitative Structure–Property Relationship

Accurate chemical property values are extremely important to process safety, industrial hygiene, and novel chemical development. Experimental measurement is the most commonly used method to determine property/toxicity values (Jiao et al. 2019a). However, the experimental setup for property measurement is costly, and most of the chemicals are highly flammable and toxic, which is extremely dangerous for conducting the experiments. For chemical mixture property predictions, mixtures have various combination patterns, making it a highly time-consuming task to measure all mixture combinations (Jiao et al. 2019b).

Quantitative structure–activity/property relationship (QSAR/QSPR) analysis involves regression and classification models and is widely used in biological and pharmaceutical science and engineering (Verma et al. 2010; Quintero et al. 2012). It also has been extensively used in chemical health and safety research recently due to its high prediction accuracy and reliability (Wang et al. 2017; Zhou et al. 2017; Wang et al. 2019). In addition, it is the area in which ML tools have been most extensively applied to assist in model development. This is because QSAR/QSPR has a well-developed data pipeline, which can serve to facilitate the development of ML-based QSAR/QSPR studies. For property predictions, the main target properties are lower flammability limit (LFL), upper flammability limit (UFL), auto-ignition temperature (AIT), and flash point (FP). There are also other properties that have been investigated such as minimum ignition energy (MIE) and self-accelerating decomposition temperature (SADT), but only limited studies are available due to the available data. More details will be discussed in Chapter 3.

1.2.2 Consequence Prediction Using Quantitative Property–Consequence Relationship

For consequence analysis, the current mainstream method is computational fluid dynamics (CFD) modeling (Jiao et al. 2019c; Mi et al. 2020; Shen et al. 2020). With the development of ML algorithms, the ANN has been widely used in consequence prediction including gas dispersion and source terms estimation. It could also be integrated with other dispersion models such as PHAST or FLACS to overcome the limitations of missing source information in emergency response cases (Ma et al. 2020).

Since 2010, DL has gained much popularity due to its excellent accuracy when trained with a large amount of data from the dispersion model. Ni et al. (2019) introduced the deep belief networks (DBNs) and convolution neural networks (CNNs) to solve the conflict between accuracy and efficiency requirements of the gas dispersion model. Qian et al. (2019) proposed a specially designed long short-term memory (LSTM) model for gas dispersion in the real environment. The dropout technique was used to prevent overfitting and to improve the generalization ability of the model. There are also studies working to implement the concept of QSAR/QSPR into consequence analysis modeling by using parameters in consequence modeling as property descriptors and make the consequence values as target variables. Sun et al. (2019) and Jiao et al. (2020b, 2021) used PHAST simulation to construct consequence databases for fire radiation distance and flammable dispersion and used them to train the model to develop a quantitative property–consequence relationship (QPCR) model, which can efficiently predict the corresponding consequence results. More details will be discussed in Chapter 4.

1.2.3 Machine Learning in Process Safety and Asset Integrity Management

Asset integrity management (AIM) plays a crucial role in ensuring process safety and risk management. AIM includes structural health of the asset and risk management (Khan et al. 2021). AIM integrates technology, process, and personnel and improves system operation, design, and reliability through a management system and effective system operation to avoid accidents. In other words, AIM combines several basic principles such as system theory, risk assessment, optimality, and sustainability. It aims to maximize economic benefits by ensuring the safety and reliability of assets, and its core goal is to actively maintain and achieve inherent safety. The risk-based AIM method includes three major parts: risk-based inspection (RBI), reliability-centered maintenance (RCM), and safety integrity levels (SILs). RBI is a systematic risk-based evaluation method, which assesses the risk by confirming the damage mechanism of an asset and the consequences caused by the failure (Vinod et al. 2014; Rachman and Ratnayake 2019). Risk is then effectively managed through targeted material selection, corrosion management, preventive testing, and process monitoring. After that, an inspection plan (e.g. maintenance method, maintenance position, maintenance time) can be determined. This can help a company minimize risk and maintenance costs. RCM refers to maintaining the inherent reliability and safety of the equipment using the fewest resources, using decision methods to determine preventive maintenance requirements of equipment. Its main tasks are to analyze the functions and failures of the system. Moreover, on-site failure data statistics, expert evaluation, and quantitative modeling are utilized to optimize the maintenance strategy of the system. Safety integrity level (SIL) is proposed in IEC 61508, which focuses on measuring the confidence that a system (e.g. safety instrumentation system (SIS)) can expect to perform its safety functions (Deshpande and Modak 2002). It is an index that measures the importance of safety instrumented functions (SIF). These three components (e.g. RBI, RCM, and SIL) can facilitate safety evaluations for different types of equipment in process systems and rank the safety of each equipment to generate a targeted maintenance plan. AIM can help to reduce the overall operation costs of the equipment and improve equipment efficiency and productivity. More details will be discussed in Chapter 5.

1.2.4   Machine Learning for Process Fault Detection and Diagnosis

Fault detection and diagnosis are crucial for safe operation of process systems. Continuous operation, and thus economic and production objectives of a plant, can be achieved by accurately detecting fault conditions and their proper diagnoses before the faults occur that lead to failures. Many fault detection and diagnosis approaches have been developed over the years, and these methods are generally classified as analytical model-based, knowledge-based, and data-driven approaches. More details will be discussed in Chapter 6.

1.2.5   Intelligent Method for Chemical Emission Source Identification

Inspired by natural animal and human olfactory, researchers proposed the concept of an "electronic nose (EN)" or "artificial olfactory system (AOS)." The multidimensional sensor array and signal process modules coupled with pattern recognition algorithms constitute the AOS. A single gas sensor may respond to various gases, but the response of one sensor to certain gases has specific features. Therefore, the multidimensional response signals captured from the sensor array can provide more information for gas recognition compared with single sensors. Moreover, it can identify the gases qualitatively and quantitatively. The cross-response of a single metal oxide semiconductor (MOS) sensor can be overcome with AOS. Many methods based on the AOS were proposed, such as principal component analysis (PCA), linear discriminant analysis (LDA), discriminant factor analysis (DFA), partial least square (PLS), principal component regression (PCR), support vector machine (SVM), and cluster analysis (CA). In addition, some ANN and deep learning models (DLM) also were used for gas recognition. More details will be discussed in Chapter 7.

1.2.6   Machine Learning and Deep Learning Applications in Medical Image Analysis

The tremendous success of ML algorithms for image recognition tasks led to a rapid rise in the potential use of ML in various medical imaging tasks, such as risk assessment, detection, diagnosis, prognosis, and therapy response, as well as in multi-omics disease discovery

(Giger 2018). Specifically, traditional methods to diagnose pulmonary diseases involve costly and invasive procedures such as X-ray screening and bronchoscopes. Thus, it is imperative and beneficial to detect the obstruction locations of peripheral lung lesions precisely with noninvasive diagnostic methods. More details will be discussed in Chapter 8.

1.2.7 Predictive Nanotoxicology: Nanoinformatics Approach to Toxicity Analysis of Nanomaterials

Engineered nanomaterials (ENMs) have been utilized in a variety of industrial applications such as cosmetics, therapeutics, electronics, manufacturing, and healthcare (Bao et al. 2013; Thiruvengadam et al. 2018; Siddiquee et al. 2019; Sahoo et al. 2021). The current body of toxicity knowledge for each ENM typically spans a multitude of studies each examining a limited cross-section of attributes in a given experimental system. Nanotechnology is therefore in need of predictive methods (e.g. fundamental first-principle models and QSARs) to identify and quantify physicochemical properties of ENMs. More details will be discussed in Chapter 9.

1.2.8 Machine Learning in Environmental Exposure Assessment

Environmental exposure assessment seeks to quantify exposure to potentially toxic environmental stressors, especially airborne pollutants. Air pollution kills millions of people each year and is a substantial threat to animals and the environment (Chuang et al. 2011; Kim et al. 2015). The long-term health consequences of pollution exposure can be difficult to infer, but understanding its impact on population health is essential for mitigating negative outcomes. The harmful and even fatal consequences of human exposure to potentially toxic stressors render experimental studies unethical, so the health effects of exposure are usually assessed retrospectively using observational data. More details will be discussed in Chapter 10.

1.2.9 Air Quality Prediction Using Machine Learning

Air pollution is a global concern due to its significant effects on human health, agriculture and ecosystems, and even climate (Myhre et al. 2013; Fuhrer et al. 2016; Cohen et al. 2017). Over the past decades, observing the atmospheric compositions and pollutants using various instrumental platforms has become an increasingly powerful tool for understanding atmospheric processes and air quality. The platforms include earth observation satellite instruments, ground-based in situ remote sensing stations, and instrumented aircrafts (Laj et al. 2009). Many species, such as particulate matter (PM), nitrogen dioxide (NO_2), sulfur dioxide (SO_2), carbon monoxide (CO), ozone (O_3), lead, and volatile organic compounds (VOCs), can be directly used in air quality studies. These studies include monitoring that is nearly in real time, source apportionment, dispersion modeling, and air quality prediction (Maykut et al. 2003; Engel-Cox et al. 2004; Cai et al. 2010; Li et al. 2011; Iskandaryan et al. 2020). More details will be discussed in Chapter 11.

1.3   Software and Tools

1.3.1   R

R is a language and operating environment for statistical analysis and ML. R is a branch of the S language that was created sometime close to 1980 and is widely used in the statistical field. R is a free and open-source software belonging to the General Public License (GNU) system and is an excellent tool for statistical calculation ML model construction (R Core Team 2013).

The use of the R language is largely aided by various R packages. To some extent, R packages are plug-ins for R. Different plug-ins meet different needs. CRAN (Comprehensive R Archive Network) has already included more than 17,000 packages of various types including many popular ML/DL packages. Some of these packages are summarized as follows:

Machine Learning	
R Package	**Description**
Arules	Mining Association Rules and Frequent Itemsets:
(Hahsler et al, 2021)	This provides the infrastructure for representing, manipulating, and analyzing transaction data and patterns (frequent itemsets and association rules). It also provides C implementations of the association mining algorithms Apriori and Eclat.
Caret	Classification and Regression Training:
(Kuhn 2021)	This consists of miscellaneous functions for training and plotting classification and regression models.
CORElearn	Classification, Regression, and Feature Evaluation:
(Robnik-Sikonja and Savicky 2021)	This contains several learning techniques for classification and regression. Predictive models include classification and regression trees with optional constructive induction and models in the leaves, random forests, k-nearest neighbors algorithm (kNN), naive Bayes, and locally weighted regression.
DataExplorer	Automate Data Exploration and Treatment:
(Cui 2020)	This automates data exploration processes for analytic tasks and predictive modeling.
dplyr	A Grammar of Data Manipulation:
(Wickham et al. 2021a)	This is a fast, consistent tool for working with data frame-like objects, both in memory and out of memory.
e1071	Misc Functions of the Department of Statistics, Probability Theory Group:
(Meyer et al. 2021)	This functions for latent class analysis, short-time Fourier transform, fuzzy clustering, support vector machines, shortest path computation, bagged clustering, naive Bayes classifier, and generalized k-nearest neighbor.
gbm	Generalized Boosted Regression Models:
(Greenwell et al. 2020)	This is an implementation of extensions to Freund and Schapire's AdaBoost algorithm and Friedman's gradient boosting machine.
ggplot2	Create Elegant Data Visualizations Using the Grammar of Graphics:
(Wickham 2016)	

(Continued)

Machine Learning	
R Package	**Description**
	This is a system for "declaratively" creating graphics based on "The Grammar of Graphics."
glmnet (Friedman et al. 2010)	Lasso and Elastic-Net Regularized Generalized Linear Models: This involves efficient procedures for fitting the entire lasso or elastic-net regularization path for linear regression, logistic and multinomial regression models, Poisson regression, Cox model, multiple-response Gaussian, and the grouped multinomial regression.
kernlab (Karatzoglou et al. 2004)	Kernel-Based Machine Learning Lab: This consists of kernel-based machine learning methods for classification, regression, clustering, novelty detection, quantile regression, and dimensionality reduction.
mboost (Hothorn et al. 2021)	Model-Based Boosting: This is a functional gradient descent algorithm (boosting) for optimizing general risk functions utilizing component-wise (penalized) least-squares estimates or regression trees as base learners for fitting generalized linear, additive, and interaction models to potentially high-dimensional data.
mice (van Buuren and Groothuis-Oudshoorn 2011)	Multivariate Imputation by Chained Equations: This consists of multivariate imputation using fully conditional specification (FCS) implemented by the MICE algorithm. Built-in imputation models are provided for continuous data (predictive mean matching, normal), binary data (logistic regression), unordered categorical data (polytomous logistic regression), and ordered categorical data (proportional odds).
mlr (Casalicchio et al. 2019)	Machine Learning in R: This is an interface for a large number of classification and regression techniques, including machine-readable parameter descriptions.
Party (Hothorn et al. 2006)	A Laboratory for Recursive Partitioning: This is a computational toolbox for recursive partitioning. The core of the package is ctree(), an implementation of conditional inference trees that embed tree-structured regression models into a well-defined theory of conditional inference procedures.
Random Forest (Liaw and Wiener 2002)	Breiman and Cutler's Random Forests for Classification and Regression: This involves classification and regression based on a forest of trees using random inputs.
ROCR (Sing et al. 2005)	Visualizing the Performance of Scoring Classifiers: ROC graphs, sensitivity/specificity curves, lift charts, and precision/recall plots are popular examples of trade-off visualizations for specific pairs of performance measures.
rpart (Therneau and Atkinson 2019)	Recursive Partitioning and Regression Trees: Recursive partitioning for classification, regression, and survival trees involve implementation of most of the functionality of the 1984 book by Breiman, Friedman, Olshen, and Stone.
tidyr	Tidy Messy Data:

Machine Learning	
R Package	**Description**
(Wickham 2021b)	This consists of tools to help to create tidy data, where each column is a variable, each row is an observation, and each cell contains a single value.
tm (Feinerer and Hornik 2020)	Text Mining Package: This is a framework for text mining applications within R.
Xgboost (Chen et al. 2021)	Extreme Gradient Boosting: Extreme Gradient Boosting is an efficient implementation of the gradient boosting framework. The package includes efficient linear model solver and tree learning algorithms. It supports various objective functions, including regression, classification, and ranking.

Deep Learning	
R Package	**Description**
deepnet (Rong 2014)	Deep learning toolkit in R: This implements some deep learning architectures and neural network algorithms, including back propagation (BP), restricted Boltzmann machine (RBM), deep belief network (DBN), Deep Autoencoder, etc.
h2o (LeDell et al. 2021)	R Interface for the "H2O" Scalable Machine Learning Platform: R interface for "H2O" is a scalable open-source machine learning platform that offers parallelized implementations of many supervised and unsupervised machine learning algorithms. It includes generalized linear models (GLM), gradient boosting machines (including XGBoost), random forests, deep neural networks (deep learning), stacked ensembles, naive Bayes, generalized additive models (GAM), Cox proportional hazards, K-means, PCA, Word2Vec, as well as a fully automatic machine learning algorithm (H2O AutoML).
keras (Allaire and Chollet 2021)	R Interface to the Keras Deep Learning Library: This provides a consistent interface to the "Keras" Deep Learning Library directly from within R.
nnet (Venables and Ripley 2002)	Feed-Forward Neural Networks and Multinomial Log-Linear Models: This is software for feed-forward neural networks with a single hidden layer and for multinomial log-linear models.
neuralnet (Fritsch et al. 2019)	Training of Neural Networks: This involves training of neural networks using backpropagation, resilient backpropagation with or without weight backtracking, or the modified globally convergent version
rnn (Quast 2016)	Recurrent Neural Network: This involves implementation of recurrent neural network architectures in native R, including Long ShortTerm Memory

(Continued)

Machine Learning	
R Package	**Description**
	(Hochreiter and Schmidhuber), Gated Recurrent Unit (Chung et al), and vanilla RNN.
Tensorflow (Allaire and Tang 2021)	R Interface to "TensorFlow": This is an interface to "TensorFlow" <https://www.tensorflow.org/>, an open-source software library for numerical computation using data flow graphs.

1.3.2 Python

Python is a cross-platform, general-purpose programming language that was developed by Guido van Rossum and first released in 1991. Python's design philosophy emphasizes code readability with its notable use of significant whitespace. It is a high-level scripting language that combines explanatory, compliable, interactive, and object-oriented features. With the continuous version updates and addition of new language features, it is becoming increasingly utilized for the development of independent and large-scale projects. It is also the default tool for beginners as well as professionals to learn and use ML/DL algorithms (Jiao et al. 2019c).

Compared to proprietary software such as MATLAB, using open-source programming languages such as Python for ML/DL model development has some important advantages. First, MATLAB is a costly proprietary software. Python, on the other hand, is free, and many open-source ML/DL and scientific computing libraries provide Python calling interfaces. In addition to some highly specialized toolboxes of MATLAB that cannot be replaced, most of the commonly used functions of MATLAB can be found in Python. Users can install Python and most of its extension libraries on any computer for free, and Python also provides a state-of-art ML/DL library that can easily complete various advanced tasks and achieve superior performance. In addition, compared to MATLAB, Python is a programming language that is easier to learn and more rigorous that can make composed code easier for users to write, read, and maintain.

Python package	Description
Caffe (Jia et al. 2014)	Caffe is a deep learning framework made with expression, speed, and modularity in mind. It is developed by Berkeley AI Research (BAIR) and by community contributors.
Keras (Chollet 2015)	Keras is a deep learning API written in Python, running in conjunction with the machine learning platform TensorFlow. It is developed with a focus on enabling fast experimentation.
Matplotlib (Hunter 2007)	Matplotlib is a comprehensive library for creating static, animated, and interactive visualizations in Python.

Python package	Description
NumPy (Harris et al. 2020)	NumPy involves a powerful N-dimensional array object as well as useful linear algebra, Fourier transform, and random number capabilities.
Pandas (McKinney 2010)	Pandas is a Python package that provides fast, flexible, and expressive data structures designed to make working with structured (tabular, multidimensional, potentially heterogeneous) and time series data both simple and intuitive.
Pytorch (Paszke et al. 2019)	PyTorch is a Python package that provides two high-level features: Tensor computation (like NumPy) with strong GPU acceleration and deep neural networks built on a tape-based autograd system.
Scikit-Learn (Pedregosa et al. 2011)	Scikit-Learn is a Python module for machine learning built on top of SciPy and is distributed under the 3-Clause BSD license.
SciPy (Virtanen et al. 2020)	SciPy is an open-source software for mathematics, science, and engineering. The SciPy library depends on NumPy, which provides convenient and fast N-dimensional array manipulation. The SciPy library is built to work with NumPy arrays, and provides many user-friendly and efficient numerical routines including for numerical integration and optimization.
Tensorflow (Abadi et al. 2016)	TensorFlow is an open-source software library for high-performance numerical computation. Its flexible architecture allows easy deployment of computation capabilities across a variety of platforms.

References

Abadi, M., Barham, P., Chen, J. et al. (2016). Tensorflow: a system for large-scale machine learning. *12th Symposium on Operating Systems Design and Implementation*, 16, pp. 265–283.

Allaire, J.J. and Chollet, F. (2021a). keras: R Interface to 'Keras'. R package version 2.4.0. https://CRAN.R-project.org/package=keras.

Allaire, J.J. and Tang, Y. (2021). tensorflow: R Interface to 'TensorFlow'. R package version 2.5.0. https://CRAN.R-project.org/package=tensorflow.

Andrieu, C., Freitas, N., Doucet, A., and Jordan, M.I. (2003). An introduction to MCMC for machine learning. *Mach. Learn.* 50: 5–43.

Bakator, M. and Radosav, D. (2008). Deep learning and medical diagnosis: a review of literature. *Multimodal Technol. Interact.* 2 (3): 47.

Bao, G., Mitragotri, S., and Tong, S. (2013). Multifunctional nanoparticles for drug delivery and molecular imaging. *Annu. Rev. Biomed. Eng.* 15: 253–282.

Cai, C., Geng, F., Tie, X. et al. (2010). Characteristics and source apportionment of VOCs measured in Shanghai, China. *Atmos. Environ.* 44 (38): 5005–5014.

Cambria, E. and White, B. (2014). Jumping NLP curves: a review of natural language processing research. *IEEE Comput. Intell. Mag.* 9 (2): 48–57.

Casalicchio, G., Bossek, J., Lang, M. et al. (2019). OpenML: an R package to connect to the machine learning platform OpenML. *Comput. Stat.* 34: 977–991.

Chaki, J., Dey, N., Shi, F., and Sherratt, R.S. (2019). Pattern mining approaches used in sensor-based biometric recognition: a review. *IEEE Sensors J.* 19 (10): 3569–3580.

Chen, T., He, T., Benesty, M. et al. (2021). xgboost: Extreme Gradient Boosting. R package version 1.4.1.1.

Chollet, F. (2015). Keras. GitHub. https://github.com/fchollet/keras.

Chong, E., Han, C., and Park, F.C. (2017). Deep learning networks for stock market analysis and prediction: methodology, data representations, and case studies. *Expert Syst. Appl.* 83: 187–205.

Chuang, K.J., Yan, Y.H., Chiu, S.Y. et al. (2011). Long-term air pollution exposure and risk factors for cardiovascular diseases among the elderly in Taiwan. *Occup. Environ. Med.* 68 (1): 64–68.

Cohen, A.J., Brauer, M., Burnett, R. et al. (2017). Estimates and 25-year trends of the global burden of disease attributable to ambient air pollution: an analysis of data from the Global Burden of Diseases Study 2015. *Lancet* 389 (10082): 1907–1918.

Cui, B. (2020). DataExplorer: Automate Data Exploration and Treatment. R package version 0.8.2. https://CRAN.R-project.org/package=DataExplorer.

Deshpande, V.S. and Modak, J.P. (2002). Application of RCM to a medium scale industry. *Reliab. Eng. Syst. Saf.* 77: 31–43.

Engel-Cox, J.A., Hoff, R.M., and Haymet, A.D. (2004). Recommendations on the use of satellite remote-sensing data for urban air quality. *J. Air Waste Manage. Assoc.* 54: 1360–1371.

Feinerer, I. and Hornik, K. (2020). tm: Text Mining Package. R package version 0.7-8. https://CRAN.R-project.org/package=tm.

Friedman, J., Hastie, T., and Tibshirani, R. (2010). Regularization paths for generalized linear models via coordinate descent. *J. Stat. Softw.* 33 (1): 1–22. https://www.jstatsoft.org/v33/i01.

Fritsch, S., Guenther, F., and Wright, M. (2019). neuralnet: Training of Neural Networks. R package version 1.44.2. https://CRAN.R-project.org/package=neuralnet.

Fuhrer, J., Val Martin, M., Mills, G. et al. (2016). Current and future ozone risks to global terrestrial biodiversity and ecosystem processes. *Ecol. Evol.* 6 (24): 8785–8799.

Giger, M. (2018). Machine learning in medical imaging. *J. Am. Coll. Radiol.* 15: 512–520.

Greenwell, B., Boehmke, B., Cunningham, J., and GBM developers (2020). gbm: Generalized Boosted Regression Models. R package version 2.1.8. https://CRAN.R-project.org/package=gbm.

Hahsler, M., Buchta, C., Gruen, B., and Hornil, K. (2021). arules: Mining Association Rules and Frequent Itemsets. R package version 1.6-8. https://CRAN.R-project.org/package=arules

Harris, C.R., Millman, K.J., van der Walt, S.J. et al. (2020). Array programming with NumPy. *Nature* 585: 357–362.

Hinton, G.E., Osindero, S., and Teh, Y. (2006). A fast learning algorithm for deep belief nets. *Neural Comput.* 18: 1527–1554.

Hothorn, T., Hornik, K., and Zeileis, A. (2006). Unbiased recursive partitioning: a conditional inference framework. *J. Comput. Graph. Stat.* 15 (3): 651–674.

Hothorn, T., Buehlmann, P., Kneib, T. et al. (2021). mboost: Model-Based Boosting, R package version 2.9-5, https://CRAN.R-project.org/package=mboost.

Hunter, J.D. (2007). Matplotlib: a 2D graphics environment. *Comput. Sci. Eng.* 9 (3): 90–95.

Iskandaryan, D., Ramos, F., and Trilles, S. (2020). Air quality prediction in smart cities using machine learning technologies based on sensor data: a review. *Appl. Sci.* 10 (7): 2401.

James, G., Witten, D., Hastie, T., and Tibshirani, R. (2017). *An Introduction to Statistical Learning: with Applications in R*. New York: Springer.

Jia, Y., Shelhamer, E., Donahue, J. et al. (2014). Caffe: Convolutional architecture for fast feature embedding. *Proceedings of ACM Multimedia*, pages 675–678.

Jiao, Z., Escobar-Hernandez, H.U., Parker, T., and Wang, Q. (2019a). Review of recent developments of quantitative structure-property relationship models on fire and explosion-related properties. *Process Saf. Environ. Prot.* 129: 280–290.

Jiao, Z., Yuan, S., Zhang, Z., and Wang, Q. (2019b). Machine learning prediction of hydrocarbon mixture lower flammability limits using quantitative structure-property relationship models. *Process. Saf. Prog.* 39 (2): e12103.

Jiao, Z., Yuan, S., Ji, C. et al. (2019c). Optimization of dilution ventilation layout design in confined environments using Computational Fluid Dynamics (CFD). *J. Loss Prev. Process Ind.* 60: 195–202.

Jiao, Z., Hu, P., Xu, H., and Wang, Q. (2020a). Machine learning and deep learning in chemical health and safety: a systematic review of techniques and applications. *ACS J. Chem. Health Saf.* 27 (6): 316–334.

Jiao, Z., Sun, Y., Hong, Y. et al. (2020b). Development of flammable dispersion quantitative property-consequence relationship (QPCR) models using extreme gradient boosting. *Ind. Eng. Chem. Res.* 59 (33): 15109–15118.

Jiao, Z., Ji, C., Sun, Y. et al. (2021). Deep learning based quantitative property-consequence relationship (QPCR) models for toxic dispersion prediction. *Process Saf. Environ. Prot.* 152: 352–360.

Jordan, M.I. and Mitchell, T.M. (2015). Machine learning: Trends, perspectives, and prospects. *Science* 349 (6245): 255–260.

Karatzoglou, A., Smola, A., Hornik, K., and Zeileis, A. (2004). kernlab - an S4 package for kernel methods in R. *J. Stat. Softw.* 11 (9): 1–20. http://www.jstatsoft.org/v11/i09.

Khan, F., Yarveisy, R., and Abbassi, R. (2021). Risk-based pipeline integrity management: a road map for the resilient pipelines. *J. Pipeline Sci. Eng.* 1 (1): 74–87.

Kim, K.H., Kabir, E., and Kabir, S. (2015). A review on the human health impact of airborne particulate matter. *Environ. Int.* 74: 136–143.

Kuhn, M. (2021). caret: Classification and Regression Training. R package version 6.0-88. https://CRAN.R-project.org/package=caret.

Laj, P., Klausen, J., Bilde, M. et al. (2009). Measuring atmospheric composition change. *Atmos. Environ.* 43: 5351–5414.

Lecun, Y., Bengio, Y., and Hinton, G. (2015). Deep learning. *Nature* 521 (7553): 436–444.

LeDell, E., Gill, N., Aiello, S. et al. (2021). h2o: R Interface for the 'H2O' Scalable Machine Learning Platform. R package version 3.32.1.3. https://CRAN.R-project.org/package=h2o.

Lee, Y., Buchanan, B.G., Mattison, D.M. et al. (1995). Learning rules to predict rodent carcinogenicity of non-genotoxic chemicals. *Mutat. Res. Fundam. Mol. Mech. Mutagen.* 328: 127–149.

Li, C., Hsu, N.C., and Tsay, S.-C. (2011). A study on the potential applications of satellite data in air quality monitoring and forecasting. *Atmos. Environ.* 45: 3663–3675.

Liaw, A. and Wiener, M. (2002). Classification and regression by RandomForest. *R News.* 2 (3): 18–22.

Ma, D., Gao, J., Zhang, Z. et al. (2020). Locating the gas leakage source in the atmosphere using the dispersion wave method. *J. Loss Prev. Process Ind.* 63: 104031.

Maykut, N.N., Lewtas, J., Kim, E., and Larson, T.V. (2003). Source apportionment of PM2. 5 at an urban IMPROVE site in Seattle Washington. *Environ. Sci. Technol.* 37 (22): 5135–5142.

McCulloch, W.S. and Pitts, W. (1943). A logical calculus of the ideas immanent in nervous activity. *Bull. Math. Biol.* 5: 115–133.

McKinney, W. (2010). Data structures for statistical computing in python. *Proceedings of the 9th Python in Science Conference* Vol. 445, pp. 51–56.

Meyer, D., Dimitriadou, E., Hornik, K. et al. (2021). e1071: Misc Functions of the Department of Statistics, Probability Theory Group (Formerly: E1071), TU Wien. R package version 1.7-8. https://CRAN.R-project.org/package=e1071.

Mi, H., Liu, Y., Jiao, Z. et al. (2020). A numerical study on the optimization of ventilation mode during emergency of cable fire in utility tunnel. *Tunn. Undergr. Space Technol.* 100: 103403.

Mitchell, T.M. (1999). Machine learning and data mining. *Commun. ACM* 42 (11): 30–36.

Modi, K. and Dayma, R. (2017). Review on fraud detection methods in credit card transactions. *2017 International Conference on Intelligent Computing and Control (I2C2)*.

Myhre, G., Shindell, D., and Pongratz, J. (2013). Anthropogenic and natural radiative forcing. In: *Climate Change 2013: The Physical Science Basis. Contribution of Working Group I to the Fifth Assessment Report of the Intergovernmental Panel on Climate Change* (ed. T.F. Stocker, D. Qin, G.-K. Plattner, et al.). Cambridge, United Kingdom and New York, NY, USA: Cambridge University Press.

Nassif, A.B., Shahin, I., Attili, I. et al. (2019). Speech recognition using deep Neural networks: a systematic review. *IEEE Access.* 7: 19143–19165.

Ni, J., Yang, H., Yao, J. et al. (2019). Toxic gas dispersion prediction for point source emission using deep learning method. *Hum. Ecol. Risk Assess. Int. J.* 26: 1–14.

Pan, Y., Jiang, J., Wang, R., and Cao, H. (2008). Advantages of support vector machine in QSPR studies for predicting auto-ignition temperatures of organic compounds. *Chemom. Intell. Lab. Syst.* 92 (2): 169–178.

Pan, Y., Jiang, J., Wang, R. et al. (2009). A novel QSPR model for prediction of lower flammability limits of organic compounds based on support vector machine. *J. Hazard. Mater.* 168 (2–3): 962–969.

Paszke, A., Gross, S., Massa, F. et al. (2019). PyTorch: an imperative style, high-performance deep learning library. *Adv. Neural Inf. Proces. Syst.* 32: 8024–8035. Curran Associates, Inc. http://papers.neurips.cc/paper/9015-pytorch-an-imperative-style-high-performance-deep-learning-library.pdf.

Pedregosa, F., Varoquaux, G., Gramfort, A. et al. (2011). Scikit-learn: machine learning in Python. *J. Mach. Learn. Res.* 12 (Ooctober): 2825–2830.

Pierson, H.A. and Gashler, M.S. (2017). Deep learning in robotics: a review of recent research. *Adv. Robot.* 31 (16): 821–835.

Qian, F., Chen, L., Li, J. et al. (2019). Direct prediction of the toxic gas diffusion rule in a real environment based on LSTM. *Int. J. Environ. Res. Public Health* 16: 2133.

Quast, B.A. (2016). rnn: a Recurrent Neural Network in R. Working Papers. https://qua.st/rnn/.

Quintero, F.A., Patel, S.J., Muñoz, F., and Mannan, M.S. (2012). Review of existing QSAR/QSPR models developed for properties used in hazardous chemicals classification system. *Ind. Eng. Chem. Res.* 51 (49): 16101–16115.

R Core Team (2013). *R: A Language and Environment for Statistical Computing*. Vienna, Austria: R Core Team.

Rachman, A. and Ratnayake, R.M. (2019). Machine learning approach for risk-based inspection screening assessment. *Reliab. Eng. Syst. Saf.* 185: 518–532.

Robertson, G. and Watson, I. (2015). A review of real-time strategy game AI. *AI Mag.* 35 (4): 75.

Robnik-Sikonja, M. and Savicky, P. (2021). CORElearn: Classification, Regression and Feature Evaluation. R package version 1.56.0. https://CRAN.R-project.org/package=CORElearn.

Rong, X. (2014). deepnet: deep learning toolkit in R. R package version 0.2. https://CRAN.R-project.org/package=deepnet.

Rosenblatt, F. (1958). The perceptron: a probabilistic model for information storage and organization in the brain. *Psychol. Rev.* 65: 386–408.

Rumelhart, D., Hinton, G., and Williams, R. (1986). Learning representations by back-propagating errors. *Nature* 323: 533–536.

Sahoo, M., Vishwakarma, S., Panigrahi, C., and Kumar, J. (2021). Nanotechnology: current applications and future scope in food. *Food Front.* 2 (1): 3–22.

Samuel, A.L. (1959). Some studies in machine learning using the game of checkers. *IBM J. Res. Dev.* 3 (3): 210–229.

Shen, R., Jiao, Z., Parker, T. et al. (2020). Recent application of computational fluid dynamics (CFD) in process safety and loss prevention: a review. *J. Loss Prev. Process Ind.* 67: 104252.

Siddiquee, S., Melvin, G.J.H., and Rahman, M. (2019). *Nanotechnology: Applications in Energy, Drug and Food*. Springer.

Sing, T., Sander, O., Beerenwinkel, N., and Lengauer, T. (2005). ROCR: visualizing classifier performance in R. *Bioinformatics* 21 (20): 7881. http://rocr.bioinf.mpi-sb.mpg.de.

Sun, Y., Wang, J., Zhu, W. et al. (2019). Development of consequent models for three categories of fire through artificial neural networks. *Ind. Eng. Chem. Res.* 59 (1): 464–474.

Therneau, T. and Atkinson, B. (2019). rpart: Recursive Partitioning and Regression Trees. R package version 4.1-15. https://CRAN.R-project.org/package=rpart.

Thiruvengadam, M., Rajakumar, G., and Chung, I.M. (2018). Nanotechnology: current uses and future applications in the food industry. *3 Biotech* 8 (1): 74.

Van Buuren, S. and Groothuis-Oudshoorn, K. (2011). mice: Multivariate imputation by chained equations in R. *J. Stat. Softw.* 45 (3): 1–67. https://www.jstatsoft.org/v45/i03/.

Venables, W.N. and Ripley, B.D. (2002). *Modern Applied Statistics with S*, 4e. New York: Springer ISBN 0-387-95457-0.

Verma, J., Khedkar, V., and Coutinho, E. (2010). 3D-QSAR in drug design - a review. *Curr. Top. Med. Chem.* 10: 95–115.

Vinod, G., Sharma, P.K., Santosh, T.V. et al. (2014). New approach for risk-based inspection of H2S based process plants. *Ann. Nucl. Energy* 66: 13–19.

Virtanen, P., Gommers, R., Oliphant, T.E. et al. (2020). SciPy 1.0: fundamental algorithms for scientific computing in Python. *Nat. Methods* 17: 261–272. https://doi.org/10.1038/s41592-019-0686-2.

Voulodimos, A., Doulamis, N., Doulamis, A., and Protopapadakis, E. (2018). Deep learning for computer vision: a brief review. *Comput. Intell. Neurosci.* 2018: 1–13.

Wang, B., Yi, H., Xu, K., and Wang, Q. (2017). Prediction of the self-accelerating decomposition temperature of organic peroxides using QSPR models. *J. Therm. Anal. Calorim.* 128 (1): 399–406.

Wang, B., Xu, K., and Wang, Q. (2019). Prediction of upper flammability limits for fuel mixtures using quantitative structure-property relationship models. *Chem. Eng. Commun.* 206 (2): 247–253.

Wickham, H. (2016). *Ggplot2: Elegant Graphics for Data Analysis.* New York: Springer-Verlag.

Wickham. H. (2021b). tidyr: Tidy Messy Data. R package version 1.1.3. https://CRAN.R-project.org/package=tidyr.

Wickham, H., François, R., Henry, L., and Müller, K. (2021a). dplyr: A Grammar of Data Manipulation. R package version 1.0.7. https://CRAN.R-project.org/package=dplyr.

Witten, I.H., Frank, E., Hall, M.A., and Pal, C.J. (2017). *Data Mining: Practical Machine Learning Tools and Techniques.* Amsterdam: Morgan Kaufmann.

Zhou, L., Wang, B., Jiang, J. et al. (2017). Quantitative structure-property relationship (QSPR) study for predicting gas-liquid critical temperatures of organic compounds. *Thermochim. Acta* 655: 112–116.

2

Machine Learning Fundamentals

Yan Yan

School of Electrical Engineering and Computer Science, Washington State University, Pullman, WA, USA

Machine learning covers a large group of computational methods built upon past data and experience, by which we expect to perform prediction on unseen data. This chapter serves as a gentle introduction to machine learning fundamentals. It covers key concepts of machine learning tasks, methods, and applications. It mainly answers four questions: (i) What is learning? (ii) What can be learned? (iii) How to learn? What condition? (iv) How to optimize the parameters? This chapter is highly inspired by seminal textbooks on machine learning (Shalev Shwartz and Ben-David 2014; Mohri et al. 2018).

2.1 What Is Learning?

In the context of machine learning, learning is generally a procedure that one can extract useful information from past experience, and then convert it into expertise that can be used as a predictor in future. Machine learning can be viewed as an automated learning process that is usually implemented by computer programs. Particularly, in machine learning, a training dataset plays the role of past experience containing useful information, while the machine learning model learned on this training dataset serves as the expertise that helps us make prediction. In this way, machine learning studies the approaches for designing and building algorithms and methods that extract information from the training dataset, convert to expertise, and produce accurate prediction on unseen data.

One may name a long list of real-world examples of machine learning. For example, spam classification is a classical application of machine learning, which aims to identify those junk emails massively sent out in bulk to a recipient list. It helps automatically filter out spam emails by analyzing the content so that our inbox may not be overwhelmed. Image classification is another popular application of machine learning. It reads a new image and can tell what objects this image contains. Object detection, another very commonly seen use case, has a further requirement localizing the object with a bounding box in addition to identifying its presence in images or videos. Machine translation targets the conversion from a sequence of symbols in one language to a sequence of symbols in another language, such as from English to German or from French to English. This particular task of natural language processing can also be implemented by building and training a machine learning. When one would like to capture the shared properties of common data and detect the abnormal data in the future, anomaly detection is the one that machine learning can help. One application of such task is credit card fraud detection.

We may continue this list of machine learning tasks for the areas of autonomous driving, gaming, healthcare, software development, and so forth. In the next sections, we summarize some common applications and tasks in machine learning in a little bit more details. We also roughly group machine learning models into five categories and highlight their connection and difference.

2.1.1 Machine Learning Applications and Examples

One may find a large variety of applications of machine learning in daily life. Even though they may have very different data formats, problem demands, and requirements, one can still cast them into machine learning models. We give a gentle introduction to several most used applications and areas.

- Computer vision aims to generate high-level understanding from visual data, including images and videos, and covers various use cases. Image classification, for example, is to classify an image to a particular class. Generally, the classification criterion is to identify whether a target object is contained in the image by inputting an image and outputting a vector that represents the presence of the object of interest. To do so, each element of the output vector presents the likelihood of the presence of one particular object, which is later used to make the decision of presence. In such task, the spatial information is sometimes not critical – machine learning, especially deep neural network models, can be invariant to image translation, which means that models can recognize the object of interest regardless some appearance and location variation. This property is also called invariance.

  Object detection is another popular example, which, in addition to decide whether the object of interest is present in the image, also requires to localize the target object in the image using a bounding box. Object detection is often more difficult than image classification due to the additional localization requirement. Typically, the output of an object detector includes (i) the likelihood of the presence of objects and (ii) the predicted bounding box that covers the object properly (not a too large or too small coverage) under a certain measure.

  Image segmentation, unlike image classification and object detection, is to label each pixel in the image (e.g. pixel labeling) that generates finer localization prediction. This requires to decide the label of each pixel – which particular object each pixel is associated with. In video domain, action recognition is an important application that requires to additionally consider temporal information to classify an action in video clips. We may skip the full list of use cases such as pose estimation, face recognition, and pedestrian detection.

- Natural language processing aims to gain high-level understanding from documents or text data. Machine translation is to translate texts from one human language to another one (e.g. from English to French). Question answering is to provide an answer to a question, where both the answer and question are in human language. Text summarization is to provide a short text summary for a long document such as research papers or news articles. Other examples of natural language processing include natural language generation, semantic parsing, sentiment analysis, etc.

- Speech processing is to understand and interpret the semantic meaning of human speech data so that is able to provide intellectual services. Examples include virtual assistants on smart phones, such as Siri and Alexa. Speech recognition is one of the most popular tasks in speech processing, which focuses on the translation from audio data to text data with some additional integration of grammar, syntax, and structure information.

2.1.2 Machine Learning Tasks

One may summarize some simple properties of machine learning models shared by the preceding applications. First, machine learning models aim to learn a mapping from the input space to the output space, e.g. $f: S_{input} \to S_{output}$, where S_{input} and S_{output} denote the input space and output space, respectively.

In the context of image classification, the input space is the image space, e.g. the set of all possible images with 32×32 resolution denoted by $S_{input} = \mathbb{R}^{32 \times 32}$, while the output space is a certain vector space. If we consider a 10-class image classification problem, recalling that the output vector represents likelihood of the presence of objects of interest, then the output space can be $S_{output} = [0, 1]^{10}$. We highlight that the aforementioned input space and output space can be further precisely defined according to the real condition: the image space can be precisely $S_{input} = \{0,255\}^{32 \times 32}$ for the grayscale image format, which takes a value from $\{0, 1, ..., 255\}$ on each pixel; the output space can be precisely $S_{output} = \{v \in [0, 1]^{10} : v^\top 1 = 1\}$, which satisfies the requirement of probability. Moreover, instead of predicting the likelihood of each class as illustrated earlier, real applications usually require to predict a hard label, in which we may slightly modify S_{output} as $S_{output} = \{v \in \{0, 1\}^{10} : v^\top 1 = 1\}$, leading to only one label as the determination.

If considering object detection as an example, the mapping to be learned can be written as $f: S_{input} \to S_{output}$ where particularly, $S_{output} = \{v \in [0, 1]^{10} : v^\top 1 = 1\} \times \{[x, y, w, h] \mid x, y \geq 0, w, h > 0, x + w \leq 32, y + h \leq 32\}$. The second component in S_{output} defines a bounding box by specifying the coordinate of initial vertex (x, y) and its width w as well as heigt h. For semantic segmentation, similarly, the input space can be re-defined as $S_{output} = \{1, 2, ..., K\}^{32 \times 32}$, where K is the number of classes we are concerned with. It implies that our output prediction should assign a class label on each pixel.

Earlier we take image classification, object detection, and semantic segmentation as examples to shed light on different machine learning tasks on a particular scenario (32×32 images). Next we summarize two major types of machine learning tasks regarding the output prediction in a relatively more general way.

- *Classification* usually deals with integer predictions, and is the most frequently considered case in machine learning area. The reason why we need integer predictions is that we can use them to indicate whether or not a particular object/class is present in an input data sample. To this end, for binary classification where there are only two different classes, we need to set $S_{\text{output}} = \{0, 1\}$, or $\{-1, +1\}$. If the prediction on a data sample is 0 or -1, then it means this data sample is predicted as a negative class. When the prediction is 1, it indicates this data sample is a positive class. We may also have two more complicated cases, e.g. multiclass classification and multilabel classification, where there are more than two classes to be determined on each input data sample, say K classes for $K > 2$. In multiclass classification, specifically, the output space is defined as $S_{\text{output}} = \{1, ..., K\}$, which requires to predict only *one class* label for each data sample. In multilabel classification, the output space is defined as $\{0, 1\}^K$, where a prediction is a *one-hot vector* of length K. The i-th element of the *one-hot vector* indicates whether the i-th class is present or not, so it handles the case where multiple classes are present simultaneously on one single data sample.

- Regression deals with real values as predictions, which generally can be written as $f:$ $S_{\text{input}} \rightarrow S_{\text{output}}$, where $S_{\text{input}} = \mathbb{R}^d$ and $S_{\text{output}} = \mathbb{R}^n$. Specifically, when $n = 1$, e.g. $S_{\text{output}} = \mathbb{R}$, the prediction function f is a real-valued function. When $n > 1$, the prediction function f is a vector-valued function. In the example of semantic segmentation, the second component of the output (e.g. $\{ [x, y, w, h] \mid x, y \geq 0, w, h > 0, x + w \leq 32, y + h \leq 32 \}$ for generating bounding boxes) is a typical regression task, which requires to output vector values associated with the coordinates on an image.

There are of course more different machine learning tasks in addition to the preceding two. For example, *ranking* tasks deal with the order prediction given a set of data samples, say $\{z_1, ..., z_n\}$. In this case, the input space $S_{\text{input}} = \{\{z_1, ..., z_n\}\}$, while the output space S_{output} is all possible permutation of this dataset.

2.2 Concepts of Machine Learning

Before we formally introduce machine learning foundations, we first list some key concepts and their definitions.

- *Feature*. Feature is the descriptor of a data sample. The raw data samples from real applications may not be easily described in the digital format or processed by machine learning models. For example, images can be stored in many approaches in computers, including grayscale (then they are matrices), RGB color model (then they are tensors), etc. However, if there is no special mechanism such as convolutional layers used in convolutional neural networks, most traditional machine learning models can only take a vector as input, rather than a matrix or tensor. Therefore, we may need a representation feature of each data sample for machine learning models
- *Label*. Because a machine learning model learns a mapping function $f: S_{\text{input}} \rightarrow S_{\text{output}}$ that reflects the relation between the input data and output response, for each input data sample, it requires a label as response to provide the information of such relation. Typically, if label supervision is completely available, we usually group feature and label together as a data sample, e.g. $z = (x, y)$, where x is the representation feature and y is the response label.

- *Training data.* To allow a machine learning model to learn the relation between the feature and label, one needs a sufficiently large number of data samples that carry such information. This set of data samples are called training set. Typically, we may denote $S_{\text{train}} = (z_1, z_2, ..., z_n)$, where $z_i = (x_i, y_i)$ for $1 \leq i \leq n$ is a feature-label pair.

- *Hypothesis set.* After preparing a training set, one can consider how to learn a good machine learning model, or more specifically learn a good mapping function f. The mapping function $f: S_{\text{input}} \rightarrow S_{\text{output}}$ can also be called a hypothesis. Usually, the learning process requires to select a hypothesis from a predefined hypothesis set due to a number of concerns. For example, we may actively choose a family of functions to approximate the relation between the feature and label on the training set. This chosen family of functions can be linear functions or quadratic functions, depending on our real requirements. Thus, the hypothesis set can be defined accordingly, e.g. a set containing all possible linear functions or quadratic functions. Another example of our concern is that we may need to control the complexity of the hypothesis (see Section 2.7 later for more discussion), so we use regularization or create a constraint on the hypothesis set. Different concerns and choices on the hypothesis set can be present simultaneously, leading to a narrowed-down hypothesis set finally.

- *Model parameters.* The selected hypothesis set includes many hypotheses, each of which is determined by a set of parameters. Take one-dimensional linear functions as an example, and consider a general structure of linear functions $f(x) = a * x + b$, where a and b are two parameters in linear functions. If we determine the values of a parameter pair of (a, b), then we can determine a linear function. The same situation is present for other function families and hypothesis set – there are a set of parameters associated with the hypothesis set, and their values determine the hypothesis.

- *Learning process.* Given a hypothesis set, it is natural to see that some hypotheses can perform better than others, so we need to select a sufficiently good one to approximate the relation between feature and labels. This selection can be done by selecting the values of model parameters, e.g. (a, b) in linear functions. The process for determining the values of model parameters is the learning process. In practice, optimization algorithms can be critical to learning process, including zeroth order gradient methods, first order gradient methods, etc. (see Section 2.9 for examples). One of the mostly used optimization algorithms is stochastic gradient descent (SGD) method, which is an iterative algorithm, updating parameters iteratively until convergence. There are also many new variants of SGD designed for machine learning problems.

- *Hyperparameters.* There may be some parameters used in machine learning models that do not need to be updated in the learning process (usually predefined before the learning process), but can have significant impact on the hypothesis selection as well as the final performance. Those parameters are called hyperparameters. For example, the regularization term (see Section 2.7 for more discussion) can be weighted by multiplying a scalar, which is the hyperparameter of regularization. Since it controls the weight of regularization and thus the hypothesis selection, we may observe a significant influence on performance by assigning different values.

- *Validation data.* Due to the significance of hyperparameters, it is reasonable to figure out what proper values we should assign to hyperparameters during the learning process. A common way is to predefine a set of candidate values for the hyperparameter and then perform validation. To this end, typically, a validation set, e.g. $\{z_1, ..., z_{n_{\text{validate}}}\}$, is required and should be *independent* from the training set. After setting a particular candidate value to the hyperparameter and learning model parameters, we can evaluate this learned model on the validation set using a certain performance measure. Repeating such *setting-learning-evaluation* process for each of the predefined candidate values, we thus understand which candidate values can be better choices.

- *Testing data.* After the learning process and validating the hyperparameters, we finalize our machine learning model. The final step is to get to know the real performance of the learned model. Similar to validation, we need a testing set, e.g. $\{z_1, ..., z_{n_{test}}\}$, which is independent from the training set. Then we may evaluate the learned model on this testing set by computing the target performance measure. In classification, accuracy (or error rate) is the most used performance metric. In some cases where class imbalance happens, F-measure, Area Under Curve (AUC), etc., can be better options rather than accuracy. In regression, mean square error (MSE), mean absolute error (MAE), etc., are commonly used.

2.3 Machine Learning Paradigms

Depending on how the label supervision may be available, e.g. how y_i is available in the training set S_{train}, we may roughly group machine learning models into several categories. Note that some of the following machine learning paradigms are not necessarily disjoint.

- *Supervised learning* is a classical machine learning paradigm, where complete information of the feature-label pair (x_i, y_i) is available in the training set S_{train}.
- *Weakly supervised learning* may have a broader coverage of real applications. In this learning paradigm, incomplete information of the feature-label pair (x_i, y_i) is available. We may further group them into several subtypes.

In *semi-supervised learning* (SSL), a popular machine learning problem setting, only a small portion of training data samples have complete feature-label pairs, denoted by $S_{labeled} = \left\{ (x_1, y_1), (x_2, y_2), ..., \left(x_{n_l}, y_{n_l}\right) \right\}$. Here n_l denotes the number of such samples, or labeled data. For the rest part of training set, we have only feature information instead of the associated labels, denoted by $S_{unlabeled} = \{x_{n_l + 1}, x_{n_l + 2}, ..., x_{n_l + n_u}\}$, where n_u is the number of such unlabeled data samples. The reason for considering SSL setting is that, in some applications, it can be difficult to collect high quality labels, so the number of feature-label pairs can be limited. The majority part of collected data samples contain features only. Some well-known applications include medical image analysis as in Irvin et al. (2019). For example, it is time consuming to annotate X-ray images for screening, diagnosis, and management of diseases.

In *unsupervised learning*, the scarcity issue of label supervision can be much worse – there is no label information at all in the training set, which means $S_{train} = \{x_1, x_2, ..., x_n\}$, but still our target is to learn a prediction function that can differentiate data samples from different underlying classes. Even if we cannot access to the true label information, there are some great techniques we can make use of. For example, we could gradually include pseudo labels into the training set (Rizve et al., 2021). By pseudo labels, we refer to as the predicted labels, but we use them as they are the true labels. Another research line is contrastive learning (Chuang et al., 2020), which is built on an assumption that we can augment (generate) training data by applying some slight variation on non-essential features, e.g. translating images or other appearance change (Chen et al., 2020). If such non-essential variation does not impact the underlying discriminative features, then the augmented data sample should be assigned by the same label as the original data sample. In this way, some reasonable labels are available – the original sample and augmented sample can be regarded as a positive pair, while different samples are regarded as negative pairs. Then a contrastive loss function can be used to push away negative pairs and pull close positive pairs simultaneously.

Real applications may also include a lot of more complicated cases, rather than the two settings mentioned earlier. SSL and unsupervised learning assume for each sample, whether there is a true label y_i associated with the feature x_i. In practice, the true label y_i can be more complicated than a single class label – recall the example of object detection in Section 2.1.1, where the true label includes two components, the class label, and the coordinate of bounding box. In a supervised setting, both class labels and coordinates of bounding boxes are required on each data sample. If the training set contains only class labels without bounding boxes, then for each data sample, we have partial supervision information. Such case can also be in weakly supervised learning setting – we have only class labels in training data, and aim to predict both class labels and bounding boxes (Wei et al., 2018).

Again, we highlight that the preceding list cannot cover all topics in machine learning. There are some other learning paradigms depending on how label supervision is available, some of which are very practical in many applications. Reinforcement learning, for example, is a well-known learning paradigm, which does not have explicit labels. Instead, rewards from the environment are the key information driving the learning process. We will avoid the over-complete list of machine learning paradigms and focus on our main topics. In the next subsection, we introduce a well-known general learning framework, *probably approximately correct learning*, which establishes the foundation of most machine learning models.

2.4 Probably Approximately Correct Learning

Probably approximately correct (PAC) learning is a general machine learning framework that answers several fundamental questions: (i) What can be learned by machine learning models? (ii) What conditions are required to successfully learn them? (iii) How many examples are needed to learn them? To answer these questions, we first need to define what machine learning models learn. In previous subsection, we mention that machine learning is to learn a mapping function from an input space to an output space, so it is natural to start the analysis from the underlying mapping function.

Classical analysis considers two types of the underlying function. First, in the deterministic setting, we assume that the underlying mapping is a deterministic function. In the second setting, we assume that the underlying mapping is characterized by a joint distribution over the feature and label, which introduces randomness into the mapping. The deterministic underlying mapping is easy to understand, as formalized in the following Section 2.4.1, but the stochastic setting may need more discussion. A vivid example of stochastic setting in real world is the papaya classification problem analyzed in Shalev-Shwartz and Ben-David (2014). Suppose this problem aims to classify papayas into two classes, tasty and not-tasty. The considered features to describe papaya samples include color, softness, etc. In many cases, we can correctly classify papayas using these two features, but we can also possibly find two papayas with the same color and softness that belong to different classes, which describes the impact of randomness. As a result, stochastic setting can be more realistic and complicated compared with the deterministic one.

In the following subsections, we first discuss deterministic case in Section 2.4.1. Under this particular setting, we further consider two subcases, e.g. consistent and inconsistent setting (see their definitions later), respectively. Then in Section 2.4.2, we mainly discuss stochastic case.

2.4.1 Deterministic Setting

Suppose there exists an underlying mapping function $c : X \rightarrow Y$ that takes a feature $x \in X$ as input and predicts a label $y \in Y$ as output, where X and Y denote the feature space and label space, respectively. For simplicity, throughout this chapter, we consider a binary classification problem, which makes $Y = \{0, 1\}$ (or $-1, +1$). More generally, as we define a hypothesis set H for f, we also define a hypothesis set C and assume that C contains all underlying hypothesis that we would like to learn. The underlying mapping function c is not accessible directly, but we can still collect a training set that carries information of the mapping function. Let $S_{\text{train}} = ((x_1, y_1), ..., (x_n, y_n))$ be the collected training set, where each $x_i \sim D$ is an independent and identically distributed (*iid*) random variable drawn from the underlying distribution D defined over X. Note that their labels are also determined by the underlying mapping function c, e.g. $y_i = c(x_i)$.

After preparing this training set, we can define a hypothesis set H (for example, linear functions) and select a hypothesis $f \in H$ such that f shares very similar behavior with $c \in C$. To quantify how close f is to c, we need some new measures. The first measure we would like to introduce is *generalization error*, or *risk*.

Definition 2.1 (Generalization Error for Deterministic Setting)
Suppose there exists a hypothesis $f \in H$, a target hypothesis $c \in C$, and an underlying distribution D over X. The *generalization error* or *risk* of the hypothesis f is defined as follows:

$$R(f) := \mathbb{P}_{x \sim D}[f(x) \neq c(x)] = \mathbb{E}_{x \sim D}\left[\mathbb{I}_{f(x) \neq c(x)} \right] \tag{2.1}$$

where $\mathbb{I}_{\mathcal{E}}$ is the indicator function of an event $\mathcal{E}$, e.g. $\mathbb{I}_{\mathcal{E}} = 1$ if $\mathcal{E}$ is true, and $\mathbb{I}_{\mathcal{E}} = 0$ if $\mathcal{E}$ is false.

One thing to highlight is that the preceding generalization error involves all data samples x that is drawn from D, so it measures the error or risk of the hypothesis f in the sense of population level. Moreover, we observe that D is the underlying distribution and thus cannot be accessed, which means that generalization error cannot be directly computed. To make performance measure feasible, we need the following *empirical risk*.

Definition 2.2   (Empirical Error for Deterministic Setting)
Suppose there exists a hypothesis $f \in H$, a target hypothesis $c \in C$, and a set of data, $S = (x_1, ..., x_n)$, which are *iid* samples drawn from the underlying distribution D over X. The empirical error or empirical risk of the hypothesis f is defined as follows:

$$\hat{R}_S(f) := \frac{1}{n}\sum_{i=1}^{n} \mathbb{I}_{f(x)\neq c(x)} \tag{2.2}$$

The key difference between empirical error and generalization error is that empirical error is evaluated on the empirical data S, which is available and thus can be directly computed.

After introducing the generalization error $R(f)$, we are ready to answer the first two questions in the beginning of this subsection: what can be learned and what conditions are required to learn successfully.

Definition 2.3   (PAC Learnability)
A target hypothesis class C is PAC learnable if there exists an algorithm A and a polynomial function $n_H(\cdot, \cdot)$ such that for any $\epsilon > 0$, $\delta > 0$, for all distributions D on X and for any target hypothesis $c \in C$, the following inequality holds for any sample size $n \geq n_H(1/\epsilon, 1/\delta)$:

$$\mathop{\mathbb{P}}_{S \sim D^n}[R(f_S) \leq c] \geq 1 - \delta$$

where f_S denote a hypothesis learned on S.

Remark 2.1   It can be seen from the definition that PAC learning framework includes two parameters, ϵ and δ. First, δ is a confidence parameter that controls the probability of the event $R(f_S) \leq \epsilon$. When δ is very small, this event happens with high probability at least $1 - \delta$. Second, ϵ controls the correctness of f_S. Note that since all labels are generated by the underlying mapping function c, which is a deterministic function, c is the optimal hypothesis for the generalization error R, e.g. $R(c) = 0$. Therefore, ϵ indeed controls the difference between $R(c)$ and $R(f_S)$, as the upper bound of errors of f_S. To sum up, the preceding definition provides a learning framework that, with high *Probability* at least $1 - \delta$, the learned model is *Approximately Correct* (at most ϵ error).

Given the two parameters ϵ and δ controlling somewhat the quality of a learned model f_S on the training set S, there is one term left in the aforementioned definition, the polynomial function $n_H(1/c, 1/\delta)$, which depends on H and serves as the lower bound of n, the number of samples included in training data S. A natural question can be: can we identify the structure of the polynomial function $n_H(1/\epsilon, 1/\delta)$? The answer is affirmative.

Next, we show two results for the structure of $n_H(1/\epsilon, 1/\delta)$ under two different settings, *consistent* and *inconsistent* cases. For both settings in this section, we first assume that H is a *finite hypothesis set*, which means that the cardinality of H is finite, e.g. $|H| < \infty$.

Remark 2.2   Could H be *finite* in practice? The answer is affirmative, if we use the "discretization trick" discussed in Remark 4.1 of Shalev-Shwartz and Ben-David (2014).

Consider a machine learning model parameterized by a d-dimensional vector (e.g. a linear regressor in Section 2.9.1). If we build this linear regression model using computer, where each element of the d-dimensional vector is represented by a float point with 64 bits, so the hypothesis class H contains at most 2^{64d} hypotheses. In this case, $\log(|H|) = 64d$.

In the first setting, *consistent case*, we assume that the underlying target hypothesis c is contained by the predefined hypothesis set H, e.g. $c \in H$. It implies that there is at least one hypothesis in H that does not make any mistake in its prediction – for example, $c \in H$ itself is one. In this case of $f_S = c$, we can guarantee that $R(f_S) = 0$ and $\hat{R}_S(f_S) = 0$ (simply using the definition of generalization error in Definition 2.1 and empirical error in Definition 2.2). As a result of the preceding analysis, it is possible for us to find such best hypothesis if we learn with H.

In the second setting, *inconsistent case*, we assume that the underlying target hypothesis c is not contained by the predefined hypothesis set H, e.g. $c \notin H$. Consequently, some useful properties available in the consistent case cannot hold now, e.g. $R(f_S) = 0$ and $\hat{R}_S(f_S) = 0$. This makes PAC learnability more difficult compared with the consistent case, which can be verified from the comparison between results of two settings later.

The following analysis holds for the consistent setting.

Theorem 2.1 (*Theorem 2.5 in Mohri et al. (2018), learning bound for deterministic setting, consistent case) Let H be a finite set of functions mapping from X to Y and A be an algorithm that for any target hypothesis $c \in H$ and iid sample S returns a consistent hypothesis f_S: $\hat{R}_S(f_S) = 0$. Then for any $\epsilon, \delta > 0$, the inequality $\mathbb{P}_{S \sim D^n}[R(f_S) \leq \epsilon] \geq 1 - \delta$ holds if:*

$$n \geq \frac{\log(|H|/\delta)}{\epsilon}$$

This sample complexity result admits the following equivalent statement as a generalization bound: for any $\epsilon, \delta > 0$ with probability at least $1 - \delta$,

$$R(f_S) \leq \frac{\log(|H|/\delta)}{n}$$

Remark 2.3 Theorem 2.1 reveals two quantities and their insights. (i) A lower bound of sample complexity required to achieve PAC learnability of a target hypothesis c, e.g. $\frac{\log(|H|/\delta)}{\epsilon}$, which is dominated by a polynomial in $1/\epsilon$ and $1/\delta$. This quantity is associated with the polynomial function $n_H(1/\epsilon, 1/\delta)$ in Definition 2.3 of PAC learnability and gives the least sample size to make PAC learnability of c hold when considering consistent case. (ii) An upper bound of the generalization error of the PAC learned hypothesis $\frac{\log(|H|/\delta)}{n}$. This result gives a straightforward conclusion for controlling the upper bound of $R(f_S)$. If n increases, then this upper bound decreases as the same growth rate. For example, if we increase the sample size n from 1000 to 1,000,000 without changing anything else, then the upper bound decreases 1000 time compared with the original quantity. If $|H|$ or $1/\delta$ increases at a certain growth rate, then $R(f_S)$ increases logarithmically in this growth rate.

One reminder here is that logarithmic function grows significantly slower than the linear function, so changing n earlier introduces a more significant impact to $R(f_S)$ compared with changing $|H|$ and δ at same order of amount.

Above we introduce the sample complexity and generalization error for the PAC learned hypothesis in the context of consistent setting, which gives a promising result. Next we introduce the result for inconsistent setting.

Theorem 2.2   *(Theorem 2.13 in Mohri et al. (2018), learning bound for deterministic setting, inconsistent case) Let H be a finite hypothesis set. Then for any $\delta > 0$, with probability at least $1 - \delta$, the following inequality holds:*

$$\forall f \in H, R(f) \le \hat{R}_S(f) + \sqrt{\frac{\log(2|H|/\delta)}{2n}}$$

Remark 2.4   Since inconsistent case does not guarantee $R(f_S) = 0$ and $\hat{R}_S(f_S) = 0$, the preceding theorem upper bounds the gap between $R(f)$ and $\hat{R}_S(f)$ for any $f \in H$. As Theorem 2.1, this gap includes dependency on ϵ, δ, and n, but in a different order, e.g. a square root function outside. Suppose again that we increase n from 1000 to 1,000,000 and both the dependencies on δ and $|H|$ are dominated by that on n due to the logarithmic growth rate. In the consistent case, we can decrease the learning bound around 1000 times, while in the inconsistent case, we can only decrease the learning bound around $\sqrt{1000} \approx 31.6$ times, which can be much less than the former. This comparison verifies that the inconsistent case can be more difficult to learn a PAC model than the consistent case.

2.4.2   Stochastic Setting

In this subsection, we consider the more complicated stochastic setting. In the previous setting, we assume the underlying mapping can be done by a deterministic function c, where given an input data sample x, c performs a mapping to $y = c(x)$. This setting is deterministic because y is uniquely determined as long as x is uniquely determined – there is no exception. However, in reality, deterministic mapping can be very rare.

We again use the example of papaya classification problem from Chapter 2 of Shalev-Shwartz and Ben-David (2014) to elaborate the limit of deterministic setting. Suppose we would like to predict whether a papaya is tasty or not with two features: (i) color, which can be dark green, orange, red to dark brown; (ii) softness, which can be rock hard to musty. Clearly both features roughly indicate the stage of the development of papayas, given that maturity significantly impacts the taste, so these features can be used for our classification problem. For example, the orange color and moderate softness of a papaya indicates that it is fully mature and may be tasty.

Whereas the aforementioned relation between features and taste does not always hold or is not deterministic, sometimes two papayas with same color and softness can have different taste, which is the mostly seen case in reality, so it is critical to extend and generalize the deterministic setting to cover the stochastic setting. To this end, we change how the mapping is performed. Instead of assuming the underlying mapping is described by a deterministic function c for a given $x \sim D_X$, we now consider a joint distribution $D_{X \times Y}$ defined over $X \times Y$. For an instantiation of random variable (X, Y), e.g. (x, y), its joint probability can be written as:

$$\mathbb{P}[X = x, Y = y] = \mathbb{P}[X = x] \cdot \mathbb{P}[Y = y \mid X = x]$$

In this way, randomness of relation between features and labels can be introduced.

Accordingly, we then need to make necessary modification on the definitions of risk and empirical risk from Definition 2.1 and 2.2.

Definition 2.4 **(Generalization Error for Stochastic Setting)**
Suppose there exists a hypothesis $f \in H$, a joint distribution D defined over $X \times Y$, and an underlying distribution D over X. The *generalization error* or *risk* of the hypothesis f is defined as follows:

$$R(f) := \mathbb{P}_{(x,y) \sim D}[f(x) \neq y] = \mathbb{E}_{(x,y) \sim D}\left[\mathbb{I}_{f(x) \neq y}\right] \qquad (2.3)$$

Definition 2.5 **(Empirical Error for Stochastic Setting)**
Suppose there exists a hypothesis $f \in H$, and a set of data, $S = ((x_1, y_1), ..., (x_n, y_n))$ that are *iid* samples drawn from the underlying distribution D over $X \times Y$. The *empirical error* or *empirical risk* of the hypothesis f is defined as follows:

$$\hat{R}_S(f) := \frac{1}{n} \sum_{i=1}^{n} \mathbb{I}_{f(x) \neq y} \qquad (2.4)$$

Then the generalization of PAC learning from deterministic setting to stochastic setting can be summarized as follows.

Definition 2.6 **(Agnostic PAC Learnability)**
Let H be a hypothesis set. A is an agnostic PAC-learning algorithm if there exists a polynomial function $n_H(\cdot, \cdot)$ such that for any $\epsilon > 0$ and $\delta > 0$, for all distributions D over $X \times Y$, the following inequality holds for any sample size $n \geq n_H(1/\epsilon, 1/\delta)$:

$$\mathbb{P}_{S \sim D^n}\left[R(f_S) - \min_{f \in H} R(f) \leq \epsilon\right] \geq 1 - \delta$$

Remark 2.5   If there is no randomness in the joint distribution over $X \times Y$, then it reduces to the deterministic function, e.g. a mapping $c : X \rightarrow Y$ with probability 1. On the other hand, if we consider the randomness from the feature-label joint distribution, then we generalize PAC learnability to cover a much more significant variety of cases in reality. Moreover, as in the inconsistent case (Theorem 2.2) of PAC learnability (Definition 2.3), in the stochastic setting, agnostic PAC learnability does not guarantee an arbitrarily small absolute error. Instead, the target is to achieve an arbitrarily small gap between $R(f_S)$ and $\min_{f \in H} R(f)$.

After generalizing the notion of PAC learnability to stochastic setting, a natural question is how to bound $n_H(1/\epsilon, 1/\delta)$, as in the consistent case (Theorem 2.1) and inconsistent case

(Theorem 2.2) of PAC learning. In the context of finite hypothesis class, we can derive similar results as in Theorem 2.1 and 2.2. For example, Corollary 4.6 in Shalev-Shwartz and Ben-David (2014) shows the learning bound $O(\log(|H|/\delta)/\epsilon^2)$ (same with Theorem 2.2 for inconsistent deterministic setting) with probability at least $1 - \delta$ in this case of finite hypothesis class. In the following sections, instead, we introduce the learning bounds for infinite hypothesis class, e.g. without the condition $|H| < \infty$.

2.5 Estimation and Approximation

Recall that PAC learning provides the answers to questions about (i) what can be learned, (ii) under what conditions, and (iii) how many data samples needed to learn. These analyses mainly focus on the fundamental principles of machine learning. Then the next question is how to design an algorithmic way to find the hypothesis f from the considered hypothesis class H under PAC learning principle? This problem can be regarded as *model selection* problems, e.g. selecting one hypothesis from the entire hypothesis class H. Then a selection criterion is required. In this chapter, we study what can be a good criterion for the model selection problem.

To proceed our analysis, we need the following new concepts of *Bayes error* and *Bayes hypothesis*.

Definition 2.7 (Bayes Error and Bayes Hypothesis)
Suppose $\mathscr{D}$ is the underlying joint distribution defined over $X \times Y$. Bayes error is defined as follows:

$$R^* = \inf_f R(f)$$

where f is any measurable function.

Those hypotheses that achieve the Bayes error are Bayes hypotheses or Bayes classifier, e.g. f_{Bayes} is a Bayes hypothesis if $R(f_{\text{Bayes}}) = R^*$.

Note that in the definition of Bayes error and Bayes hypothesis, the infimum is taken over all measurable hypotheses f, rather than some considered hypothesis class H. Recall that hypothesis set H can be narrowed down purposely by selecting a particular family of functions to learn the underlying distribution D, such as linear functions and quadratic functions. The Bayes hypothesis can be considered as the optimal hypothesis from all possible hypotheses, beyond the predefined H. It is usually impossible to know what structure the optimal Bayes hypothesis is, but we aim to make the risk of our empirical hypothesis close to Bayes error, e.g. minimizing the gap, or *excess risk*:

$$\epsilon_{\text{excess}}(f) := R(f) - R^* \tag{2.5}$$

However, there are two main challenges to analyze this quantity. First, R is in the sense of pupulation and usually not accessible, since the underlying distribution is unknown. Second, R^* is also not available, since it is difficult to know what Bayes hypothesis is.

To enable such analysis for excess risk of a hypothesis f, one way is to decompose the preceding excess risk as follows:

$$\epsilon_{\text{excess}}(f) = R(f) - R^* = \underbrace{(R(f) - \inf_{f \in H} R(f))}_{=:\epsilon_{\text{est}}(f)} + \underbrace{(\inf_{f \in H} R(f) - R^*)}_{=:\epsilon_{\text{approx}}(f)} \tag{2.6}$$

A further inspection on ϵ_{approx} indicates that the selection of H is the only factor controlling this quantity. To see why, we can write down the full definition of R^* as follows:

$$\epsilon_{\text{approx}} = \inf_{f \in H} R(f) - \inf_{f:\text{all measurable}} R(f)$$

H is a subset of the set of all measurable hypotheses – if H is larger, e.g. H covers more measurable hypotheses, then their gap ϵ_{approx} is smaller. If H is smaller, e.g. H covers less measurable hypotheses, then their gap ϵ_{approx} is larger (see Section 2.7 for more details).

On the other hand, ϵ_{est} depends on which hypothesis f we choose from H assuming that H is fixed, since $\inf_{f \in H} R(f)$ only depends on H and R. Now minimizing ϵ_{est} reduces to minimizing $R(f)$ over $f \in H$. One thing to highlight is that ϵ_{est} is the "Approximately Correct" quantity required by PAC learnability, as in Definition 2.6.

In the next two subsections, we introduce empirical risk minimization (ERM) and regularization, two techniques for controlling ϵ_{est} and ϵ_{approx}, respectively, leading to a trade-off between the two components.

2.6 Empirical Risk Minimization

2.6.1 Empirical Risk Minimizer

Recall that we have introduced *empirical error* in Definition 2.5. Given that the underlying distribution D is not available, we cannot directly select a hypothesis that minimizes the generalization error. Instead, it is reasonable to find a hypothesis that minimizes the empirical error. We denote such hypothesis by *empirical risk minimizer* or *ERM solution* as follows:

$$f_S^{\text{ERM}} := \arg\min_{f \in H} \hat{R}_S(f) \tag{2.7}$$

Then we need to analyze the estimation error by plugging in f_S^{ERM}:

$$\epsilon_{\text{est}}(f_S^{\text{ERM}}) = R(f_S^{\text{ERM}}) - \inf_{f \in H} R(f) \tag{2.8}$$

The following result takes the first step to build the connection between the estimation error and the generalization bound of f, e.g. $R(f) - \hat{R}(f)$.

Proposition 2.1 (Proposition 4.1, Mohri et al. (2018))
For any sample S, the following inequality holds for the hypothesis returned by ERM:

$$\mathbb{P}\left[R(f_S^{\text{ERM}}) - \inf_{f \in H} R(f) > \epsilon\right] \leq \mathbb{P}\left[\sup_{f \in H} |R(f) - \hat{R}_S(f)| > \frac{\epsilon}{2}\right] \tag{2.9}$$

In the following subsection, we introduce a useful tool, VC-dimension, to help us upper bound the aforementioned generalization bound in (2.9).

2.6.2 VC-dimension Generalization Bound

VC-dimension, or Vapnik–Chervonenkis dimension, is a useful technique to analyze the learning bounds for *infinite* hypothesis set (different from the finite hypothesis set in previous subsection). Particularly, it reveals that it is possible to efficiently learn with finite sample size under infinite hypothesis set. In this subsection, we first introduce two new concepts, growth function and VC-dimension, and then provide the result of VC-dimension generalization bound for deriving the bound of the estimation error in (2.9).

Definition 2.8 (Growth Function)
Given a hypothesis class H, a family of functions taking values in $\{-1, +1\}$, its growth function $\Pi_H : \mathbb{N} \to \mathbb{N}$ is defined as follows:

$$\forall n \in \mathbb{N}, \Pi_H(n) = \max_{\{x_1, \ldots, x_n\} \subset X} |\{(f(x_1), \ldots, f(x_n)) : f \in H\}|$$

Let's decompose and interpret the preceding definition for growth function. For a given n, there can be a set of n data samples, where we consider the feature part only, e.g. $\{x_1, \ldots, x_n\}$. Applying f to this set, we generate a set of prediction $\{f(x_1), \ldots, f(x_n)\}$, each of which takes a value in $\{-1, +1\}$. Going through all hypotheses f contained in H, there can be many different sets of prediction. For example, if $n = 3$, then there are $8\ (2^3 = 8)$ sets of prediction, e.g. $\{-1,-1,-1\}$, $\{-1,-1,+1\}$, $\{-1,+1,-1\}$, $\{+1,-1,-1\}$, $\{-1,+1,+1\}$, $\{+1,+1,-1\}$, $\{+1,-1,+1\}$, $\{+1,+1,+1\}$. Each set of prediction is called a *dichotomy*. When H is rich enough, the total number of dichotomies can reach 2^n for a dataset with size n. When H is restricted, it is possible to have less dichotomies. The *growth function* then is used to count the number of dichotomies that a hypothesis set H can cover for a given sample size n. This concept can be viewed as a measure for the richness of H.

Definition 2.9 (VC-dimension)
The VC-dimension of a hypothesis set H is the size of the largest set that can be shattered by H:

$$\text{VCdim}(H) = \max\{n : \Pi_H(n) = 2^n\}$$

After we decompose the definition of growth function (Definition 2.8), it is not difficult to understand the VC-dimension. The growth function returns the maximum number of dichotomies for a given H and sample size n. The largest possible number of dichotomies is 2^n. This is the case where we assume that n is given. If we scan different values of n and find the largest value of n that allows H to achieve 2^n dichotomies, then that value of n is VC-dimension of H. When we check some certain value of n and find that $\Pi_H(n) < 2^n$, we can decrease the value of n, until $\Pi_H(n) = 2^n$ holds.

Listing the preceding two critical concepts, we can proceed to the VC-dimension generalization bound.

Theorem 2.3 *(VC-dimension generalization bound, Corollary 3.19, Mohri et al. (2018)) Let H be a family of functions taking values in $\{-1, +1\}$ with VC-dimension d. Then for any $\delta > 0$, with probability at least $1 - \delta$, the following holds for all $f \in H$:*

$$R(f) \leq \hat{R}_S(f) + \sqrt{\frac{2d_{vc} \log(en/d_{vc})}{n}} + \sqrt{\frac{\log(1/\delta)}{2n}} \tag{2.10}$$

Remark 2.6 Generalization bound above is used to bound the gap between $R(f)$ and $\hat{R}_S(f)$ for the same hypothesis f. Note that it holds for any $f \in H$, so it is simple to plug into (2.9) regardless the supremum operation sup over H. Let us first simplify the generalization bound in (2.10):

$$R(f) - \hat{R}_S(f) \leq O\left(\sqrt{\frac{d_{vc} \log(n/d_{vc})}{n}}\right) = O\left(\sqrt{\frac{\log(n/d_{vc})}{n/d_{vc}}}\right)$$

If we further hide the logarithmic factor using $\tilde{O}$ in the preceding equation, then we have $R(f) - \hat{R}_S(f) \leq \tilde{O}\left(\sqrt{d_{vc}/n}\right)$. There thus is a preliminary conclusion – if d_{vc} is too large (or H is too rich and covers too many hypotheses), say $d_{vc} = O(n)$, then $R(f) - \hat{R}_S(f) = O(1)$, which is too large. In contrast, if d_{vc} is relatively small, say $O(1)$, then $R(f) - \hat{R}_S(f) = O\left(\sqrt{1/n}\right)$, which is a good result.

For any $f \in H$, to make $R(f) - \hat{R}_S(f) \leq \epsilon$, as required by (2.9), we have $\epsilon = O\left(\sqrt{\frac{\log(n/d_{vc})}{n/d_{vc}}}\right)$. Combining with Proposition 2.1, we have with probability at least $1 - \delta$:

$$\epsilon_{est}(f) \leq O\left(\sqrt{\frac{\log(n/d_{vc})}{n/d_{vc}}}\right) \tag{2.11}$$

Moreover, the conclusion still holds: (i) the richer H is, the larger d_{vc} is; (ii) the larger d_{vc} is, the larger $\epsilon_{est}(f)$ is. It indicates that a too complicated H will increase the estimation error.

2.6.3 General Loss Functions

So far, we only consider 0-1 loss in R and $\hat{R}_S$ as the performance measure. However, 0-1 loss can be difficult to handle in practice, since optimization over 0-1 loss is NP-hard, and the analysis of agnostic PAC learning for infinite hypothesis class in Section 2.6.2 requires to derive the ERM solution f_S^{ERM} (see also Proposition 2.1 and (2.8)). A common way to deal with such computational challenge is to use a surrogate convex loss function that can upper bound the 0-1 loss.

To this end, we can generalize the 0-1 loss to a loss function with the structure as:

$$\ell(-yf(x))$$

such that $\ell : \mathbb{R} \to \mathbb{R}$ is a convex and non-decreasing function with $\mathbb{I}_{u \geq 0} \leq \ell(u)$ for any $u \in \mathbb{R}$. This structure covers many commonly used loss functions in machine learning. For example, for square loss, we have $\ell(u) = (u + 1)^2$; for hinge loss, we have $\ell(u) = \max(0, 1 + u)$; for logistic loss, we have $\ell(u) = \log(1 + \exp(u))$; for exponential loss, we can derive $\ell(u) = \exp(u)$; for squared hinge loss, we have $\ell(u) = \max(0, 1 + u)^2$.

Let $\mathcal{L}(f) = \mathbb{E}_{(x,y) \sim D}[\ell(-yf(x))]$ be the expected loss over D using loss ℓ. Denote by $f^*(x) = \mathbb{P}[y = +1 \mid x] - \frac{1}{2}$ the Bayes scoring function. Define $L(x, u) = \mathbb{P}[y = +1 \mid x] \cdot \ell(-u) + (1 - \mathbb{P}[y = +1 \mid x]) \cdot \ell(u)$ and $f^*(x) = \arg\min_u L(x, u)$ the optimal score for given x. Moreover, let $\mathcal{L}^* = \mathbb{E}_{(x,y) \sim D}[\ell(-yf^*(x))]$ denote the expected loss function value of $f^*(x)$ over D. The following result gives the relation between the excess error and loss function bound.

Theorem 2.4 *(Theorem 4.7 in Mohri et al. (2018)) Let ℓ be a convex and non-decreasing function. Assume that there exists $s \geq 1$ and $c > 0$ such that the following holds for all $x \in X$:*

$$|f^*(x)|^S = \left| \mathbb{P}[y = +1 \mid x] - \frac{1}{2} \right|^S \leq c^S \left(L(x, 0) - \min_u L(x, u) \right) \tag{2.12}$$

Then for any hypothesis f, the excess risk of f is bounded as follows:

$$R(f) - R^* \leq 2c(\mathcal{L}(f) - \mathcal{L}^*)^{\frac{1}{s}}$$

The preceding result shows that the loss function bound $(\mathcal{L}(f) - \mathcal{L}^*)$ can upper bound the excess risk. If we can minimize $(\mathcal{L}(f) - \mathcal{L}^*)$ (or $\mathcal{L}(f)$) over f, then we can guarantee a small value of the excess risk. It motivates and establishes theoretical foundation of derivation of ERM solution using a convex surrogate loss, which is widely used in machine learning community. There are several special cases of the value of c, s in Theorem 2.4. For hinge loss, $s = 1, c = 1/2$. For logistic loss, $s = 2, c = 1/\sqrt{2}$.

2.7 Regularization

2.7.1 Regularized Loss Minimization

Clearly, ERM algorithms do not influence the distance between f^* and f_{Bayes}, e.g. approximation error; they only work on estimation error. To control the approximation error, we need to set up the hypothesis class that is used for learning. Regularization is one of the mostly used way, which is usually combined with loss function in the ERM learning paradigm, resulting in the regularized loss minimization (RLM) problem:

$$f_S^{\text{RLM}} = \arg\min_{f \in H} \hat{R}_S(f) + \lambda R(f) \tag{2.13}$$

where λ is the hyper-parameter of regularization that controls the weight of regularization in the preceding objective function. The main goal of $R(f)$ is to restrict the complexity of f.

There are some frequently used regularization functions. Suppose that $f(x) = w^\top x$ is a linear function, parameterized by the weight vector $w \in \mathbb{R}^d$. L2 regularization instantiates $R(w) = \|w\|_2^2$, which is also called Tikhonov regularization. To see why this formulation can be used to restrict the complexity of w, let us write down the complete form as

$R(w) = \|w - \mathbf{0}\|_2^2$. If $w = \mathbf{0}$, then $R(w) = 0$, leading to no penalty in the objective function when deriving the ERM hypothesis in (2.13). On the other hand, if $w \neq \mathbf{0}$, $R(w)$ adds penalty to the ERM objective function. As a result, R avoids excessively complicated weight vector w that is far away from $\mathbf{0}$. For example, suppose that there are two weight vectors, say w_1 and w_2, achieving the same empirical risk with the same loss function and furthermore w_1 is closer to the origin, resulting in less $L2$ regularization penalty. Then w_2 is more complicated and thus far away from the origin, leading larger $L2$ regularization penalty. Thus, w_1 achieves smaller regularized loss value and is a better solution.

$L1$ regularization instantiates $R(w) = \|w\|_1$. The functionality of avoiding excessively complicated hypothesis can be similar to $L2$ regularization. $L1$ norm usually induces sparsity of the weight vector.

2.7.2 Constrained and Regularized Problem

In this subsection, we show the equivalence between constrained and regularized problem, in order to make it clear that regularization has a significant impact on the approximation error in (2.6) by controlling the richness of H. To this end, consider the following two problems (the following notations, including f, h, x, b, etc., are only used in this subsection):

$$(CP): \min_x f(x), \text{ s.t. } h(x) \leq b \tag{2.14}$$

$$(RP): \min_x f(x) + \lambda h(x) \tag{2.15}$$

where we denote them by (CP) and (RP) for the constrained problem and the regularized problem, respectively. We can interpret the preceding two problems by instantiating $h(x) = \|x\|_2^2$. For (CP), it serves as a $L2$ ball constraint, e.g. $\|x\|_2^2 \leq b$, with the center at the origin and radius $\sqrt{b}$. For (RP), it serves as regularization with hyper-parameter λ.

Target. Our general idea is to verify that, under some mild conditions, for each constraint bound b in (CP), there exists a regularization parameter λ and its corresponding (RP), such that both problems share the same minimizer.

Proposition 2.2 Suppose there are two problems (CP) and (RP) defined in (2.14) and (2.15), respectively. The minimizer of (CP) with a constraint bound b is also the minimizer of (RP) with a particular regularization parameter λ.

Proof. (of Proposition 2.2)

We start with x^* as the minimizer of (CP) with a given b:

$$x^* \in \arg\min_{h(x) \leq b} f(x),$$

which indicates $h(x^*) \leq b$ by feasibility.

The Lagrangian function of this (CP) can be written as follows:

$$L(x, \alpha) = f(x) + \alpha(h(x) - b)$$

where α is the Lagrangian multiplier. Then we can derive its dual problem with respect to α by:

$$g(\alpha) = \min_x L(x, \alpha)$$

and the corresponding dual problem as follows:

$$\max_{\alpha \geq 0} g(\alpha)$$

where we can denote $\alpha^* \in \arg\max_{\alpha \geq 0} g(\alpha)$ as one maximizer.

If the strong duality holds, then for the minimizer of the constrained problem x^*, we have $f(x^*) = g(\alpha^*)$:

$$f(x^*) = g(\alpha^*) = \min_x f(x) + \alpha^*(h(x) - b) \leq f(x^*) + \alpha^*(h(x^*) - b) \leq f(x^*)$$

where the last inequality is due to $\alpha^* \geq 0$ (Lagrangian multiplier) and $h(x^*) - b \leq 0$ (feasibility). The aforementioned inequalities should all be equality due to $f(x^*) = f(x^*)$, leading to $\alpha^*(h(x^*) - b) = 0$.

On the other hand, using the equality derived from above ones $f(x^*) = \min_x f(x) + \alpha^*$ $(h(x) - b)$, we know that $x^* \in \arg\min_x f(x) + \alpha^*(h(x) - b) = \arg\min_x f(x) + \alpha^* h(x)$, e.g. x^* is also the minimizer of the (RP) with α^* as the regularization parameter.

Taking a closer inspection on $\alpha^*(h(x^*) - b) = 0$, we have two cases. When $\alpha^* = 0$, the constraint is inactive, and the corresponding regularization parameter is 0. When $h(x^*) - b = 0$, the constraint is active, and the regularization parameter is set to α^* so that (CP) with b as constraint bound shares the same minimizer as (RP) with α^* as the regularization parameter.

After showing the equivalence between the constrained problem and regularized problem, it is easy to see why regularization is one way to adjust the hypothesis class H. If we have a heavy constraint, or a large regularization parameter, then H should be less rich and covers less hypotheses, which decreases the approximation error ϵ_{approx}. In contrast, if we have a light constraint, or a small regularization parameter, then H should be more rich and covers more hypotheses, which increases the approximation error ϵ_{approx}.

2.7.3 Trade-off Between Estimation and Approximation Error

Combining the analysis in Sections 2.6 and 2.7, it is easy to see there exists a trade-off between the estimation error and approximation error regarding the determination of H. If we have an excessively rich H, it can increase the estimation error, as shown in Section 2.6.1, and decrease the approximation error, as shown in Section 2.7. Figure 2.1 illustrates this trade-off by considering the different values of λ, which correspond to different constraints on H, showed by circles in figure.

Specifically, λ_1, λ_2, λ_3 are three regularization hyper-paremters with $\lambda_1 > \lambda_2 > \lambda_3$. Section 2.7.2 shows the equivalence between regularized and constrained problem, so the domain size associated with λ_1 is the smallest and the domain size associated with λ_3 is the largest. Given fixed current hypothesis f and Bayes hypothesis f_{Bayes}, tuning λ changes H, thus also changes $f^* = \arg\min_{f \in H} R(f)$, e.g. the population-level optimal solution in considered hypothesis class H.

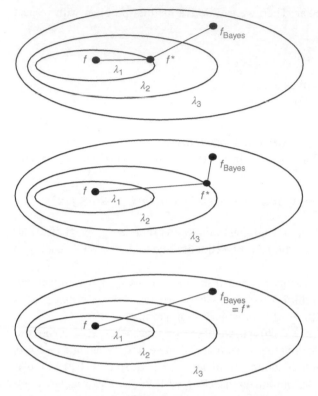

Figure 2.1 Trade-off between estimation and approximation error. Left: we consider the smallest hypothesis class associated with λ_1. Middle: we consider the hypothesis class associated with λ_2. Right: we consider the largest hypothesis class that covers f_{Bayes}, leading to $\epsilon_{\text{approx}} = 0$, where we reach the smallest ϵ_{excess}. However, larger H always means larger computational burden. It also involves the trade-off of excess risk against computational cost.

2.8 Maximum Likelihood Principle

In Section 2.6.3, we briefly mention several loss functions such as square loss and hinge loss for building a machine learning model. There can be lots of options for loss functions depending on the demands of applications. In this section, we introduce *maximum likelihood principle*, a general framework for learning the underlying conditional distribution (introduced later). It covers a large variety of machine learning loss functions, especially in many modern machine learning models such as deep neural networks (Goodfellow et al. 2016).

We first formally define the problem that maximum likelihood principle is used for density estimation. Let $S = \{x_1, ..., x_n\} \sim D^n$ be a set of random variables that are drawn iid from an underlying distribution D. Given a set of distributions, denoted by P, we aim to select one distribution from P that can best align with the observed realization of the set of random variables S. To this end, a good criterion is required to measure the closeness of the selected distribution from P to the underlying distribution D. Maximum likelihood principle provides such criterion to measure the distance between the modeled distribution and D.

2.8.1 Maximum Likelihood Estimation

Since the underlying distribution D is not available, we can only measure the closeness using the observed S. A maximum likelihood estimator can be identified by solving the following problem:

$$p_{\text{ML}} = \arg\max_{p \in P} \Pi_{i=1}^{n} p(x_i) \tag{2.16}$$

where $\Pi_{i=1}^{n} p(x_i) = p(x_1) \cdot p(x_2) \cdots p(x_n)$, meaning the probability for observing $x_1, ..., x_n$ simultaneouly with the modeled distribution p.

For an easier way to solve the aforementioned problem, a log-likelihood formulation is usually used as follows:

$$p_{\text{ML}} = \arg\max_{p \in P} \sum_{i=1}^{n} \log(p(x_i))$$

which gives the exact same solution with (2.16).

Maximum likelihood estimation is also usually related to relative entropy (Kullback–Leibler divergence or KL divergence for short) and cross entropy. Next, we introduce these two concepts and show their connection to maximum likelihood principle.

Definition 2.10   (KL Divergence for Discrete Probability Distributions)
For two discrete probability distributions, P and Q defined over space X, the KL divergence or relative entropy from Q to P is defined as follows:

$$D_{\text{KL}}(P \mid\mid Q) = \sum_{x \in X} P(x) \log(P(x)/Q(x)) \tag{2.17}$$

KL divergence can be viewed as a general distance function measuring the difference between two distributions P and Q. In the context of maximum likelihood estimation, we use KL divergence to measure the closeness between the modeled distribution p to be learned and the empirical distribution p_{emp}, where $p_{\text{emp}}(x_i) = 1/n$ for $i = 1, ..., n$. In this way, we can have a measure of closeness between p and p_{emp}. To see why we need to consider p_{emp}, we can track the information flow as follows:

$$\underbrace{p_{\text{underlying}}}_{\text{unavailable}} \xrightarrow{\text{observe}} \underbrace{p_{\text{emp}}}_{\text{available}} \xleftarrow{\text{approximate}} \underbrace{\text{PML}}_{\text{to learn}}$$

If we would like to make p approximates p_{emp}, simply minimizing the KL divergence between p and p_{emp} over p can be a reasonable approach:

$$p_{\text{KL}} = \arg\min_{p \in P} D_{\text{KL}}\left(p_{\text{emp}} \mid\mid p\right)$$

The final solution of minimizing the preceding KL divergence between p and p_{emp} is equivalent to that of maximizing likelihood, e.g. $p_{\text{KL}} = p_{\text{ML}}$. To see this, we can start the following development on the KL divergence of p_{emp} to p:

$$D_{\mathrm{KL}}\left(p_{\mathrm{emp}} \middle\| p\right) = \sum_{x \in X} p_{\mathrm{emp}}(x) \log\left(p_{\mathrm{emp}}(x)\right) - \sum_{x \in X} p_{\mathrm{emp}}(x) \log\left(p(x)\right)$$

$$= \sum_{x \in X} p_{\mathrm{emp}}(x) \log\left(p_{\mathrm{emp}}(x)\right) - \sum_{x \in X} \sum_{i=1}^{n} \mathbb{I}_{x=x_i} \cdot \frac{1}{n} \cdot \log\left(p(x)\right) \qquad (2.18)$$

$$= \sum_{x \in X} p_{\mathrm{emp}}(x) \log\left(p_{\mathrm{emp}}(x)\right) - \sum_{i=1}^{n} \frac{1}{n} \cdot \log\left(p(x)\right)$$

When minimizing the aforementioned $D_{\mathrm{KL}}(p_{\mathrm{emp}} \| p)$ over $p \in P$ since only the second term involves p, we have the following relation:

$$p_{\mathrm{KL}} = \arg\min_{p \in P} D_{\mathrm{KL}}\left(p_{\mathrm{emp}} \middle\| p\right) = \arg\min_{p \in P} - \sum_{i=1}^{n} \frac{1}{n} \cdot \log\left(p(x)\right)$$

$$= \arg\max_{p \in P} \sum_{i=1}^{n} \log\left(p(x)\right) = p_{\mathrm{ML}} \qquad (2.19)$$

Thus, we show the equivalence between KL-divergence minimization and maximum likelihood.

2.8.2 Cross Entropy Minimization

Cross entropy is often used as the measure of the quality of learning models during training Goodfellow et al. (2016). Next we give its definition and indicate its connection to KL divergence and maximum likelihood principle.

Definition 2.11 (Cross Entropy)
For two discrete probability distributions P and Q defined over space X, the cross entropy of P relative to Q is defined as:

$$H(P, Q) = -\sum_{x \in X} P(x) \log\left(Q(x)\right)$$

Let us consider the relation between $H(P, Q)$ and $D_{\mathrm{KL}}(P, Q)$ by some simple development, e.g. a more general way to decompose KL divergence compared with (2.18), as follows:

$$D_{\mathrm{KL}}\left(P \middle\| Q\right) = \sum_{x \in X} P(x) \log\left(P(x)/Q(x)\right)$$

$$= \sum_{x \in X} P(x) \left(\log\left(P(x)\right) - \log\left(Q(x)\right)\right)$$

$$= \sum_{x \in X} P(x) \log\left(P(x)\right) - \sum_{x \in X} P(x) \log\left(Q(x)\right)$$

$$= \sum_{x \in X} P(x) \log\left(P(x)\right) + H(P, Q)$$

The negative of first term in the last expression, e.g. $-\sum_{x \in X} P(x) \, \log(P(x))$, is called *entropy* of P, denoted by $H(P)$. As a result, for two distributions P and Q, we have the relation of entropy of P, cross entropy of P to Q, KL divergence of P to Q as:

$$D_{\mathrm{KL}}(P, Q) = -H(P) + H(P, Q)$$

A highlight here is that both cross entropy and KL divergence are asymmetric. When P is fixed and Q is a variable to be determined, we have:

$$\arg\min_{Q} D_{\mathrm{KL}}(P, Q) = \arg\min_{Q} H(P, Q)$$

Again, if we set $P = p_{\mathrm{emp}}$ and $Q = p$ that is the model to be learned, then minimizing $D_{\mathrm{KL}}(p_{\mathrm{emp}} \parallel p)$ over p is equivalent to minimizing $H(p_{\mathrm{emp}}, p)$ over p. As a result, we show the equivalence of KL divergence minimization, cross entropy minimization, and maximum likelihood maximization.

2.9 Optimization

After understanding what can be learned (e.g. PAC learning in Section 2.4) and how to use training data to construct a problem to learn (e.g. ERM in Section 2.6), we need the last main step, *optimization*, which is a process determining the parameters in machine learning model. In the context of ERM in Section 2.6, optimization algorithms aim to derive $f_S^{\mathrm{ERM}} = \arg\min_{f \in H} \hat{R}_S(f)$, e.g. the empirical risk minimizer. Even though we can establish an ERM problem for learning the model, we may not derive the exact ERM solution f_S^{ERM} in most practical cases. Instead, we can only achieve an approximation to f_S^{ERM} with a gap ϵ_{opt}, e.g. $\hat{R}_S(f) - \hat{R}_S(f_S^{\mathrm{ERM}}) \leq \epsilon_{\mathrm{opt}}$, where $\epsilon_{\mathrm{opt}} > 0$. This ϵ_{opt} can be called optimization error, or suboptimality.

In this case, the analysis of $\epsilon_{\mathrm{est}}(f_S^{\mathrm{ERM}})$ using Proposition 2.1 and Theorem 2.3 are insufficient, since f_S^{ERM} is not available. Instead, for an optimization solution f such that $\hat{R}_S(f) - \hat{R}_S(f_S^{\mathrm{ERM}}) \leq \epsilon_{\mathrm{opt}}$, we need to analyze $\epsilon_{\mathrm{est}}(f) = R(f) - \inf_{f \in H} R(f)$. As a result, we can elaborate the estimation error as follows:

$$\epsilon_{\mathrm{est}}(f) = R(f) - \inf_{f \in H} R(f) = \underbrace{R(f) - \hat{R}_S(f)}_{\text{Theorem 2.3}} + \underbrace{\hat{R}_S(f) - \hat{R}_S(f_S^{\mathrm{ERM}})}_{\text{optimization error}} + \underbrace{\hat{R}_S(f_S^{\mathrm{ERM}}) - R(f_S^{\mathrm{ERM}})}_{\text{Theorem 2.3}}$$
$$+ \underbrace{R(f_S^{\mathrm{ERM}}) - \inf_{f \in H} R(f)}_{\text{Proposition 2.1, \quad Theorem 2.3}}$$

$$(2.20)$$

Theorem 2.3 gives an upper bound of the difference between $R(f)$ and $\hat{R}_S(f)$ for any $f \in H$ in $\tilde{O}\left(\sqrt{d/n}\right)$, as shown in Remark 2.6. Therefore, the first, third, and last terms in (2.20) can be bounded by Theorem 2.3 (or together with Proposition 2.1).

Due to the presence of such generalization bound in $\tilde{O}\left(\sqrt{d_{vc}/n}\right)$, even if we can only derive a solution with ϵ_{opt} optimization error, the overall quantity of ϵ_{est} still stays in $\tilde{O}\left(\sqrt{d_{vc}/n}\right)$, as long as we can guarantee:

$$\epsilon_{opt} \leq \tilde{O}\left(\sqrt{d_{vc}/n}\right) \tag{2.21}$$

This justifies the reason why an approximation solution f that is sufficiently close to f_S^{ERM} can also work for machine learning models satisfying agnostic PAC learning.

In the following subsections, we consider a simple example of linear regression with square loss as an illustration of optimization algorithms. We show three different optimization algorithms to solve the concerned problem, the closed-form solution, gradient descent (GD), and SGD, respectively.

2.9.1 Linear Regression: An Example

We consider an RLM problem (recall (2.13) in Section 2.7.1) that constructs a linear model, uses square loss on training samples as performance measure, and applies Tikhonov regularization (see Section 2.7.1). Particularly, suppose the training set $S = ((x_1, y_1), ..., (x_n, y_n))$ and $x_i \in \mathbb{R}^d$ where d denotes the dimensionality of features. A linear model is represented by $f(x) = w^\top x$ parameterized by w. A square loss is denoted by $\ell_{\text{square}}(f, (x, y)) = (f(x) - y)^2 = (w^\top x - y)^2$. Tikhonov regularization, or $L2$ regularization, can be written as $R(w) = \|w\|_2^2$. Then the resulting machine learning model can be learned as follows:

$$w^* = \underset{w \in \mathbb{R}^d}{\arg\min} \left\{ F(w) := \frac{1}{n} \sum_{i=1}^{n} \left(w^\top x_i - y_i\right)^2 + \lambda \|w\|_2^2 \right\} \tag{2.22}$$

Optimization is the process that identifies or approximates the value of w^*. There can be multiple approaches to solving the aforementioned optimization problem. Next, we first introduce a closed-form solution by using simple linear algebra.

2.9.2 Closed-form Solution

A closed-form solution is a single equation that solves a problem exactly (e.g. $\epsilon_{opt} = 0$). Given the problem in (2.22), we have a closed-form solution. To this end, we calculate the gradient of $F(w)$ as follows:

$$\nabla F(w) = \frac{2}{n} \sum_{i=1}^{n} (w^\top x_i - y_i) x_i + 2\lambda w = 2\left(\lambda \mathbf{I}_n + \frac{1}{n} \sum_{i=1}^{n} x_i x_i^\top \right) w - \frac{1}{n} \sum_{i=1}^{n} y_i x_i \tag{2.23}$$

where $\mathbf{I}_n$ denotes an identity matrix of $n \times n$.

Since $F(w)$ is a convex function, setting $\nabla F(w) = 0$, we can derive the optimal solution w^* as follows:

$$w^* = \left(\lambda \mathbf{I}_n + \frac{1}{n} \sum_{i=1}^{n} x_i x_i^\top \right)^{-1} \frac{1}{2n} \sum_{i=1}^{n} y_i x_i \tag{2.24}$$

For some simple problems, such as the preceding RLM problem with linear model structure, it is possible to have a closed-form solution, which is easy to derive. However, the computational cost can be very high. For example, (2.24) involves an inversion of a matrix $\lambda \mathbf{I}_n + \frac{1}{n}\sum_{i=1}^{n} x_i x_i^{\top}$, which is of size $n \times n$. The computational complexity for matrix inversion of size n is generally $O(n^3)$ (some optimized algorithms can achieve better complexity such as $O(n^{2.373})$, but we only use the general one as baseline). When n is large, e.g. 1 million training samples like ImageNet in Deng et al. (2009), this complexity leads to unaffordable computational cost.

On the other hand, the preceding example of regularized linear regression only involves simple and special form of objective function (linear model, square loss function, and $L2$ regularization). If the problem involves a more general and complicated objective function, a closed-form solution may not be accessible. For example, support vector machines (SVMs) with hinge loss function do not have closed-form solutions Cortes and Vapnik (1995). Other common models, such as logistic regression (for binary classification), softmax regression (for multiclass classification), and deep neural network models, do not have closed-form solutions. Consequently, in order to solve those general problems, optimization algorithms are necessary.

In the following two subsections, we particularly introduce two simple and commonly used methods in machine learning, GD and SGD.

2.9.3 Gradient Descent

Gradient descent (GD) is an iterative algorithm that has only easy updates and affordable computational cost at each iteration. In GD, updates based on gradients are iteratively performed until some stopping condition is satisfied to ensure $\epsilon_{\text{opt}} \leq \tilde{O}\left(\sqrt{d_{\text{vc}}/n}\right)$, as in (2.21). In the following, we first discuss how to update iteratively using gradients, and then consider the appropriate stopping condition for GD in the sense of convergence rate analysis.

First of all, we introduce the useful notations of gradients. For differentiable function F, we denote its gradients at w by $\nabla F(w)$ (see Section 7.D in Rockafellar and Wets (2009) for detailed definition of gradients). For non-differentiable function F, we assume its subgradients are available, denoted by $\partial F(w)$ (see Definition 8.3 in Section 8.B of Rockafellar and Wets (2009) for detailed definition of subgradients and Exercise 8.8 of Rockafellar and Wets (2009) for relation between gradients and subgradients).

GD updates. To apply GD to Problem (2.22), the update at the t-th iteration can be done as follows:

$$w_t = w_{t-1} - \eta \nabla F(w_{t-1}), \text{ for } 1 \leq t \leq T \tag{2.25}$$

where T is the total number of iterations and w_t is an initialized solution.

GD convergence. Suppose w^* is the optimal solution. The convergence rate analysis is to build a connection between ϵ_{opt} and the number of iterations T. Denote such connection by $\epsilon(T)$, which implies a function of T (so depends on T) as follows:

$$F(w_T) - F(w^*) \leq \epsilon(T)$$

To establish a theoretical result of convergence analysis, we need some assumptions for F. Next we list two conditions that are both satisfied by Problem (2.22) and the key to the analysis.

Definition 2.12 (Strongly Convex Function)

A function $f: X \rightarrow \mathbb{R}$ is μ-strongly convex if the following inequality holds for any $x, x' \in X$:

$$\partial f(x')^\top (x - x') + \frac{\mu}{2} \| x - x' \|_2^2 \leq f(x) - f(x')$$

Definition 2.13 (Smooth Function)

A function $f: X \rightarrow \mathbb{R}$ is L-smooth if the following inequality holds for any $x, x' \in X$:

$$\nabla f(x')^\top (x - x') + \frac{L}{2} \| x - x' \|_2^2 \geq f(x) - f(x')$$

For our considered problem in (2.22), the objective function F satisfies both μ-strongly convex and L-smoothness, with $\mu = 2\lambda$ and $L = 2 \| \lambda \mathbf{I}_n + \frac{1}{n} \sum_{i=1}^n x_i x_i^\top \|_2$ (spectral norm). For details on how to derive those two values, see Theorem 2.1.11 and Lemma 2.2.2 in Nesterov et al. (2018), respectively.

Theorem 2.5 *(Convergence rate of strongly convex problems, Theorem 2.1.14, Nesterov (2003)) If F is μ-strongly convex and L-smooth, with a step size $0 < \eta < \frac{2}{\mu + L}$, then GD makes the following inequality hold:*

$$F(w_T) - F^* \leq \frac{L}{2} \left(\frac{L/\mu - 1}{L/\mu + 1} \right)^{2T} \| w_0 - w^* \|^2 \tag{2.26}$$

Remark 2.7 By setting $\alpha = \left(\frac{L/\mu - 1}{L/\mu + 1} \right)^2 < 1$ and $B_0 = L \| w_0 - w^* \|^2 / 2$, the preceding result can be simplified to $F(w_T) - F^* = B_0 \alpha^T$. Note that w^*, L, and μ depends on the objective function F; $\| w_0 - w^* \|$ depends on the initialization. As a result, we can regard both α and B_0 as pre-fixed constants, leaving only T a tunable parameter for the consideration of stopping condition. Particularly, to ensure $F(w_T) - F^* \leq \epsilon_{\text{opt}}$, it requires $B_0 \alpha^T \leq \epsilon_{\text{opt}}$, or equivalently, setting $T \geq \log_\alpha(\epsilon_{\text{opt}}/B_0) = \log_{1/\alpha}(B_0/\epsilon_{\text{opt}})$.

Compared with the closed-form solution of Problem 2.22, GD does not involve matrix inversion, so avoids the $O(n^3)$ computational complexity. Now we analyze the overall computational complexity of GD by assuming some optimization error ϵ_{opt}. We can take two examples of $\epsilon_{\text{opt}} = O(1/n)$ and $\epsilon_{\text{opt}} = O(1/\sqrt{n})$ in the following analysis (both are smaller than $\widetilde{O}\left(\sqrt{d_{\text{vc}}/n} \right)$).

In the first case of $\epsilon_{\text{opt}} = O(1/n)$, as in Remark 2.7, we can set $T \geq \log_{1/\alpha}(B_0/\epsilon_{\text{opt}}) = O(\log(1/\epsilon_{\text{opt}})) = O(\log(n))$. Moreover, note that for each iteration of GD in (2.25), we need to calculate $\nabla F(\backslash w_{t-1})$, which requires to go over all n data samples involved in F, leading to $O(n)$ computational complexity per iteration. Therefore, the overall computational complexity is $O(n \log(n))$, which is dominated by $O(n^3)$ required by closed-form solution.

Next, suppose that we require $\epsilon_{\text{opt}} = O(1/\sqrt{n})$. Using similar development, we set $T \geq O(\log(\sqrt{n}))$. Given that GD requires $O(n)$ per iteration for going over n samples in F, the overall computational complexity is $O(n\log(1/\sqrt{n}))$, which is also dominated by $O(n^3)$. Here we show that GD can be a more efficient way than closed-form solutions to satisfy the requirement of $O(1/n)$ or $O(1/\sqrt{n})$.

2.9.4 Stochastic Gradient Descent

Stochastic gradient descent (SGD) is one of the most used optimization algorithms for training modern machine learning models such as neural networks. When the training data is too large (too large n) to load data into memory entirely, SGD can be very helpful. As GD, SGD is an iterative optimization algorithm, whose update relies on stochastic gradients, instead of gradients, and can be written as follows:

$$w_t = w_{t-1} - \alpha_t \hat{\partial} F(w_{t-1}), \text{ for } 1 \leq t \leq T \tag{2.27}$$

where $\hat{\partial} F(w)$ is the stochastic subgradient of $F(w)$, e.g. $\mathbb{E}\left[\hat{\partial} F(w)\right] \in \partial F(w)$. For example, for the differentiable F defined in (2.22), subgradients reduce to gradients (see Exercise 8.8 in Rockafellar and Wets (2009)), and we derive the gradient of (2.23) as $\nabla F(w)$, while its stochastic gradient $\hat{\nabla} F$ can be written as follows:

$$\hat{\nabla} F(w) = 2\left(\lambda \mathbf{I}_n + x_i x_i^\top\right)w - y_i x_i$$

where i is a randomly sampled index from $1, \ldots, n$. It can be verified that $\mathbb{E}\left[\hat{\nabla} F(w)\right] = \nabla F(w)$.

In the following, we elaborate the convergence rate of SGD algorithm to derive computational complexity of SGD, as GD in Section 2.9.3.

Theorem 2.6 (*Optimal convergence rate of last-iterate SGD for strongly convex problems, Theorem 2.2, Jain et al. (2019)*) *Let F be μ-strongly convex and stochastic gradient be G-bounded, e.g. $\| \partial F(w)\| \leq G$ for every w. The sequence of step sizes is determined by $\alpha_t = 2^{-i}/(\mu t), \forall\, T_i < t < T_{i+1}, 0 \leq i \leq k$, where $T_i := T - \lceil T \cdot 2^{-i}\rceil$ and $k := \inf\{i : T \cdot 2^{-i} \leq 1\}$. By running SGD update in (2.27) with step size α_t, the following inequality holds:*

$$\mathbb{E}[F(w_T)] - F(w^*) \leq \frac{130G^2}{\mu T} \tag{2.28}$$

Furthermore, for all $0 < \delta < 1/e$, the following inequality holds with probability at least $1 - \delta$:

$$F(w_T) - F(w^*) \leq O\left(G^2 \log(1/\delta)/(\mu T)\right)$$

Remark 2.8 The preceding result applies a special design of step size α_t for the t-th iteration in a stagewise fashion. Specifically, k is the total number of stages and $T_0 = 0$, $T_1 = T/2$, $T_2 = 3T/4$,. During the i-th stage, the step size α_t is fixed to $2^{-i}/(\mu t)$. Theorem 2.6 provides convergence results for both expected bound and high probability bound. They can both be simplified to $O(1/T)$ by hiding constants to reflect only the factor from T. As the analysis of

GD in Remark 2.7, to make the optimization error bounded by $O(1/n)$ (resp. $O(1/\sqrt{n})$), we set $T = O(n)$ (resp. $O(\sqrt{n})$). On the other hand, the per-iteration computational complexity is in constant order $O(1)$, leading to $O(n)$ (resp. $O(\sqrt{n})$) total computational complexity.

Above we only use GD algorithm from Nesterov (2003) and SGD algorithm from Jain et al. (2019). There are many variants of GD and SGD methods. For example, in Hazan and Kale (2014), the final solution is the average over all solutions from the first iteration to the last iteration and also derives $O(1/(\mu T))$ convergence rate (with kind of different condition as in Theorem 2.6). The takeaway message is that we can satisfy agnostic PAC learning and successfully learn a machine learning model, as long as making the solution from optimization algorithms f sufficiently close to the ERM solution f_S^{ERM}.

References

Chen, T., Kornblith, S., Norouzi, M., and Hinton, G. (2020). A simple framework for contrastive learning of visual representations. International Conference on Machine Learning, pages 1597–1607. PMLR.

Chuang, C-Y., Robinson, J., Lin, Y-C., et al. (2020). Debiased contrastive learning. NeurIPS.

Cortes, C. and Vapnik, V. (1995). Support-vector networks. *Machine Learning* 20 (3): 273–297.

Deng, J., Dong, W., Socher, R., et al. (2009). Imagenet: a large-scale hierarchical image database. 2009 IEEE conference on computer vision and pattern recognition, pages 248–255. IEEE.

Goodfellow, I., Bengio, Y., and Courville, A. (2016). *Deep Learning*. MIT press.

Hazan, E. and Kale, S. (2014). Beyond the regret minimization barrier: optimal algorithms for stochastic strongly convex optimization. *The Journal of Machine Learning Research* 15 (1): 2489–2512.

Irvin, J., Rajpurkar, P., Ko, M., et al. (2019). Chexpert: a large chest radiograph dataset with uncertainty labels and expert comparison. Proceedings of the AAAI conference on artificial intelligence, volume 33, pages 590–597.

Jain, P., Nagaraj, D., and Netrapalli, P. (2019). Making the last iterate of sgd information theoretically optimal. Conference on Learning Theory, pages 1752–1755. PMLR.

Mohri, M., Rostamizadeh, A., and Talwalkar, A. (2018). *Foundations of Machine Learning*. MIT press.

Nesterov, Y. (2003). *Introductory Lectures on Convex Optimization: A Basic Course*, vol. 87. Springer Science & Business Media.

Nesterov, Y. (2018). *Lectures on Convex Optimization*, vol. 137. Springer.

Rizve, M.N., Duarte, K., Rawat, Y.S., and Shah, M. (2021). In defense of pseudo-labeling: an uncertainty-aware pseudo-label selection framework for semi-supervised learning. International Conference on Learning Representations.

Rockafellar, R.T. and Wets, R.J.-B. (2009). *Variational Analysis*, vol. 317. Springer Science & Business Media.

Shalev-Shwartz, S. and Ben-David, S. (2014). *Understanding machine learning: From theory to algorithms*. Cambridge University Press.

Wei, Y., Shen, Z., Cheng, B., et al. (2018). Ts2c: Tight box mining with surrounding segmentation context for weakly supervised object detection. Proceedings of the European Conference on Computer Vision (ECCV), pages 434–450.

3

Flammability Characteristics Prediction Using QSPR Modeling

Yong Pan and Juncheng Jiang

College of Safety Science and Engineering, Nanjing Tech University, Nanjing, China

3.1 Introduction

3.1.1 Flammability Characteristics

With the development of our society, the number and diversity of hazardous chemicals involved in our lives is increasing. Although the chemical industry puts great power to their development, hazardous chemicals are a danger to us. One of the main hazards of chemicals is flammability characteristics.

This section focuses on the combustion/flammability characteristics of chemical substances. The main physical and chemical parameters include flash point (FP), auto-ignition temperature (AIT), heat of combustion, minimum ignition energy, upper limit of combustion, lower limit of combustion, self-accelerating decomposition temperature, impact sensitivity, and electrostatic sensitivity. The FP, AIT, and heat of combustion mainly characterize the combustibility of chemical substances. The explosion limit, impact sensitivity, and electrostatic sensitivity mainly characterize the explosive properties of chemical substances. These parameters are important indicators of the hazardous properties and can characterize the probability and severity of the risk of combustion and explosion during the production, storage, and transport of hazardous chemical substances.

Due to the tremendous amount of chemicals, it is time and cost consuming to obtain these property data from experiments. However, quantitative structure–property relationship (QSPR) as a method for studying the properties of substances has the advantage of being efficient, risk-free, low cost, and so on. These qualities make it a perfect choice for studying the flammability characteristics of chemical substances. The following section describes the systematic study of QSPR modeling on the quantitative relationships between these indicators and molecular structure to achieve an in-depth and systematic understanding of the hazardous properties of chemical substances.

Machine Learning in Chemical Safety and Health: Fundamentals with Applications, First Edition.
Edited by Qingsheng Wang and Changjie Cai.
© 2023 John Wiley & Sons Ltd. Published 2023 by John Wiley & Sons Ltd.

3.1.2 QSPR Application

3.1.2.1 Concept of QSPR

QSPR refers to the intrinsic quantitative relationship between the molecular structure and properties of compounds. It is an effective method that enables the prediction of physico-chemical properties of compounds based on molecular structure. The theoretical basis of QSPR studies is the physicochemical properties of compounds attributed to their chemical structures. The molecule is the basic unit of a substance, and structural information such as the characteristics of the molecular inner-structure and the way in which the molecules are combined determines the properties of the compound.

QSPR research aims to explore the intrinsic quantitative relationship between structure and properties of a substance based on the fundamental principle that the structure of a substance determines its properties and the properties of a substance reflect its structure. All that is required for QSPR modeling are the structural information of the molecule and the experimental data of its properties. Through statistical analysis of the various structural parameters and experimental data of molecule properties, a quantitative relationship between the structure parameters and the properties of the compound from the experiments can be established. Once a reliable QSPR model was established, it can be used to predict various properties of new or not-yet synthesized compounds. QSPR studies allow for the discovery and identification of the structural factors that determine the various properties of a compound. In other words, it leads to a deep understanding of the influence of the microstructure of a compound on various macroscopic properties at the molecular level, which can provide guidance for molecular design and prevent the potential hazards to our daily life.

3.1.2.2 Trends and Characteristics of QSPR

QSPR study is constantly evolving to become more and more sophisticated and reliable, indicating that it has great potential for improvements in the near future. Looking over the development history of QSPR study, it reveals three significant trends:

1) Goal-oriented and application-oriented: In the field of safety science and engineering, QSPR research is mainly focused on the objective of mechanistic explanation or model prediction of organic hazard characteristics (Gao et al. 2020; Jiang et al. 2020).
2) Multidisciplinary: QSPR research covers multidisciplinary research fields, such as cheminformatics (chemometrics, computational chemistry), physical chemistry, biochemistry, toxicology, combustion science, explosion dynamics, and computer science mathematics, which reflect more and more multidisciplinary integration (Liu and Long 2009; Jiao et al. 2019; Awfa et al. 2021).

    The molecular structure characterization was developed from empirical molecular structure descriptors such as hydrophobicity constants and electronic effect constants, which were measured experimentally in the early days. Nowadays theoretical molecular structure descriptors, such as topological parameters and quantum chemical parameters are widely used. The development reflects the integration of mathematics, molecular topology, quantum chemistry, and computer numerical computation, which makes the characterization of molecular structure more detailed and comprehensive and provides a good basis for the successful establishment of QSPR models.

For the model establishment, it was developed from the linear regression analysis techniques at the beginning, then shifted to the comprehensive utilization of various multivariate analysis methods. Recently, nonlinear modeling techniques such as artificial neural networks (ANNs), support vector machines (SVM), and random trees have been applied for QSPR study. Combinatorial algorithms such as GA-MLR (genetic algorithm-multiple linear regression) and GA-SVM have also been developed based on the single algorithms. The application of these methods reflects the intersection of multiple disciplines and promotes the continuous improvement of modeling techniques.

3) Intelligence: With the development of computer technology, QSPR application software with various features has been appearing. The total number of the software is conservatively estimated to be over 200. The achievements of mathematics, computer graphics science, and other related sciences strongly contribute to the continuous development of QSPR research in direction from descriptive to inferential, from qualitative to quantitative, and from macro-state to micro-structure (Begam and Kumar 2016).

3.2 Flowchart for Flammability Characteristics Prediction

The theoretical basis of QSPR is that a chemical's characteristics depend on its molecular structure. A typical solution for QSPR problem is shown in Figure 3.1.

As can be seen from Figure 3.1, although there is an inherent relationship between molecular structure and property, there is a lack of a prior empirical method to directly describe this relationship. Therefore, an indirect approach is needed to solve this problem, namely, to establish a relationship model between quantitative structure and property.

In this section, according to the basic principles of QSPR research, based on the newly established combinatorial algorithms such as GA-MLR and GA-SVM, QSPR study process is set up for the flammability characteristics of organic compounds. The process includes: (i) sample set; (ii) structure input and molecular simulation; (iii) calculation of molecular descriptors; (iv) preliminary screening of molecular descriptors; (v) descriptor selection and modeling; (vi) model evaluation and validation; (vii) mechanism explanation of the model. Figure 3.2 shows QSPR study process, and each step is described next.

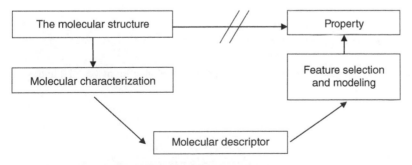

Figure 3.1 Flowchart of solution for QSPR study.

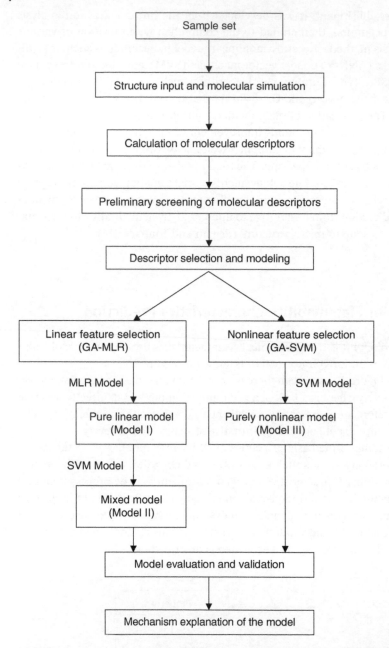

Figure 3.2 Flowchart of QSPR study process based on combinatorial algorithms GA-MLR and GA-SVM.

3.2.1 Dataset Preparation

A reliable prediction model must be based on reliable experimental samples. The premise of QSPR research is to select a data series of organic compound and their properties to constitute the corresponding experimental sample set according to certain statistical and structural criteria. The construction of sample set includes two aspects: compound selection and experimental data collection.

The criteria for compound selection should follow statistical randomness, structural representativeness and comprehensiveness, and data availability. Therefore, as many compounds as possible should be collected and reasonable distribution of compound structure types should be kept.

Then, the corresponding data of physicochemical properties are collected for the selected compounds. The data selection principle is that it must be reliable and standardized. There are three main ways of data collection: experimental measurement, scientific and technical literature search, and related database search. Experimental data must be obtained from experiments that strictly comply with relevant experimental specifications. The data obtained from scientific literature must conform to the original literature. In order to reduce the data error and obtain the QSPR model with better performance, it is better to use the property data under the same experimental methods and conditions from the same laboratory or through quality evaluation and screening of data with reliable evaluation methods according to uniform standards. If the property data are from different researchers or have not been evaluated, the property data must be analyzed and preliminarily screened before modeling to ensure the reliability of the sample data.

At present, many authoritative organizations or academic institutions have established hazardous chemical databases, and most of the databases provide free online search functions. The reliability of the data is assessed when they are included in most of these databases by checking if the data were obtained in the same way under the closest experimental conditions. The data are constantly updated as needed. For example, the main data of the International Chemical Safety Cards (ICSCs) database co-edited by the International Program on Chemical Safety (IPCS) and the European Union (EU) have been updated since 2000. It has proved to be more reliable than earlier published data and has been used as a major source of reliable data by the United Nations Environment Program (UNEP), the International Labor Organization (ILO), the World Health Organization (WHO), and the EU.

Therefore, in order to eliminate the influence on the prediction model from the differences between experimental data to the maximum extent, the main authoritative database can be selected as the main source of flammability characteristics to ensure the accuracy and authority of the experimental data. The main databases are as follows:

1) DIPPR® database of the American Institute of Chemical Engineers (AIChE)
2) International Chemical Safety Cards (ICSCs) of the International Program on Chemical Safety (IPCS)
3) Hazardous Chemical Database of Oxford University, United Kingdom (Chemical and Other Safety Information)
4) Hazardous Chemical Database of the University of Akron (The Chemical Database)

3.2.2 Structure Input and Molecular Simulation

One of the first task in QSPR research is the description of molecular structure; that is, selecting different molecular descriptors to reflect the structural characteristics of molecules. The calculation of molecular descriptors must be based on the stable two-dimensional or three-dimensional molecular model to obtain various microstructure information such as atomic composition, inter-atomic connection, atomic space coordinates, and atomic charge.

At present, researchers mainly use molecular simulation techniques based on molecular mechanics and quantum mechanical methods to obtain the optimal molecular conformation model.

1) Molecular mechanics method

The method of molecular mechanics is also called the force field method, which takes the minimum energy as the objective function and optimizes the initial molecular model through the calculation of molecular force field. Molecular mechanics considers a molecule as a "field" of interactions between its atoms. Atomic interactions fall into two categories: bonding and non-bonding interactions. The former includes bond stretching, bond angle stretching, bond torsion, dihedral angle stretching, etc. The latter includes electrostatic action and van der Waals action. These intramolecular interactions are expressed as functions of intramolecular coordinates in energy terms. These functions contain a series of parameters obtained by fitting experimental data of compounds. Because of the detailed classification of atomic types, these parameters are generally portable and can be widely used in molecular mechanics calculations for other substances.

Compared with quantum mechanics, the molecular mechanics method is simple and fast, especially for the simulation of large molecules. Among them, the force fields commonly used for small molecules include MM2/MM3/MM4, TINKER, UFF, MOMEC, COSMOS, etc., and those for biomacromolecules include AMBER, GROMOS, OPLS, ECEPP, MMFF, etc.

2) Quantum mechanical method

The theory of quantum mechanics regards molecules as combinations of basic particles (nucleus and electrons) with energy and motion states. The motion state of each particle is represented by a wave function. The particle energy and wave function satisfy the Schrödinger equation. Given the initial molecular geometry structure, the optimal molecular model can be obtained by quantum mechanical calculation according to the minimum molecular energy standard.

The core of the quantum mechanics method is to solve the Schrödinger equation. Solving the Schrödinger equation can determine the electronic structure and energy of the molecule, and then other molecular properties such as atomic net charge, dipole moment, polarizability, orbital energy, and local ionization potential can be calculated. The commonly used methods for solving Schrödinger equation include molecular orbital theory, semi-empirical molecular orbital theory, ab initio calculation theory, and density functional theory.

Molecular orbital (MO) theory is derived from the Hückel molecular orbital (HMO) theory proposed by Erich Huckel in 1930. It is a hand calculation method with large approximation degree. In 1963, the extended HMO appeared, and the assumptions of HMO were treated more closely to reality.

The semi-empirical molecular orbital theory calculates electron interactions by mean field theory, self-consistent field theory, and Hartree–Fock theory and compensates for the deviations caused by approximations with adjustable parameters. Based on different assumptions about the center of the atom in the integration process, a series of gradually improved semi-empirical molecular orbital methods were developed, such as CNDO, INDO, and NDDO in the 1960s; MNDO appeared in 1977 and MNDO/H, AM1, and PM3 after modification were developed in 1980 and 1990s; in the 1990s, MNDO/d, AM1(d), AM1, PM3-tm, and PM3(d) methods were obtained based on the introduction of d-orbital system.

The ab initio method starts with the self-consistent field theory and solves the Schrödinger equation by accurately calculating all integrals. This method takes a long time and consumes a lot of computing resources, but the results are accurate and can provide more detailed molecular structure information.

Density functional theory does not calculate molecular orbitals. It relates the energy of isolated electrons to the real energy of molecules and expresses it as a function of electron density. Local density approximations, generalized gradient approximations, and hybrid approximations are functional form assumptions.

3.2.3 Calculation of Molecular Descriptors

Molecular descriptors are the basis of the QSPR model. In principle, the molecular structure descriptors that can explain the properties should be selected. If one can judge in advance which molecular structure descriptors are relevant to the studied properties based on the mechanism or experience, the relevant molecular descriptors can be targeted. However, for most physical and chemical properties, it is often impossible to determine in advance which molecular structure descriptors are closely related to the properties. In this case, as many molecular descriptors as possible should be selected to avoid missing important molecular structure information.

As a lot of molecular descriptors have been developed, there are already a variety of chemical software for calculation. For example, CODESSA software developed by the Katritzky group at the University of Florida can calculate multiple types of descriptors such as molecular composition, geometry, topology, electrostatic forces, quantum chemistry, and thermodynamics. ADAPT, developed by group of Jurs at Pennsylvania State University, can calculate topological, geometric, electrical, and combinatorial descriptors of molecules. In addition, there are some other important computing software, such as MOLCONN-Z software, which is mainly used to calculate molecular topology descriptors; Power MV software, mainly for calculation of composition, molecular fingerprint, BCUT, and other descriptors; and MOE software, which is mainly applied to calculate descriptors such as topology, physical properties, and chemical bonds.

DRAGON software is another widely used software for calculating molecular descriptors. The software was designed and developed by Todeschini et al. from the Stoichiometry and

Table 3.1 The categories and numbers of descriptors calculated by DRAGON 7.

Descriptor type	Descriptor number	Descriptor type	Descriptor number
Constitutional	47	RDF descriptors	210
Ring descriptors	32	3D-MoRSE descriptors	224
Topological indices	75	WHIM descriptors	114
Walk and path counts	46	GETAWAY descriptors	273
Connectivity indices	37	Randic molecular profiles	41
Information indices	50	Functional groups count	154
2D matrix-based descriptors	607	Atom-centered fragments	115
2D autocorrelations	213	Atom-type E-state indices	172
Burden eigenvalues	96	CATS 2D	150
P-VSA-like descriptors	55	2D Atom pairs	1596
ETA indices	23	3D Atom pairs	36
Edge adjacency indices	324	Charge descriptors	15
Geometrical descriptors	38	Molecular properties	20
3D matrix-based descriptors	99	Drug-like indices	28
3D autocorrelations	80	CATS 3D	300

quantitative structure–activity relationship (QSAR) research team in Milan, Italy. It includes multiple functions such as molecule data reading, computing, mapping, and correlation analysis. It can calculate a large number of molecular structure descriptors including topology, composition, electrical properties, and other multi-type structural parameters of molecules that represent the spatial structure of 0–3 dimensional molecules and cover information such as atoms, chemical bond types, connectivity, charge distribution, and atomic space coordinates. Due to the advantages of diverse descriptors and simple operation, DRAGON software has been widely used in QSPR research and accepted by experts in various aspects.

At present, DRAGON software has been updated to version 7.0, which can be obtained from the internet for free and can calculate 5270 kinds of molecular structure descriptors in 18 categories (Table 3.1).

3.2.4 Preliminary Screening of Molecular Descriptors

It can be seen from the Section 3.2.3 that 5270 molecular descriptors can be calculated for each organic compound. For such a large number of descriptors, on the one hand, the molecular structure can be characterized comprehensively, and important molecular structure information will not be missed. On the other hand, it also increases the difficulty of data analysis and model building. Too many variables not only will increase the complexity of calculation but can also lead to "accidental correlation" in modeling, making the model not robust. In addition, many descriptors have high autocorrelation (collinearity) with each other, which will cause internal redundant information. Too many variables will also make it difficult to explain the mechanism of the model.

Therefore, from a statistical and chemical point of view, it is desirable to use as few variables as possible to represent as much structural information as possible. For the purpose, it is necessary to pretreat variables before establishing mathematical models, so as to get rid of descriptors with poor correlation or irrelevant descriptors, so that the basic information is included in the reduced descriptor subset. On this basis, the appropriate variable screening method is used to select the best combination of descriptors to establish the corresponding QSPR model.

The screening steps are listed as follows:

1) Structural parameters that are constants or approximate constants (with little variation) for all compounds are deleted as they cannot effectively characterize the structural differences.
2) For any two structural parameters with a high correlation coefficient, one of them is also deleted because of collinearity, as the prediction results will be worse if the two parameters are introduced into the model at the same time.
3) The parameters with a value of "0" for most samples are also deleted because they do not contain sufficient structural information.
4) Structural parameters that have a poor correlation with the target properties can also be considered for deletion.

After pre-screening, the number of descriptors can be reduced to a certain extent, but it still cannot meet the needs of QSPR modeling, so an effective variable selection method is needed for further screening.

3.2.5 Descriptor Selection and Modeling

For variable selection, the simplest but also most complicated method is the all possible regression method, which is used to establish a regression model with a combination of all subsets of the independent variable and the dependent variable, and then choose the best model. However, for the problem containing M independent variables, there is a $2^m - 1$ combination way, which means $2^m - 1$ regression models will be generated. Furthermore, with the increase of the number of independent variables, the number of models to be screened will increase exponentially. When $m > 100$, it is difficult to filter such a large number of models effectively.

To solve this problem, researchers have proposed many different variable selection methods. Among them, genetic algorithm (GA) is the most widely used as an efficient global optimization probabilistic search method. Based on GA, this chapter developed and improved combinatorial algorithms GA-MLR and GA-SVM by combining GA with MLR method and SVM technology, respectively. On the one hand, these methods can realize the optimization and screening of many molecular structural parameters and find out the structural parameters most closely related to the studied hazardous characteristics. On the other hand, it can quickly establish an efficient QSPR prediction model with the selected structural parameters, so it has a broad application prospect in QSPR research.

For the QSPR study on the flammability and explosion characteristics of substances in this chapter, the improved GA-MLR algorithm is first used to simulate the linear relationship between the flammability and explosion characteristics and their molecular structure, a set of optimal structural parameters (molecular descriptors) that can best characterize the linear relationship found out, and the optimal linear QSPR prediction model is established accordingly. Obviously, this model is a pure linear model based on the linear variable selection method (GA-MLR) and linear modeling method (MLR) and can effectively characterize the linear relationship between the flammability characteristics and molecular structure, which is defined as Model I (Figure 3.2).

Then, in order to study the possible nonlinear relationship between the characteristics and the molecular structure, a new QSPR model was established by using the nonlinear SVM method. The molecular descriptor selected in Model I was used as the input variable and the corresponding flammability characteristics as the output variable. Based on the linear variable selection method (GA-MLR) and nonlinear modeling method (SVM), this model is a hybrid model (GA+SVM), defined as Model II. In general, due to the strong data fitting ability of SVM method, the prediction accuracy of Model II is generally better than that of Model I when the input variables are the same.

However, we know that the best descriptor screened by the linear variable selection method GA-MLR is normally not the best choice for nonlinear fitting, and the nonlinear relationship between the characteristics and its molecular structure can be more reasonably characterized only by the descriptors screened with the nonlinear method. Therefore, the newly established GA-SVM algorithm was used to simulate the nonlinear relationship between the flammability and explosion characteristics of the organic compound and its molecular structure, aiming to find the best set of structural parameters (molecular descriptors) that can best characterize the nonlinear relationship, and to establish the corresponding optimal nonlinear QSPR prediction model. Based on the nonlinear variable selection method (GA-SVM) and nonlinear modeling method (SVM), this model is a pure nonlinear model, which can effectively characterize the nonlinear relationship between the characteristics and their molecular structure, defined as Model III.

By establishing the QSPR models with the preceding three different properties, this chapter attempts to conduct a more comprehensive study on the intrinsic quantitative relationship between the flammability and explosion characteristics and the molecular structure of substances and provides the corresponding prediction models for selection. The main features of the three models are summarized in Table 3.2.

Table 3.2 Features of the three model types.

Model	Variable selection	Modeling	Model property	Application method
I	Linear	Linear	Pure linear	GA-MLR
II	Linear	Nonlinear	Hybrid	GA+SVM
III	Nonlinear	Nonlinear	Pure nonlinear	GA-SVM

3.2.6 Model Validation

In 2002, at an international conference in Setubal, Portugal, researchers presented several rules for the validity of the QSAR/QSPR model, known as the "Setubal Principles." In November 2004, the Organization for Economic Cooperation and Development (OECD) further amended these rules and officially named them "OECD Principles." In February 2007, the OECD announced guidance on the verification and validation QSAR model.

According to OECD Principles, a QSAR/QSPR model must meet five conditions for the regulatory purposes. Article 4 specifies that there must be appropriate measures of goodness-of-fit, robustness, and predictivity, which means that before a QSAR/QSPR model is applied, these capabilities must be evaluated and verified comprehensively. And this is also the most-revised article from the "Setubal Principles" to the "OECD Principles."

Existing studies have shown that traditional QSPR studies on the hazardous characteristics of organic compounds often evaluate one or two of the "fitting ability, stability, and predictivity" of the model, which is lacking in terms of comprehensive and effective verification of the model. Therefore, based on the "OECD Principles," this chapter intends to comprehensively evaluate and validate the established QSPR model from three aspects: model fitting ability, stability, and predictivity.

3.2.6.1 Model Fitting Ability Evaluation

The evaluation of model fitting ability is mainly used to indicate the ability of the model to explain the changes of the training set. The following statistical evaluation indicators are described to evaluate the model fitting ability:

1) Correlation coefficient (R)

   The correlation coefficient represents the closeness of the linear relationship between the predicted value and the target value, and its value ranges from 0 to 1. If the value is closer to 1, the linear relationship is stronger, and vice versa.

2) Determination coefficient (R^2)

   The determination coefficient, also known as the multiple correlation coefficient, is an important index to determine the goodness of fit of linear regression, which is defined as:

$$R^2 = \frac{\text{Explained variation}}{\text{Total variation}} = \frac{\sum (\hat{y}_i - \bar{y})^2}{\sum (y_i - \bar{y})^2} = \frac{\sum_{i=1}^{n}(Y_i - \overline{Y})^2 - \sum_{i=1}^{n}(Y_i - \hat{Y})^2}{\sum_{i=1}^{n}(Y_i - \overline{Y})^2}$$

$$(3.1)$$

As can be seen from the preceding equation, the coefficient of determination is equal to the proportion of the sum of squares of the regression in the total sum of squares, so it represents the percentage of dependent variation that can be explained by the regression model. For example, $R^2 = 0.825$ indicates that 82.5% of the dependent variable variation is caused by the independent variable. $R^2 = 1$ indicates that all observation points fall on the regression equation. $R^2 = 0$ indicates that there is no linear relationship between the independent variable and the dependent variable.

3) Root mean square error (RMSE), average absolute error (AAE), absolute percentage error (APE), and standard error (SE)

RMSE represents the dispersion degree of random error, which is defined as:

$$RMSE = \sqrt{\frac{\sum_{i=1}^{n}\left(y_{i,\text{pred}} - y_{i,\text{obs}}\right)^2}{n}} \tag{3.2}$$

The AAE represents the difference between the predicted value and the target value, defined as:

$$AAE = \frac{\sum_{i=1}^{n}\left|y_{i,\text{pred}} - y_{i,\text{obs}}\right|}{n} \tag{3.3}$$

The APE is defined as

$$APE = \frac{\sum_{i=1}^{n}\left|\left(\left(y_{i,\text{pred}} - y_{i,\text{obs}}\right)/y_{i,\text{obs}}\right)\right|}{n} \tag{3.4}$$

SE refers to the SE of the fitting value, which is defined as:

$$SE = \sqrt{\frac{\sum_{i=1}^{n}\left(y_{i,\text{pred}} - y_{i,\text{obs}}\right)^2}{n-1}} \tag{3.5}$$

In the preceding equations, n is the number of experimental samples, and $y_{i,\text{pred}}$ and $y_{i,\text{obs}}$ are the predicted value and target value of samples, respectively. These parameters are commonly used to measure model accuracy. They depend on the range and distribution of dependent data and are affected by "delocalization points."

4) F-test

F-test is a method to test whether the linear relationship between independent variables and dependent variables is significant, which is applicable to models established based on the MLR method. It compares the sum of squares of regression deviations with the sum of squares of residual deviations to check if the difference between them is significant. If it is significant, it indicates a linear relationship between the two variables, and vice versa.

The preceding evaluation indices can show the fitting ability of the model well. However, for the two kinds of problems that often appear in QSPR research—"underfitting" and "overfitting"—they cannot be effectively used. The "underfitting" means that the model does not fully reveal the variable information contained in the sample set, resulting in lower predictivity of the model. This kind of problem often occurs in models obtained by using linear modeling methods. The "overfitting" means that the fitting degree of the model is higher than the variability of the combination of property data and descriptors due to the fitting of error information. Such problems often occur in models obtained by using nonlinear modeling methods. Because these two kinds of problems are closely related to the stability of the model, the identification of these problems often needs to be solved by the stability analysis of the model.

3.2.6.2 Model Stability Analysis

The model stability analysis is actually an analysis of the "instability" of the model. The "instability" of the model means that the model is relatively influenced by some individual compounds or a subset of compounds in the training set. If the predicted value of the substance exceeds the confidence interval of the model, the model will be unstable.

At present, there is little literature on direct quantitative analysis of model instability. Model instability is more commonly studied by internal validation, as any internal validation technique can evaluate the instability of the model to some extent. Therefore, the following internal validation techniques are used to study the stability of the model:

1) "Leave-many-out"(LMO) cross validation

"Leave-many-out" (LMO) cross validation is a common internal validation technique. In this method, n samples in the initial training set are divided into $G(=n/m)$ subsets with the size of m on average, then m data points are removed each time, and the remaining n-m samples are used as the training set to remodel and validate the test set composed of m samples. After G times of calculation, the cross-validation coefficient Q^2_{CV} is obtained to represent the stability and internal predictivity of the model. Generally, if Q^2_{CV} is greater than 0.5, the model is stable. If it is greater than 0.9, the stability of the model is excellent. The calculation equation of Q^2_{CV} is as follows:

$$Q^2_{CV} - 1 - \frac{\sum_{i=1}^{training}(y_i - \hat{y}_i)^2}{\sum_{i=1}^{training}(y_i - \bar{y})^2} \tag{3.6}$$

where y_i $\hat{y}_i$, and $\bar{y}$ represent the experimental value, predicted value, and average experimental value of the training sample, respectively.

2) "Leave-one-out" (LOO) cross validation

"Leave-one-out" (LOO) cross validation is a special "Leave-many-out" cross-validation method. Its process is similar to "Leave-many-out" cross-validation method except that $m = 1$. Since the "Leave-one-out" method uses all the sample data, it is also the most economical method. Although some researchers pointed out that the results of cross validation of "Leave-one-out" often overestimate the predictivity of the model, it is still indispensable in QSPR research, especially for studying systems with small samples.

3) Y-randomness test

The Y-randomness test is also a statistical method widely used to characterize the stability of models. It randomly scrambles the dependent variable Y of the original sample set and combines it with the original independent variable to form a new sample set to establish the model. Then the aforementioned process is repeated 50~100 times, and the results are compared with the results of the original sample set. If the performance of the prediction model in the original sample set is significantly better than those in the new sample set, it is considered that there is a real QSPR relationship in the original sample data, the model is stable, and there is no "accidental correlation" phenomenon. Otherwise, it indicates that the original model is not acceptable.

4) Residual diagram analysis

Residual diagram analysis method refers to drawing the scatter diagram of residuals with the predicted value as the horizontal axis and the residual as the vertical axis in the rectangular coordinate system. If the scattered points show obvious regularity, the model is considered to have autocorrelation, nonlinear, or non-constant variance problems. If the scatter points are randomly distributed, the model is considered to be appropriate, and no systematic error is generated in the process of model establishment.

Internal validation is an essential step in the modeling process of QSPR, but good internal validation results only indicate that the model has high stability or strong internal predictivity and cannot guarantee the model has predictivity for external samples. Therefore, it is necessary to evaluate the extrapolation of the model.

3.2.6.3 Model Predictivity Evaluation

The predictivity of QSPR models is based on the goodness of fit and stability of the model. The most effective way to evaluate the predictivity of the model is to conduct external validation, that is, to use independent sample sets not involved in the modeling as test sets to evaluate the predictivity of the model for unknown compounds. The methodology is as follows: The original sample set is randomly divided into two subsets, a training set and a test set, according to a certain ratio. The training set is used for variable selection and modeling, while the test set is used for external validation of the model. The results of external validation can not only reflect the generalization ability of the model but also reflect the real predictivity of the model to the external samples that are not involved in the modeling.

As for the proportion of samples in the training set and those in the test set, there is still no clear conclusion. In most studies, the sample number of the test set is much smaller than that of the training set. Recent research by Gramatica indicates that the validation of the predictivity of a QSPR model cannot be performed only with a small number of compounds (≤ 5), but must be based on a sufficiently large set of external tests ($\geq 20\%$ of the sample set) to avoid the occurrence of "accidental correlation."

The external predictivity of the model can be measured by the cross-validation coefficient Q^2_{ext} between the predicted value of the test set sample and the target value:

$$Q^2_{ext} = 1 - \frac{\sum_{i=1}^{prediction}(y_i - \hat{y}_i)^2}{\sum_{i=1}^{prediction}(y_i - \bar{y}_{tr})^2} \tag{3.7}$$

where y_i and $\hat{y}_i$, respectively, represent the experimental value and predicted value of the test set sample, and $\bar{y}_{tr}$ represents the average value of the experimental value of the training set.

In addition, the previous statistical evaluation indices used to evaluate fitting ability of the model, such as R^2, RMSE, AAE, APE, and SE, can be used to evaluate predictivity. Usually, the indices of the test set are a little worse than those of the training set, but if the former is far worse than the latter, then the model likely has an "overfitting" problem. In addition, there is no correlation between Q^2_{ext} and the fitting coefficient R^2 of the test set. A high Q^2_{ext} value is only a necessary condition for the model with high predictivity.

3.2.7 Model Mechanism Explanation

The mechanism explanation of the model is the most important in the QSPR study for the hazardous characteristics of chemicals, but there are few studies reported. The traditional prediction models based on the empirical correlation method and the group contribution method (GCM) only pay attention to the prediction function of the model, but cannot explain the mechanism of the model. Researchers are accustomed to using some empirical parameters to quantitatively describe the structure of compounds, and only focus on good prediction performance of the model. In fact, a successful QSPR model is helpful to explain the mechanism reflected in the model. On the one hand, the mechanism explanation can discover and determine the molecular structure factors that play a decisive role in the hazardous characteristics of substances, explore the influence of these factors on the characteristics, and realize the influence of the microstructure of substances on the macro hazardous characteristics at the molecular level. On the other hand, the elucidated results of the structure–activity relationship can provide directions for the design, screening, or prediction of compounds with any property. Therefore, it is very important and necessary to explain the mechanism of the established QSPR model.

In this chapter, the QSPR method is used to predict flammability characteristics, and at the same time, the mechanism of the model is explained from the perspective of molecular structure, so as to reveal and describe the intrinsic quantitative relationship between the flammability characteristics of substances and their molecular structure through the following two aspects:

1) Select molecular structure descriptors with clear physical and chemical significance to establish the model, such as the commonly used basic property descriptors, such as molecular weight, topological descriptors, geometric descriptors, and electrical descriptors, in order to facilitate the explanation of the model mechanism.
2) Adopt appropriate research methods to evaluate the importance of descriptors in each model, so as to clarify the influence degree of the corresponding structural factors on the target properties.

3.2.8 Summary of QSPR Process

In this chapter, based on the basic principles and QSPR process, the corresponding QSPR study system for flammability and explosion characteristics was established by applying the newly established QSPR algorithms such as GA-MLR and GA-SVM. The features of the system are:

1) Three QSPR models with different properties were established by using linear GA-MLR algorithm, hybrid GA+SVM algorithm, and nonlinear GA-SVM algorithm, respectively. The quantitative relationship between the flammability characteristics and molecular structure was studied comprehensively.
2) From the three aspects of model fitting ability, stability, and predictivity, the prediction models are evaluated and validated comprehensively and effectively.
3) On the basis of evaluation and validation of the models, the mechanism of each model is explained, and the inherent quantitative relationship between the flammability

characteristics of an organic compound and its molecular structure is revealed and described, so as to master the influence of molecular structure on the characteristics.

3.3 QSPR Review for Flammability Characteristics

3.3.1 Flammability Limits

The flammability limit is one of the most widely studied properties in the field of QSPR analysis. Most flammability limit data are obtained from the Institute for Design of Physical Property Relations (DIPPR) database. This database is publicly available and contains property data for over 2200 pure chemical compounds and is managed by the American Institute of Chemical Engineers (AIChE) (Thomson 1996). Table 3.3 shows the quantity of data available for flammability limits in the DIPPR-801 dataset.

3.3.1.1 LFLT and LFL

Data on the lower flammability limit (LFL) for pure compounds are relatively easy to obtain, and the feasible ranges for LFL and lower flammability limit temperature (LFLT) are usually much lower than for upper flammability limit (UFL).

The LFLT is defined as the temperature at which the concentration of a saturated vapor/air mixture equals the lower limit of flammability at atmospheric pressure (Thomson 1996). Gharagheizi used a GA to select 6 of the 1664 available molecular descriptors and used these descriptors to analyze LFLT data for 1171 pure compounds (Gharagheizi 2009d). The model had a training R^2 value of 0.945 and a tested RMSE of 15.406, indicating that the prediction accuracy of the MLR-based QSPR model was within acceptable limits. In another work (Gharagheizi 2009b), an ANN method was applied to 1429 pure compound LFLT data. The model had a training R^2 value of 0.981 and a test RMSE of 13. Lazzús performed an ANN-based QSPR analysis by integrating a particle swarm optimization (PSO) descriptor screening approach (Lazzús 2011c). In this work, 418 pure compounds were used to build the model with a training R^2 value of 0.9952. This model showed significantly more accurate results compared to previous models. Gharagheizi et al. introduced the corresponding state method (CSM) in LFLT prediction (Gharagheizi et al. 2013), where the prediction model was constructed based on five other relevant physical properties, and the gene expression programming (GEP) method was used to determine the regression model. The prediction model was generated based on 1480 pure chemical LFLT data and had a training R^2 value of

Table 3.3 Data available from DIPPR-801 dataset.

Property	No. of available data	Data type	
		Experimental	Other
Upper flammability limit (UFL)	1598	400	1198
Lower flammability limit (LFL)	1716	470	1246

0.992 and a test RMSE of 7.9, which is better than the QSPR-based model. However, the parameters available for state-based models are often limited compared to the large number of chemical descriptors, which limits its application.

The LFL is defined as the minimum concentration of combustible gas or vapor that can propagate a flame through a homogeneous mixture of the component and air under a specified test condition.

Suzuki developed a model based on 112 sample data by using multiple regression analysis with R^2 of 0.910 and RMSE of 0.23 (Suzuki 1994). It was demonstrated that in some organic compounds there is a relationship between the combustion limit and the standard enthalpy of combustion. They gave an estimation scheme for the LFLs of compounds containing carbon, hydrogen, oxygen, nitrogen, and sulfur elements.

Suzuki and Ishida have proposed a new method for predicting flammability limits (Suzuki and Ishida 1995). They developed a neural network model based on MLR and neural network methods for chemical structures. They compared the prediction model based on a three-layer neural network and a backpropagation (BP) algorithm with the model on multiple regression analysis. The comparison results show that their neural network has better generality and reliability.

Hshieh developed an empirical formula based on the atomic contribution method to predict the lower limits of combustion for organic and silicone compounds from the net heat of combustion (Hshieh 1999). The SE and mean APE of the formulae for predicting LFLs were 0.22 and 9.3 vol%, respectively.

Gharagheizi developed a predictive model for the LFL based on data for 1056 pure compounds (Gharagheizi 2008). The mean absolute error, the square of the correlation coefficient, and the root mean square of the error of the model were 7.68%, 0.9698, and 0.084, respectively.

Gharagheizi developed an ANN-based QSPR model using a filtering process based on group contributions (Gharagheizi 2009a), obtaining a training R^2 value of 0.99 and a test RMSE of 0.084. This model was based on 1057 pure chemical compounds possessing 105 functional groups. Pan et al. performed MLR regression on 354 pure LFL data based on a GA implemented descriptor filtering method and obtained a model training R^2 value of 0.966 (Pan et al. 2009a). They also introduced an SVM regression method in LFL prediction (Pan et al. 2009b). By comparing SVM, MLR, and ANN-based QSPR models, the results showed that SVM had the highest training R^2 value of 0.979 and the lowest test RMSE of 0.076, indicating that the SVM regression method has a stronger predictivity. Lazzús introduced the PSO descriptive filtering method for the first time to develop an ANN-based QSPR model for LFL prediction (Lazzús 2011c). In this model, 418 pure compounds were used and a training R^2 value of 0.9876 was obtained. The result showed a significant improvement in the predictivity of the model compared to previous work, and it facilitated the use of PSO for molecular descriptor screening in another work (Bagheri et al. 2012b). The Adaptive Neuro-Fuzzy Interface System (ANFIS) method was used to generate regression models suitable for LFL values, giving a training R^2 value of 0.929 and a minimum test RMSE of 0.153. This showed a significant improvement in predictivity compared to the MLR-based model, which had a training R^2 value of 0.906 and a minimum test RMSE of 0.335. Albahri used 543 LFL data points from the American Petroleum Institute Technical Data Book and the DIPPR-801 database to construct an ANN-based QSPR model (Albahri 2013). In this work, structural group filtering methods were also used to achieve simpler model. The model in this work has a training R^2 value of 0.9998, which gives it particularly high predictivity compared to other work.

It should be noted that these models are based on large amount of data, of which most of the LFL data in the DIPPR database are predicted values or values that have not been further validated. According to the work of Chen et al. (2017), only 27.39% of the LFL data were verified as experimental measurements. As predicted value-based models did not show high validity in predicting LFL values, they re-evaluated the MLR-based model of Gharagheizi, Pan, and Bagheri by getting rid of predicted and unvalidated LFL data. However, the results showed that the predictivity was significantly lowered by just using the experimental data. Then, they proposed a new four-variable MLR model that outperformed the previous model with a training R^2 value of 0.9256.

Compared with the pure chemical, studies on QSPR-based LFL prediction models for chemical mixtures are few. This may be due to the fact that QSPR modeling is considered to be based on a big data approach that requires a large amount of data for regression modeling. Wang et al. obtained 86 mixture data from one reference to construct a QSPR model based on MLR (Wang et al. 2018a); the result showed good accuracy with a training R^2 value of 0.923 and a test RMSE of 0.048. Pan et al. further extended the dataset by obtaining 181 mixture data from multiple sources (Pan et al. 2019). They obtained a total of 12 multilinear models based on different mathematical formulations. The model with the best performance is the six-parameter model issued from the norm of the molar contribution formula. After validation and comparison, the results demonstrated the robustness, validity, and satisfactory predictivity of the proposed model. Although the model is not sufficiently accurate for individual data points, the model provides a good measure of prediction for the LFL of mixtures.

3.3.1.2 UFLT and UFL

The upper flammability limit temperature (UFLT) or upper explosion point is defined as the liquid temperature at which the concentration of vapor in thermodynamic equilibrium with the liquid equals the UFL (Vanderstraeten et al. 1997). Although Lazzús performed an ANN-based QSPR analysis by integrating PSO descriptor screening methods, research on UFLT is limited (Lazzús 2011b). 418 pure compound UFLT data were used to build this model, yielding a training R^2 value of 0.9949. In addition, Gharagheizi used CSM to build a prediction model based on 1462 pure chemical compound UFLT data (Gharagheizi et al. 2012). The model had a training R^2 value of 0.990 and a test RMSE of 8.8, which indicates its good predictivity for UFLT of pure chemical compounds. However, additional QSPR models for UFLTs should be further developed to improve the predictivity.

The UFL is defined as the maximum concentration of combustible material capable of propagating a flame in a homogeneous mixture of substances and gaseous oxidants under specified test conditions. The UFL typically has a larger range than the LFL, which makes it more difficult to obtain an accurate model.

High and Danner proposed a GCM for predicting the UFL of pure compounds (High and Danner 1987). The method was applied to a wide range of compounds. The theory has an AAE of 4.8 vol.% and a maximum absolute error of 88.9 vol.%. This method has a large error in the prediction results and needs to be refined.

Suzuki and Koide examined the relationship between the UFL and other thermochemical properties (Suzuki and Koide 1994). It was found that the UFLs of some organic compounds are closely related to the standard enthalpy of combustion. The general relationship equation they developed can predict the UFL from the standard enthalpy of combustion.

Pan et al. obtained UFL data from multiple reference sources for 588 pure organic compounds and proposed two different MLR-based QSPR models using GA filtering techniques (Pan et al. 2009d). The performance of these two models were compared and the most accurate model had a training R^2 value of 0.758, which was acceptable for predicting the UFL of pure organic compounds. Pan et al. also used the GA descriptor filtering method to perform MLR regression on 278 pure hydrocarbon UFL data and obtained a model with a training R^2 value of 0.904 (Pan et al. 2009a), which was significantly higher than the model previously (Gharagheizi 2009c) developed based on 865 pure compound UFL data. Gharagheizi improved the model by implementing the GCM (Gharagheizi 2010). In this work, they selected 867 pure chemical UFL data from 113 functional groups and modeled them by using ANN for regression normalization. This improved model had a training R^2 value of 0.9469 and a test RMSE of 0.882, indicating significantly better predictivity than the previous model. Lazzús developed an ANN-based QSPR model for UFL prediction with a PSO descriptor screening approach (Lazzús 2011b). 418 pure compounds were selected to develop this model and a training R^2 value of 0.9780 was obtained. Frutiger et al. investigated the Marrero/Gani based GCM for LFL prediction (Marrero and Gani 2001; Frutiger et al. 2016), where 443 experimental data were selected to develop the prediction model. Robust regression (RR) and linear error propagation methods were also used to obtain more accurate models, with results showing a training R^2 value of 0.91. Yuan et al. applied machine learning methods (Yuan et al. 2019), including k-nearest neighbor (k-NN) and random forest (RF) methods, to build QSPR models to predict the UFL of pure organic chemicals, and compared the performance of these models with MLR and SVM-based models. The result showed that the RF-based model has the best performance of all four proposed models, with a training R^2 value of 0.961 and a test RMSE of 0.136. This work suggests that it is possible to use more machine learning-based approaches to further improve existing QSPR models.

For the UFL prediction of combustible mixtures, Wang et al. generated an SVM-based QSPR model using experimental UFL data for 78 mixtures (Wang et al. 2018b), obtaining a training R^2 value of 0.950 and a test RMSE of 2.21. The consistency of the data was improved by obtaining data from references. These results indicate that the model has good predictivity for the mixed UFL. However, more experimental data are needed to further evaluate the model.

3.3.2 Flash Point

The flash point (FP) of a flammable substance is an important property of flammability and is defined as the lowest temperature at which a mixture of the substance and outside air will flash and burn immediately when in contact with a flame.

Zhokhova et al. constructed MLR and ANN models for the FPs of a number of different organic compounds using structural fragment descriptors (Zhokhova et al. 2003).

A neural network analysis of 398 compounds using 25 fragment descriptors gave the best results: $R^2 = 0.959$.

Gramatica et al. studied the properties of solvents in order to provide tools for the selection of suitable solvents (Gramatica et al. 1999). Starting with a large subset of descriptions, they developed a QSPR model based on the FPs of 136 organic solvents. A GA-variable subset selection (GA-VSS) procedure was used to obtain a six-parameter model with $R^2 = 0.813$. A dataset of 153 esters was studied for the prediction of their basic physicochemical properties. The prediction results of the class-specific QSPR model were compared with those of the US-EPIWIN model, and the results showed that the QSPR model outperformed the US-EPIWIN model.

Tetteh et al. developed a radial basis function (RBF) neural network model for simultaneous estimation of flash and boiling points based on 25 molecular functional groups and their primary molecular connectivity indices (Tetteh et al. 1999). The RBF network was trained using an orthogonality least squares (OLS) learning algorithm. The validation and test sets had flash and boiling points in the range of $10 \sim 12\,°C$ and $11 \sim 14\,°C$, respectively, in agreement with the experimental value of approximately $10\,°C$. This work is an extension of their previous study. They built an MLR model based on the entire dataset of 400 compounds ($R^2 = 0.94$, $s = 13.5\,°C$).

A general three-parameter QSPR model is provided by Katritzky et al. for the QSPR study of the FPs of 271 different compounds (Katritzky et al. 2001). It was concluded that difference in charged partial surface areas (DPSA) plays an important role in intermolecular polar interactions, while the quantum chemical descriptor is related to the reaction of the carbon atoms of the molecule during combustion reactions.

Catoire and Naudet chose 59 carbon-containing compounds to develop a unique empirical equation for estimating the FPs of most classes of organic liquids (Catoire and Naudet 2004). The equation was developed by taking into account measurement reliability, a wide temperature range (-50–$133.9\,°C$), and structural variability. At the same time, several accurate theoretical estimation methods are proposed to modify the model in the absence of experimental data for two empirical parameters. The mean absolute deviation of the developed equations from the FP prediction was $2.9\,°C$, with a maximum absolute error of $7\,°C$. They tested the equations for various organic compounds and found that in this model, polyhalogenated compounds were detected as outliers.

Stefanis et al. developed a GCM (Stefanis et al. 2004). They used primary groups describing the basic molecular structure of the compound and secondary groups based on the theory of conjugation operators to provide more structural information to distinguish the isomers, allowing the accuracy of the predictions to be improved. They define new groups, broaden the range of compounds that can be described, and the reliability of predictions is improved.

The prediction of the properties of mixtures is difficult. Vidal et al. described a study of the estimated FP of mixtures (Vidal et al. 2004). They also discussed the phenomenon in which the FP of a mixture of substances is lower than that of the individual components (Vidal et al. 2006). Moreover, their study found that the UNIFAC GCM can be used to predict the FP of binary mixtures when mixed with non-ideal liquids. Catoire et al. extended their equations for pure compounds (Catoire and Naudet 2004) to binary and ternary mixtures (Catoire et al. 2006a, 2006b) and reproduced the experimentally measured minimum flash point behavior (MFPB) phenomenon.

Pan et al. developed a neural network model based on the structure-flashpoint relationship of 92 alkanes using the group bond contribution method (Pan et al. 2007). Group bonds containing information on group properties and connectivity were used as molecular descriptors. The dataset of alkanes was divided into a training set (62), a validation set (15), and a test set (15). The parameters were adjusted by the trial-and-error method to make the optimal conditions of the neural network. The simulations were carried out using the final optimized BP neural network [9-5-1], and the results showed that the predicted FPs matched the experimental data well, with an average absolute deviation of 4.8 K and an RMSE of 6.86. The new neural network model outperformed the MLR method.

Katritzky et al. provided an update to their previous QSPR study on FPs (Katritzky et al. 2001; Katritzky et al. 2007). They collected data on 758 organic compounds from the literature published before 2004. The best model obtained has an average error of 13.9 K and characterizes the FP well. The descriptors that play a major role in the model are more related to electrostatic and hydrogen bonding interactions and molecular shape. Their ANN model also gives better statistical properties with $R^2 = 0.878$ and a mean error of 12.6 K. The QSPR model developed is predictive for a wide range of organic compounds.

Patel et al. made an important contribution to the field of process safety by using computer-aided molecular design (CAMD) methods to study the FPs of various solvents such as alcohols, hydrocarbons, amines, and ethers (Patel et al. 2010). However, the predictivity of these models was lower than those of Frutiger et al. and Dai et al. (Dai et al. 2015; Frutiger et al. 2016). Frutiger et al. analyzed 180 organic compounds with an RR model and a GCM with 927 descriptors and obtained a training R^2 value of 0.99 (Frutiger et al. 2016). In addition, Dai et al. analyzed 50 hydrocarbons, specifically for alkanes, by using ANN and MLR methods as well as GCM and obtained the highest correlation for ANN (Dai et al. 2015). The work yielded a training R^2 value of 0.9989 and a test RMSE of 0.0337.

From an analysis of the literature, the difference between the size of the dataset and the number of descriptors is apparent when more than 1651 substances are included (Bagheri et al. 2012a). It is worth noting that, as recommended by AIChE, databases such as DIPPR can be used to obtain model data when studies involve a larger number of substances. However, studies involving a smaller number of compounds and focusing on specific substances are better validated by experiments.

Other important contributions include FP studies of crude oil fractions (Liu and Liu 2010). In addition, Jia et al. investigated the positional distribution contribution method (Jia et al. 2012), where the functional group of the molecule is the only variable used to calculate FP.

QSPR studies on prediction of FPs have remained relatively constant over the last decade or so. However, based on these results, the number of QSPR literature involving FP of mixtures has increased in recent years. This is due to the large application of hydrocarbon mixtures, which suggests that we need more robust models for mixtures. Typically, mixture models will be more accurate in describing practical applications than ideal pure component models. Phoon et al. found that the predictivity of mixture QSPR models developed in previous studies was poor (Phoon et al. 2014), which further indicates that we need more accurate mixture models.

A similar trend is evident in calculation work on FP descriptors, particularly in studies involving GCMs. This is now a widely used approach that involves exploring the effect of selected molecular functional groups on the overall behavior of the molecule, as well as interactions between molecular groups, to predict properties. Important contributors to this field include Dai et al., Alibakshi, and Mirshahvalad et al. (Dai et al. 2015; Alibakshi 2018; Mirshahvalad et al. 2019). These studies used both traditional MLR and ANN methods to build models, giving high training R^2 (>0.95) values and demonstrating that GCM remains a descriptor calculation method with sufficient accuracy. Saldana et al. took a genetic function approximation (GFA), partial least squares (PLS), ANN with general regression neural network (GRNN), and SVM in a complex combination, yielding a training value of $R^2 = 0.922$ (Saldana et al. 2011).

3.3.3 Auto-ignition Temperature

Auto-ignition temperature (AIT) is the lowest temperature at which a substance can spontaneously combust in air in the absence of a spark or flame. Spontaneous combustion occurs when the heat generated by an exothermic oxidation reaction of a substance exceeds the heat released into the surrounding environment. The research to date has found that the main structural features affecting AIT are the chain length, unsaturation, branching, aromaticity, and functional groups of the compound.

The first studies of AITs were carried out in the early 1990s by Egolf and Jurs (Egolf and Jurs 1992), and after that more studies of AITs have been carried out gradually.

They selected a total of 312 different hydrocarbons, alcohols, and esters (Egolf and Jurs 1992), then correlated their chemical structural characteristics with their AIT. They proposed two different mechanisms for the AIT of hydrocarbons at high (633–848 K) and low (475–610 K) temperatures, based on the displacement of the regression line on the observed versus calculated AIT diagram. Four sets of MLR models were finally derived for which R^2 ranges from 0.88 to 0.95 and s is between 12 and 24 K. The MLR models contain between four and eight molecular descriptors. It was concluded that structural features such as radical stability, spatial strain, and molecular rigidity play a key role in the simulation of AITs. Mitchell et al. carried out a replication of the descriptors on a dataset containing 327 organic compounds (Mitchell and Jurs 1997). The AIT prediction model built using MLR and convolutional neural network (CNN) has predictive ability within the experimental error range (RMSE ≤35 °C).

Tetteh et al. modeled AIT by using RBF and BP neural networks (Tetteh et al. 1996). The model had six descriptors, two empirical subsets, and four structural subsets. The RBF and BP neural networks produced a good simulation result for a training set of 85 with 0.953 and 0.945 of R^2, respectively. But the predictions for the validation set of 148 were moderate with R^2 of 0.834 and 0.837 and mean errors of 30.1 and 29.9 °C, respectively. In a subsequent study, Tetteh et al. expanded their dataset to 232 organic compounds and used bimodal and spline interpolation to optimize the expansion parameters and number of neurons in the hidden layer of the RBF neural network (Tetteh et al. 1998).

Yoshida and Funatsu proposed a new hybrid approach (Yoshida and Funatsu 1997). They combined the GA and the quadratic partial least squares (QPLS) method as a nonlinear simulation of PLS based on data of 85 compounds from Tetteh et al. and used this to build an AIT prediction model. This method shows a significant improvement compared to the traditional QPLS method.

Kim et al. selected 72 molecular descriptors from 200 different organic compounds and used the GFA to build a very high quality MLR model for AIT with the training set $R^2 = 0.920$ and RMSE $= 25.9\,°C$ (Kim et al. 2002). The corresponding statistical parameters for the test set were $R^2 = 0.910$ and RMSE $= 29.0\,°C$.

Albahri and George used ANNs to identify structural groups within the framework of the structural group contribution (SGC) method (Albahri and George 2003). The range of applicability of the model performed well due to the selection of 470 compounds. They used this model to predict the AIT of 20 pure components with an average error of 2.6% and R^2 of 0.98.

The largest regression coefficients obtained in the known studies were from the model built by Lazzús (Lazzús 2011a). The model used 343 organic compounds including hydrocarbons from the ANN and GCM study DIPPR 801 database with 42 descriptors and obtained a training R^2 value of 0.9899 and an AAE of 10.5 K.

Another regression method is a BP method based on neural networks. This approach has been used for traditional ANNs, and Pan et al. showed that the neural network-based approach can significantly improve QSPR studies compared to traditional statistical methods (Pan et al. 2008a, 2008b).

An extensive and comprehensive work on the application of QSPR to AIT has been done by Pan et al. (Pan et al. 2008a, 2008b, 2009a, 2009c). The study included different models related to AIT, as well as a comparison of regression methods, including MLR, ANN, GA, and SVM. In addition, various descriptor calculation methods were used, including the Electronic State Index (ETSI) method, where the electronic state of a molecule is considered to be the descriptor of the model. In addition, the datasets chosen for these studies are more comprehensive compared to those used in other related studies. The sources of datasets used in these studies include: (i) International Chemical Safety Cards (ICSCs), (ii) Chemical and Other Safety Information Database of the Physical and Theoretical Chemistry Laboratory at Oxford University (United Kingdom), (iii) Chemical Database of the Department of Chemistry at the University of Akron (USA), (iv) chemical manufacturer material safety data sheet (MSDSs), and (v) *Lange's Handbook of Chemistry*. As other studies are based on data obtained from smaller, more accurate datasets and number of descriptors, the models developed are more sensitive to the analyzed compounds than models obtained using larger datasets.

3.3.4 Heat of Combustion

The heat of combustion is the amount of heat released when a substance undergoes a complete combustion reaction with oxygen. This property is particularly important in areas such as fire protection.

Hshieh developed an empirical formula for predicting the total and net heat of combustion of organosilicon compounds based on the atomic contribution method (Hshieh 1999). The average APE of the predictions was 1.4–1.5%.

Gharagheizi et al. used data of 4590 organic compounds including hydrocarbons from the DIPPR 801 database (Gharagheizi et al. 2011). They established a neural network regression model and GCM of 142 descriptors, and obtained a training R^2 of 0.99999 and a test RMSE of 12.57 kJ/mol.

In most studies on the heat of combustion, large datasets of over 500 substances were chosen as the database for the study. Only in a few studies are small amounts of data selected, for example, Yunus and Zahari chose 30 specific substances (Yunus and Zahari 2017).

A relatively novel approach to this study has also emerged (Frutiger et al. 2015). They established a new method for developing a system for GC model building, parameter estimation, and uncertainty analysis. This has been successfully applied to the development of a new GC-based heat of combustion prediction model. In particular, the new models developed by this system have a higher accuracy than existing GC models and are easier to apply than ANN models.

It is worth noting that from the QSPR study for different characteristics, the variability of the heat of combustion is relatively small in the software used to develop the model and in the comparison of other properties.

3.3.5 Minimum Ignition Energy

Minimum ignition energy (MIE) is the minimum amount of energy that can ignite a given mixture of combustible material and air or oxygen. Although previous studies have proposed mathematical models to predict MIE (Frendi and Sibulkin 1990; Lian et al. 2012), the development of QSPR prediction models for MIE requires further research.

Currently, the lack of available MIE data is the most important limitation for QSPR study. As DIPPR and other publicly available databases do not contain purely chemical MIE data, MIE data can only be found in few studies and all data are obtained from journal articles.

Wang et al. obtained experimental MIE data for 61 pure chemicals from a reference where the experiments were conducted with stoichiometric fuel-air mixtures under ambient conditions at atmospheric pressure (Wang et al. 2016b). The study used MLR and SVM methods to develop QSPR models. The results showed that the SVM-based QSPR model outperformed the MLR-based model, producing a training R^2 value of 0.837 and a test RMSE of 0.165, indicating that the model has good predictivity. As the B3LYP-based molecular optimization model led to complex calculation when applied to QSPR studies, Wang et al. used a simpler AM1-based model for molecular structure optimization (Wang et al. 2017). The MIE data were retested using the MLR approach, and the updated model produced a training R^2 value of 0.835 and a test RMSE of 0.210. Although the performance of the AM1-based optimization model was slightly lower than that of the B3LYP-based model, the results can still provide some meaningful guidance for the development of safety standards. ROMÂNĂ developed MLR and ANN-based QSPR models (Li et al. 2018), and obtained test RMSEs of 0.126 and 0.116, respectively, which also demonstrates that the QSPR-based MIE prediction models have good predictivity.

3.3.6 Gas-liquid Critical Temperature

The gas-liquid critical temperature is defined as the temperature at which a gas is no longer likely to liquefy.

In the early stages, the critical temperature was estimated from a number of property parameters that could be measured. These parameters included: boiling point, parachor, and molar refractive index. Gradually, a number of methods emerged to predict the critical temperature by calculating it from molecular descriptors (Needham et al. 1988; Kuanar et al. 1996; Klein et al. 1997).

Egolf et al. developed two independent critical temperature models based on boiling point and critical temperature data extracted from the DIPPR database by computational neural network and MLR analysis, respectively (Egolf et al. 1994). The former model was an eight-parameter MLR model containing three charged partial surface areas (CPSAs), two topological descriptors, and three structural descriptors. The latter regression model, parameterized by experimental boiling points and containing three descriptors, predicted critical temperatures for 147 different organic compounds with R^2 of 0.988 and s of 8.48. Subsequently, the MLR model with eight descriptors was improved by excluding two compounds, quinoline and hexanenitrile, and RMS error of 9.16 K and R^2 of 0.986 was obtained. After building the linear model, a nonlinear CNN model with an 8-4-1 architecture (8 input neurons for the eight descriptors, 4 hidden neurons, and 1 output neuron for a total of 41 adjustable parameters) was built with an observed RMSE of 7.33 K for the 132 training set compounds, 7.73 K for 15 cross validation, and 9.85 K for the 18 prediction set compounds.

Hall and Story used a neural network based on an atomic-type electro-topological state index to analyze critical temperatures for a group containing 165 organic compounds (Hall and Story 1996). The overall relative error was 0.97 ~ 1.17% and the mean absolute error was 4.52 K.

Katritzky et al. correlated the critical temperatures of 76 hydrocarbons and 165 structurally different compounds by the QSPR model (Katritzky et al. 1998). The one-parameter model yielded R^2 of 0.953 and the three-parameter model yielded R^2 of 0.955. The results revealed the influence of the structure on the properties of the liquid.

Shacham et al. applied the molecular similarity approach to a variety of physicochemical properties for unmeasured data (Shacham et al. 2004), and they obtained an average prediction error of 0.91 for the critical temperatures of 18 compounds.

Many different topological parameters calculated from structures have been applied to predict the critical properties of compounds (Shamsipur et al. 2004; Ni et al. 2005). Charton reviewed the nature of topological parameters in predicting the physicochemical properties of compounds, noting that topological parameters are composites representing the number of atoms, bonds, electrons, and branches (Charton 2003).

Godavarthy et al. used CODESSA PRO to develop QSPR models for critical temperatures for different datasets containing over 1230 organic compounds (Godavarthy et al. 2008). Several methods such as linear, nonlinear, and GAs were used for model development. The resulting nonlinear QSPR models were able to predict the critical temperatures of different compounds very well.

The gas-liquid critical temperatures of 692 organic compounds were collected by Zhou et al. and applied to build a QSPR model (Zhou et al. 2017a). DRAGON software was used to obtain their molecular structure information. MLR and SVM methods combined with GAs were used to build the models. Training R^2 values of 0.851 and 0.935 were obtained for each, respectively. In addition, test RMSE values of 0.038 for MLR and 0.019 for SVM were reported. The gas-liquid critical temperature of binary mixtures was also investigated by Zhou et al. (Zhou et al. 2017b). The resulting model was built to obtain a training R^2 value of 0.900 and a test RMSE of 0.030, and the model can be used to predict the critical temperature of new binary mixtures or binary mixtures of which experimental data are unknown. This study also provides an effective method for the rapid determination of the critical temperature of binary mixtures.

3.3.7 Other Properties

Wang et al. developed two new QSPR models to predict the self-accelerated decomposition temperature of organic peroxides by MLR and SVM analysis based on experimental data and molecular simulation results (Wang et al. 2016a). The results show that both models have good fit, internal robustness, and external predictivity. The MLR model is easier to apply, while the SVM model has better predictivity. This study provides a model that can be used as a reference for predicting the self-accelerated decomposition temperature of organic peroxides with unknown experimental values or new organic peroxides.

The limiting oxygen index (LOI) is the concentration of the volume fraction of oxygen at which a polymer in a mixture of oxygen and nitrogen is just able to support its combustion. It is an index that characterizes the flammability behavior of a material. Crisan et al. used the LOI as an indicator for polyphosphates that can be used as flame retardant additives in the production of polymers (Crisan et al. 2016). LOI data for 28 polyphosphates synthesized were used to develop a QSPR model based on MLR and PLS. The model had a training R^2 value of 0.976 and a test RMSE value of 0.016. Despite the high predictivity of the model, only 28 data points were involved in the development of the model, suggesting that more data are needed to further validate the model.

Parandekar et al. used the GFA regression method to develop models for three properties, heat release capacity (HRC), total heat release (THR), and char yield, as a means of assessing the combustion performance of polymers (Parandekar et al. 2015). Data from 83 polymers were used. External datasets for approximately 20 polymers were also predicted and the QSPR predictions for the external datasets were compared with the molar GCM. The results show a good relationship between polymer unit structure and flammability parameters. An understanding of the molecular level of flammability characteristics can be obtained based on the molecular descriptors in the QSPR equation. The QSPR model is shown to have good potential for studying the flammability properties of polymers.

3.4 Limitations

For current QSPR research, there are still three main limitations. The first is data limitations. The lack of data not only limits the validation of models by QSPR research but also limits the scope of application of QSPR models. Most research is limited to databases including DIPPR or American petroleum Institute (API), and there is little research outside of the databases. One reason is the difficulty of validating QSPR models for substances outside of databases, which also limits QSPR research to known databases. It is difficult to extend the application of QSPR models based on the small amount of experimental validation data available.

The second limitation is that the prediction accuracy of models based on commonly used databases is lower than that of models based on experimental validation data. There are many prediction models that are not sufficiently reliable for characteristics such as MIE because there is not enough experimental data to confirm their reliability.

Finally, there are limitations in terms of descriptors. The descriptor screening methods that are commonly used today need to be further improved in order to increase the efficiency of the screening process. The accuracy of the descriptor calculation also needs to be further improved. Furthermore, the descriptors chosen for many studies are mostly extracted from the DIPRR database, and a large number of descriptors are not relevant to the properties selected for the study.

3.5 Conclusions and Future Prospects

As society develops, the number of hazardous chemicals is increasing dramatically, which requires higher capability for loss prevention and safety control. It is therefore particularly important to understand the flammability characteristics of various hazardous chemicals.

Over the past decade, significant progress has been made in predicting the flammability properties of substances through QSPR models. A large number of QSPR models have been developed to study the flammability properties of a wide range of substances with good performance.

The current QSPR research shows that cheminformatics, quantum chemistry, statistics, and computer science have great potential for application in this field. We can further enrich the library of regression methods by applying the newly developed statistical methods to the building of QSPR models.

Many innovative machine learning regression methods including GA, ANN, and RF are widely used and have greater predictivity than traditional MLR methods, and show great potential of development in the future. It is thus clear that machine learning is a powerful tool. It acts as a cross-discipline between statistics and computer science, and machine learning has great potential in terms of both predictive accuracy and efficiency. Therefore, machine learning needs to be added to the research of QSPR, both in the development of software and theory.

Of all the flammability properties nowadays, FP, auto-ignition points, and flammability limits are the most studied, and multiple regression methods and descriptor screening methods have been thoroughly validated. Most of the models show satisfactory predictivity for the flammability characteristics.

However, data limitations and the lack of connections between model development and industrial applications need to be addressed in future QSPR research. Researchers should go further and integrate QSPR studies with process safety to better predict fire- and explosion-related properties. There is also a need to bring chemicals that are not in commonly used databases into the developed models for extensive validation as a way to ensure model validity and consistency.

References

Albahri, T.A. (2013). Prediction of the lower flammability limit percent in air of pure compounds from their molecular structures. *Fire Safety Journal* 59: 188–201. https://doi.org/10.1016/j.firesaf.2013.04.007.

Albahri, T.A. and George, R.S. (2003). Artificial neural network investigation of the structural group contribution method for predicting pure components auto ignition temperature. *Industrial & Engineering Chemistry Research* 42: 5708–5714. https://doi.org/10.1021/ie0300373.

Alibakshi, A. (2018). Strategies to develop robust neural network models: prediction of flash point as a case study. *Analytica Chimica Acta* 1026: 69–76. https://doi.org/10.1016/j.aca.2018.05.015.

Awfa, D., Ateia, M., Mendoza, D. et al. (2021). Application of quantitative structure–property relationship predictive models to water treatment: a critical review. *ACS ES&T Water* 1: 498–517. https://doi.org/10.1021/acsestwater.0c00206.

Bagheri, M., Bagheri, M., Heidari, F. et al. (2012a). Nonlinear molecular based modeling of the flash point for application in inherently safer design. *Journal of Loss Prevention in the Process Industries* 25: 40–51. https://doi.org/10.1016/j.jlp.2011.06.025.

Bagheri, M., Rajabi, M., Mirbagheri, M. et al. (2012b). BPSO-MLR and ANFIS based modeling of lower flammability limit. *Journal of Loss Prevention in the Process Industries* 25: 373–382. https://doi.org/10.1016/j.jlp.2011.10.005.

Begam, B.F. and Kumar, J.S. (2016). Computer Assisted QSAR/QSPR Approaches – A Review. *Indian Journal of Science and Technology* 9: 1–8. https://doi.org/10.17485/ijst/2016/v9i8/87901.

Catoire, L. and Naudet, V. (2004). A unique equation to estimate flash points of selected pure liquids application to the correction of probably erroneous flash point values. *Journal of Physical and Chemical Reference Data* 33: 1083–1111. https://doi.org/10.1063/1.1835321.

Catoire, L., Paulmier, S., and Naudet, V. (2006a). Estimation of closed cup flash points of combustible solvent blends. *Journal of Physical and Chemical Reference Data* 35: 9–14. https://doi.org/10.1063/1.1928236.

Catoire, L., Paulmier, S., and Naudet, V. (2006b). Experimental determination and estimation of closed cup flash points of mixtures of flammable solvents. *Process Safety Progress* 25: 33–39. https://doi.org/10.1002/prs.10112.

Charton, M. (2003). The nature of topological parameters. I. Are topological parametersfundamental properties. *Journal of Computer-Aided Molecular Design* 17: 197–209. https://doi.org/10.1023/A:1025378125128.

Chen, C.C., Lai, C.P., and Guo, Y.C. (2017). A novel model for predicting lower flammability limits using quantitative structure activity relationship approach. *Journal of Loss Prevention in the Process Industries* 49: 240–247. https://doi.org/10.1016/j.jlp.2017.07.007.

Crisan, L., Iliescu, S., and Funar-Timofei, S. (2016). Structure-flammability relationship study of phosphoester dimers by MLR and PLS. *Polimeros* 26: 129–136. https://doi.org/10.1590/0104-1428.2306.

Dai, Y.M., Liu, H., Chen, X.Q. et al. (2015). A new group contribution-based method for estimation of flash point temperature of alkanes. *Journal of Central South University* 22: 30–36. https://doi.org/10.1007/s11771-015-2491-0.

Egolf, L.M. and Jurs, P.C. (1992). Estimation of autoignition temperatures of hydrocarbons, alcohols, and esters from molecular structure. *Industrial & Engineering Chemistry Research* 31: 1798–1807. https://doi.org/10.1021/ie00007a027.

Egolf, L.M., Wessel, M.D., and Jurs, P.C. (1994). Prediction of boiling points and critical temperatures of industrially important organic compounds from molecular structure. *Journal of Chemical Information and Computer Sciences* 34: 947–956. https://doi.org/10.1021/ci00020a032.

Frendi, A. and Sibulkin, M. (1990). Dependence of Minimum Ignition Energy on Ignition Parameters. *Combustion Science and Technology* 73: 395–413. https://doi.org/10.1080/00102209008951659.

Frutiger, J., Marcarie, C., Abildskov, J. et al. (2015). A comprehensive methodology for development, parameter estimation, and uncertainty analysis of group contribution based property models—an application to the heat of combustion. *Journal of Chemical & Engineering Data* 61: 602–613. https://doi.org/10.1021/acs.jced.5b00750.

Frutiger, J., Marcarie, C., Abildskov, J. et al. (2016). Group-contribution based property estimation and uncertainty analysis for flammability-related properties. *Journal of Hazardous Materials* 318: 783–793. https://doi.org/10.1016/j.jhazmat.2016.06.018.

Gao, Y., Zhang, X., Zhang, Z. et al. (2020). Research progress of quantitative structure-property relationship (QSPR) method for predicting typical combustion and explosion characteristics of mixtures. *Safety Science and Technology* 20–26. https://doi.org/10.11731/j.

Gharagheizi, F. (2008). Quantitative structure– property relationship for prediction of the lower flammability limit of pure compounds. *Energy & Fuels* 22: 3037–3039. https://doi.org/10.1021/ef800375b.

Gharagheizi, F. (2009a). A new group contribution-based model for estimation of lower flammability limit of pure compounds. *Journal of Hazardous Materials* 170: 595–604. https://doi.org/10.1016/j.jhazmat.2009.05.023.

Gharagheizi, F. (2009b). New neural network group contribution model for estimation of lower flammability limit temperature of pure compounds. *Industrial & Engineering Chemistry Research* 48: 7406–7416. https://doi.org/10.1021/ie9003738.

Gharagheizi, F. (2009c). Prediction of upper flammability limit percent of pure compounds from their molecular structures. *Journal of Hazardous Materials* 167: 507–510. https://doi.org/10.1016/j.jhazmat.2009.01.002.

Gharagheizi, F. (2009d). A QSPR model for estimation of lower flammability limit temperature of pure compounds based on molecular structure. *Journal of Hazardous Materials* 169: 217–220. https://doi.org/10.1016/j.jhazmat.2009.03.083.

Gharagheizi, F. (2010). Chemical structure-based model for estimation of the upper flammability limit of pure compounds. *Energy & Fuels* 24: 3867–3871. https://doi.org/10.1021/ef100207x.

Gharagheizi, F., Mirkhani, S.A., and Tofangchi Mahyari, A.-R. (2011). Prediction of standard enthalpy of combustion of pure compounds using a very accurate group-contribution-based method. *Energy & Fuels* 25: 2651–2654. https://doi.org/10.1021/ef200081a.

Gharagheizi, F., Ilani-Kashkouli, P., and Mohammadi, A.H. (2012). Corresponding states method for estimation of upper flammability limit temperature of chemical compounds. *Industrial & Engineering Chemistry Research* 51: 6265–6269. https://doi.org/10.1021/ie300375k.

Gharagheizi, F., Ilani-Kashkouli, P., and Mohammadi, A.H. (2013). Estimation of lower flammability limit temperature of chemical compounds using a corresponding state method. *Fuel* 103: 899–904. https://doi.org/10.1016/j.fuel.2012.06.101.

Godavarthy, S.S., Robinson, R.L., and Gasem, K.A.M. (2008). Improved structure–property relationship models for prediction of critical properties. *Fluid Phase Equilibria* 264: 122–136. https://doi.org/10.1016/j.fluid.2007.11.003.

Gramatica, P., Navas, N., and Todeschini, R. (1999). Classification of organic solvents and modelling of their physico-chemical properties by chemometric methods using different sets of molecular descriptors. *Trends in Analytical Chemistry* 18: 461–471. https://doi.org/10.1016/S0165-9936(99)00115-6.

Hall, L.H. and Story, C. (1996). Boiling point and critical temperature of a heterogeneous data set: QSAR with atom type electrotopological state indices using artificial neural networks. *Journal of Chemical Information and Computer Sciences* 36: 1004–1014. https://doi.org/10.1021/ci960375x.

High, M.S. and Danner, R.P. (1987). Prediction of upper flammability limit by a group contribution method. *Industrial & Engineering Chemistry Research* 26: 1395–1399. https://doi.org/10.1021/ie00067a021.

Hshieh, F.Y. (1999). Predicting heats of combustion and lower flammability limits of organosilicon compounds. *Fire and Materials* 23: 79–89. https://doi.org/10.1002/(SICI)1099-1018(199903/04)23:2.

Jia, Q., Wang, Q., Ma, P. et al. (2012). Prediction of the Flash Point Temperature of Organic Compounds with the Positional Distributive Contribution Method. *Journal of Chemical & Engineering Data* 57: 3357–3367. https://doi.org/10.1021/je301070f.

Jiang, J., Duan, W., Wei, Q. et al. (2020). Development of quantitative structure-property relationship (QSPR) models for predicting the thermal hazard of ionic liquids: a review of methods and models. *Journal of Molecular Liquids* 301: 112471. https://doi.org/10.1016/j.molliq.2020.112471.

Jiao, Z., Escobar-Hernandez, H.U., Parker, T. et al. (2019). Review of recent developments of quantitative structure-property relationship models on fire and explosion-related properties. *Process Safety and Environmental Protection* 129: 280–290. https://doi.org/10.1016/j.psep.2019.06.027.

Katritzky, A.R., Mu, L., and Karelson, M. (1998). Relationships of critical temperatures to calculated molecular properties. *Journal of Chemical Information and Computer Sciences* 38: 293–299. https://doi.org/10.1021/ci970071q.

Katritzky, A.R., Petrukhin, R., Jain, R. et al. (2001). QSPR analysis of flash points. *Journal of Chemical Information and Computer Sciences* 41: 1521–1530. https://doi.org/10.1021/ci010043e.

Katritzky, A.R., Stoyanova-Slavova, I.B., Dobchev, D.A. et al. (2007). QSPR modeling of flash points: an update. *Journal of Molecular Graphics and Modelling* 26: 529–536. https://doi.org/10.1016/j.jmgm.2007.03.006.

Kim, Y.S., Lee, S.K., Kim, J.H. et al. (2002). Prediction of autoignition temperatures (AITs) for hydrocarbons and compounds containing heteroatoms by the quantitative structure–property relationship. *The Royal Society of Chemistry* 2: 2087–2092. https://doi.org/10.1039/b207203c.

Klein, D.J., Randić, M., Babić, D. et al. (1997). Hierarchical orthogonalization of descriptors. *International Journal of Quantum Chemistry* 63: 215–222. https://doi.org/10.1002/(SICI)1097-461X(1997)63:1.

Kuanar, M., Mishra, R., and Mishra, B. (1996). Optimal parametrization of structure for prediction of properties of alkanes. *Indian Journal of Chemistry* 35A: 1026–1033.

Lazzús, J.A. (2011a). Autoignition temperature prediction using an artificial neural network with particle swarm optimization. *International Journal of Thermophysics* 32: 957–973. https://doi.org/10.1007/s10765-011-0956-4.

Lazzús, J.A. (2011b). Neural network/particle swarm method to predict flammability limits in air of organic compounds. *Thermochimica Acta* 512: 150–156. https://doi.org/10.1016/j.tca.2010.09.018.

Lazzús, J.A. (2011c). Prediction of flammability limit temperatures from molecular structures using a neural network–particle swarm algorithm. *Journal of the Taiwan Institute of Chemical Engineers* 42: 447–453. https://doi.org/10.1016/j.jtice.2010.08.005.

Li, S., Wei, C., Fan, L., and Xu, J. (2018). Validated QSPR models for the prediction of minimum ignition energy. *Revue Roumaine de Chimie* 63 (2): 111–121.

Lian, P., Gao, X., and Mannan, M.S. (2012). Prediction of minimum ignition energy of aerosols using flame kernel modeling combined with flame front propagation theory. *Journal of Loss Prevention in the Process Industries* 25: 103–113. https://doi.org/10.1016/j.jlp.2011.07.006.

Liu, X. and Liu, Z. (2010). Research progress on flash point prediction. *Journal of Chemical & Engineering Data* 55: 2943–2950. https://doi.org/10.1021/je1003143.

Liu, P. and Long, W. (2009). Current mathematical methods used in QSAR/QSPR studies. *International Journal of Molecular Sciences* 10: 1978–1998. https://doi.org/10.3390/ijms10051978.

Marrero, J. and Gani, R. (2001). Group-contribution based estimation of pure component properties. *Fluid Phase Equilibria* 183: 183–208. https://doi.org/10.1016/S0378-3812(01)00431-9.

Mirshahvalad, H., Ghasemiasl, R., Raoufi, N. et al. (2019). A Neural Network QSPR Model for Accurate Prediction of Flash Point of Pure Hydrocarbons. *Molecular Informatics* 38: https://doi.org/10.1002/minf.201800094.

Mitchell, B.E. and Jurs, P.C. (1997). Prediction of autoignition temperatures of organic compounds from molecular structure. *Journal of Chemical Information and Computer Sciences* 37: 538–547. https://doi.org/10.1021/ci960175l.

Needham, D.E., Wei, I.C., and Seybold, P.G. (1988). Molecular modeling of the physical properties of alkanes. *Journal of the American Chemical Society* 110: 4186–4194. https://doi.org/10.1002/chin.198842037.

Ni, C.H., Zeng, X.Y., and Huang, H. (2005). Studies on the physical properties of alkanes using edge-adjacency information topological index. *Chinese Chemical Letters* 16: 709–710. https://doi.org/CNKI:SUN:FXKB.0.2005-05-041.

Pan, Y., Jiang, J., and Wang, Z. (2007). Quantitative structure-property relationship studies for predicting flash points of alkanes using group bond contribution method with back-propagation neural network. *Journal of Hazardous Materials* 147: 424–430. https://doi.org/10.1016/j.jhazmat.2007.01.025.

Pan, Y., Jiang, J., Wang, R. et al. (2008a). Advantages of support vector machine in QSPR studies for predicting auto-ignition temperatures of organic compounds. *Chemometrics and Intelligent Laboratory Systems* 92: 169–178. https://doi.org/10.1016/j.chemolab.2008.03.002.

Pan, Y., Jiang, J., Wang, R. et al. (2008b). Prediction of auto-ignition temperatures of hydrocarbons by neural network based on atom-type electrotopological-state indices. *Journal of Hazardous Materials* 157: 510–517. https://doi.org/10.1016/j.jhazmat.2008.01.016.

Pan, Y., Jiang, J., Ding, X. et al. (2009a). Prediction of flammability characteristics of pure hydrocarbons from molecular structures. *AICHE Journal* 56: 690–701. https://doi.org/10.1002/aic.12007.

Pan, Y., Jiang, J., Wang, R. et al. (2009b). A novel QSPR model for prediction of lower flammability limits of organic compounds based on support vector machine. *Journal of Hazardous Materials* 168: 962–969. https://doi.org/10.1016/j.jhazmat.2009.02.122.

Pan, Y., Jiang, J., Wang, R. et al. (2009c). Predicting the auto-ignition temperatures of organic compounds from molecular structure using support vector machine. *Journal of Hazardous Materials* 164: 1242–1249. https://doi.org/10.1016/j.jhazmat.2008.09.031.

Pan, Y., Jiang, J., Wang, R. et al. (2009d). Prediction of the upper flammability limits of organic compounds from molecular structures. *Industrial & Engineering Chemistry Research* 48: 5064–5069. https://doi.org/10.1021/ie900193r.

Pan, Y., Ji, X., Ding, L. et al. (2019). Prediction of lower flammability limits for binary hydrocarbon gases by quantitative structure-a property relationship approach. *Molecules* 24: 748. https://doi.org/10.3390/molecules24040748.

Parandekar, P.V., Browning, A.R., and Prakash, O. (2015). Modeling the flammability characteristics of polymers using quantitative structure–property relationships (QSPR). *Polymer Engineering & Science* 55: 1553–1559. https://doi.org/10.1002/pen.24093.

Patel, S.J., Ng, D., and Mannan, M.S. (2010). QSPR flash point prediction of solvents using topological indices for application in computer aided molecular design. *Industrial & Engineering Chemistry Research* 49: 8282–8287. https://doi.org/10.1021/ie101378h.

Phoon, L.Y., Mustaffa, A.A., Hashim, H. et al. (2014). A review of flash point prediction models for flammable liquid mixtures. *Industrial & Engineering Chemistry Research* 53: 12553–12565. https://doi.org/10.1021/ie501233g.

Saldana, D.A., Starck, L., Mougin, P. et al. (2011). Flash point and cetane number predictions for fuel compounds using quantitative structure property relationship (QSPR) methods. *Energy & Fuels* 25: 3900–3908. https://doi.org/10.1021/ef200795j.

Shacham, M., Brauner, N., Cholakov, G. et al. (2004). Property prediction by correlations based on similarity of molecular structures. *AICHE Journal* 50: 2481–2492. https://doi.org/10.1002/aic.10248.

Shamsipur, M., Ghavami, R., Hemmateenejad, B. et al. (2004). Highly correlating distance-connectivity-based topological indices. 2: Prediction of 15 properties of a large set of alkanes using a stepwise factor selection-based PCR analysis. *QSAR & Combinatorial Science* 23: 734–753. https://doi.org/10.1002/qsar.200430894.

Stefanis, E., Constantinou, L., and Panayiotou, C. (2004). A group-contribution method for predicting pure component properties of biochemical and safety interest. *Industrial & Engineering Chemistry Research* 43: 6253–6261. https://doi.org/10.1021/ie0497184.

Suzuki, T. (1994). Note: Empirical relationship between lower flammability limits and standard enthalpies of combustion of organic compounds. *Fire and Materials* 18: 333–336. https://doi.org/10.1002/fam.810180509.

Suzuki, T. and Ishida, M. (1995). Neural network techniques applied to predict flammability limits of organic compounds. *Fire and Materials* 19: 179–189. https://doi.org/10.1002/fam.810190404.

Suzuki, T. and Koide, K. (1994). Correlation between upper flammability limits and thermochemical properties of organic compounds. *Fire and Materials* 18: 393–397. https://doi.org/10.1002/fam.810180608.

Tetteh, J., Metcalfe, E., and Howells, S.L. (1996). Optimisation of radial basis and backpropagation neural networks for modelling auto-ignition temperature by quantitative-structure property relationships. *Chemometrics and Intelligent Laboratory Systems* 32: 177–191. https://doi.org/10.1016/0169-7439(95)00088-7.

Tetteh, J., Howells, S., Metcalfe, E. et al. (1998). Optimisation of radial basis function neural networks using biharmonic spline interpolation. *Chemometrics and Intelligent Laboratory Systems* 41: 17–29. https://doi.org/10.1016/S0169-7439(98)00035-5.

Tetteh, J., Suzuki, T., Metcalfe, E. et al. (1999). Quantitative structure– property relationships for the estimation of boiling point and flash point using a radial basis function neural network. *Journal of Chemical Information and Computer Sciences* 39: 491–507. https://doi.org/10.1021/ci980026y.

Thomson, G. (1996). The DIPPR® databases. *International Journal of Thermophysics* 17: 223–232. https://doi.org/10.1007/BF01448224.

Vanderstraeten, B., Tuerlinckx, D., Berghmans, J. et al. (1997). Experimental study of the pressure and temperature dependence on the upper flammability limit of methane/air mixtures. *Journal of Hazardous Materials* 56: 237–246. https://doi.org/10.1016/S0304-3894(97)00045-9.

Vidal, M., Rogers, W.J., Holste, J.C. et al. (2004). A review of estimation methods for flash points and flammability limits. *Process Safety Progress* 23: 47–55. https://doi.org/10.1002/prs.10004.

Vidal, M., Rogers, W.J., and Mannan, M.S. (2006). Prediction of minimum flash point behaviour for binary mixtures. *Process Safety and Environmental Protection* 84: 1–9. https://doi.org/10.1205/psep.05041.

Wang, B., Yi, H., Xu, K. et al. (2016a). Prediction of the self-accelerating decomposition temperature of organic peroxides using QSPR models. *Journal of Thermal Analysis and Calorimetry* 128: 399–406. https://doi.org/10.1007/s10973-016-5922-8.

Wang, B., Zhou, L., Xu, K. et al. (2016b). Prediction of minimum ignition energy from molecular structure using quantitative structure–property relationship (QSPR) models. *Industrial & Engineering Chemistry Research* 56: 47–51. https://doi.org/10.1021/acs.iecr.6b04347.

Wang, B., Zhou, L., Xu, K. et al. (2017). Fast prediction of minimum ignition energy from molecular structure using simple QSPR model. *Journal of Loss Prevention in the Process Industries* 50: 290–294. https://doi.org/10.1016/j.jlp.2017.10.010.

Wang, B., Park, H., Xu, K. et al. (2018a). Prediction of lower flammability limits of blended gases based on quantitative structure–property relationship. *Journal of Thermal Analysis and Calorimetry* 132: 1125–1130. https://doi.org/10.1007/s10973-017-6941-9.

Wang, B., Xu, K., and Wang, Q. (2018b). Prediction of upper flammability limits for fuel mixtures using quantitative structure–property relationship models. *Chemical Engineering Communications* 206: 247–253. https://doi.org/10.1080/00986445.2018.1483350.

Yoshida, H. and Funatsu, K. (1997). Optimization of the inner relation function of QPLS using genetic algorithm. *Journal of Chemical Information and Computer Sciences* 37: 1115–1121. https://doi.org/10.1021/ci970026i.

Yuan, S., Jiao, Z., Quddus, N. et al. (2019). Developing quantitative structure–property relationship models to predict the upper flammability limit using machine learning. *Industrial & Engineering Chemistry Research* 58: 3531–3537. https://doi.org/10.1021/acs.iecr.8b05938.

Yunus, N.A. and Zahari, N.N.N.N.M. (2017). Prediction of standard heat of combustion using two-step regression. *Chemical Engineering Transactions* 56: 1063–1068. https://doi.org/10.3303/CET1756178.

Zhokhova, N.I., Baskin, I.I., Palyulin, V.A. et al. (2003). Fragmental descriptors in QSPR: flash point calculations. *Russian Chemical Bulletin* 52: 1885–1892. https://doi.org/10.1023/B: RUCB.0000009629.38661.4c.

Zhou, L., Wang, B., Jiang, J. et al. (2017a). Predicting the gas-liquid critical temperature of binary mixtures based on the quantitative structure property relationship. *Chemometrics and Intelligent Laboratory Systems* 167: 190–195. https://doi.org/10.1016/j.chemolab.2017.06.009.

Zhou, L., Wang, B., Jiang, J. et al. (2017b). Quantitative structure-property relationship (QSPR) study for predicting gas-liquid critical temperatures of organic compounds. *Thermochimica Acta* 655: 112–116. https://doi.org/10.1016/j.tca.2017.06.021.

4

Consequence Prediction Using Quantitative Property–Consequence Relationship Models

Zeren Jiao and Qingsheng Wang

Artie McFerrin Department of Chemical Engineering, Texas A&M University, College Station, TX, USA

4.1 Introduction

Consequence prediction includes prediction of fire, explosion, and dispersion consequences, which is an essential component of process safety. An accurate and efficient consequence prediction model can facilitate the development of emergency response, mitigation, and intervention plans, as well as design phase risk assessment and evaluation. The release of flammable chemicals can result in fire or explosion in the presence of an ignition source, while jet fire and fireballs are caused by the instant ignition of flammable chemicals and a delayed ignition may cause an explosion or flash fire. The consequences of fire and explosion are enormous, including the Piper Alpha offshore oil platform incident that resulted in 167 fatalities and was caused by a leak of highly flammable hydrocarbons and subsequent fire and catastrophic explosion (Paté-Cornell 1993). Incidental release and dispersion of toxic chemicals may also lead to serious short-term and long-term consequences to humanity as well as to the environment. The Union Carbide methyl isocyanate (MIC) release incident in Bhopal, India, killed at least 3787 people and injured 57,000 people immediately, and the morbidity and premature death rate increased significantly afterward (Broughton 2005). Laws and regulations have become more stringent in recent years involving the storage and use of flammable and toxic chemicals. Additionally, the emergency response and process safety protocols have developed significantly in recent decades. However, in spite of this, fire, explosion, and toxic release incidents continue to occur. Consequence prediction, as a key step in both emergency response and risk assessment, is crucial for obtaining the effect estimation of fire, explosion, and dispersion scenarios to vicinity equipment, plants, and communities. It is also capable of providing a baseline for subsequence mitigation planning and optimization for process design. Therefore, tremendous efforts have been taken to develop efficient and reliable consequence prediction tools. Machine learning, with its highly accurate and adaptive nature, has very broad application prospects.

Machine Learning in Chemical Safety and Health: Fundamentals with Applications, First Edition.
Edited by Qingsheng Wang and Changjie Cai.
© 2023 John Wiley & Sons Ltd. Published 2023 by John Wiley & Sons Ltd.

In this chapter, different conventional consequence prediction methods are introduced and discussed first, and recent studies of implementing machine learning and deep learning in consequence prediction are summarized. Furthermore, a novel machine learning and deep learning-based consequence prediction model as well as a quantitative property–consequence prediction method are discussed in detail, and the challenges and future directions are presented at the end.

4.2 Conventional Consequence Prediction Methods

For consequence prediction, there are three principal methods, which are shown in Figure 4.1. These include an empirical method, which utilizes equations and computation graphs; a CFD method, which uses numerical analysis of fluid flow with the aid of computers to solve Navier–Stokes equations; and an integral method, which combines the advantages of empirical and CFD methods (Shen et al. 2020). Each method possesses unique advantages and disadvantages.

4.2.1 Empirical Method

Compared to property prediction, consequence prediction is more complex due to its macroscale nature, which has many more factors that can influence the prediction performance. Using dispersion as an example, there are various parameters that can affect the dispersion consequence prediction: ground terrain, weather conditions, and atmospheric stability. For dispersion, the most widely used prediction models are Pasquill–Gifford and Britter–McQuaid models as they are able to predict downwind plume distances using tables, derived equations, and computation graphs. However, the Pasquill–Gifford dispersion model is only applicable to neutrally buoyant gas dispersions, with a total of 15 cases that need to be differentiated for application. The Britter–McQuaid model, on the other hand, can only be implemented for dense gas dispersion. The calculation procedures for these two models are complex and are unable to account for the influence of terrain and obstacles. For explosion, there are two main empirical methods, the TNT equivalency

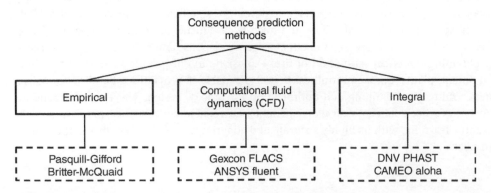

Figure 4.1 Three different types of consequence prediction methods.

method and TNO multi-energy method. The TNT equivalency method is based on the assumption that the chemical vapor cloud explosion behaves like a TNT explosion. However, this is usually not the case and this method tends to overpredict the explosion overpressure near the explosion source but underpredict as the distance increases. The TNO multi-energy method is a semi-empirical method that will identify explosion confined spaces with an assigned degree of confinement and then calculate the explosion overpressure. As it involves both equations and computational graphs, the TNO multi-energy method can provide more accurate explosion predictions compared to the TNT equivalency method. However, it is relatively difficult to choose an appropriate severity factor based on confinement as only limited guidance is available. There is also a Baker–Strehlow method that is similar to the multi-energy method. All of the methods are able to provide acceptable predictions in certain cases but lack generalized accuracy and efficiency (Crowl and Louvar 2019, pp. 224–247).

4.2.2 Computational Fluid Dynamics (CFD) Method

Computational fluid dynamics (CFD) is another commonly used method for conducting consequence analysis and prediction (Shen et al. 2020). A large number of efforts have been made for investigating and understanding the fluid dynamics and interactions in different complex scenarios (Jiao et al. 2019b; Mi et al. 2020; Shen et al. 2020).

The mainstream software packages for CFD simulation in safety are ANSYS Fluent and CFX, GEXCON FLACS, and open-sourced OpenFOAM. All of them are highly capable of conducting numerical simulations of fire, explosion, and dispersion. One key advantage of CFD-based consequence analysis is that CFD is comprehensively capable of taking obstacles and terrain into consideration, which can provide the most accurate result among all three methods. Additionally, unlike empirical methods, which can only calculate transient state consequences, CFD can also conduct continuous simulations, which can provide detailed results during the course of the incident.

However, CFD simulations also require the most professional knowledge. The geometry setup is cumbersome and time-consuming and grid independency analysis is necessary anytime a new grid is formed. Furthermore, the parameters, boundary conditions, and model adjustments also require a high level of fluid dynamics and transport knowledge that makes it more difficult for broad applications. In addition, the CFD method is also extremely resource demanding, as a complex simulation may require days of calculation with aid of the supercomputer, making it impossible to implement during an emergency response.

On the other hand, due to the high accuracy of CFD for consequence prediction, it is suitable for generating a database for the machine learning-based model, which can compensate for its disadvantages of being resource demanding and time-consuming. One of the most significant challenges for machine learning implementation in consequence analysis is the difficulty in collecting sufficient data, as field experiment data is very rare. CFD simulation models validated against experiments will be far more practical for machine learning model dataset collection. However, many complex consequences still cannot be accurately simulated using CFD, such as boiling liquid expanding vapor explosions (BLEVE), deflagration to detonation transitions (DDT), and dust explosions. Furthermore, with the increase in cases and scenarios, the shortcomings of being computationally resource demanding and time-consuming will be further highlighted.

4.2.3 Integral Method

Integral models are lumped-parameter models that are usually developed for one or two-dimensional consequence prediction. This takes account of physical phenomena to determine the parameters that are further tuned and optimized using field test data. Integral methods do not require high-level knowledge compared with CFD, but their accuracies are highly dependent on the field test dataset they used for model parameter tuning (Derudi et al. 2014). Popular integral prediction methods such as HEavy GAs Dispersion from Area Sources (HEGADIS), National Center for Atmospheric Research (NCAR), Dispersion of Releases Involving Flammables or Toxics (DRIFT), and Unified Dispersion Model (UDM) are developed using various field test datasets. They are able to overcome the disadvantages of empirical and CFD methods while providing relatively accurate prediction results with much lower computational costs. As an example, PHAST is one of the most popular process hazard analysis software and has a dispersion prediction tool that uses UDM and 3D explosion modules to simulate the complete course of incident scenarios. This spans from initial leakage to far-field dispersion and subsequent explosion, and it is able to model the rainout and subsequent vaporization, fire radiation distance, and vapor cloud explosion. The PHAST consequence prediction module has been widely validated against experimental results and field test data for different scenarios with very high predictive accuracies. Sensitivity analysis has also been conducted for PHAST dispersion prediction, which shows its capabilities in consequence prediction (Jiao et al. 2021).

Since the integral method is able to provide fast consequence prediction with high accuracy, it is also an ideal data collection tool for machine learning-based consequence prediction model development. However, the experimental setups of the field tests for integral models mostly do not involve obstacles, which makes the data collected from integral models limited to open field conditions, which limits the model applications. The most accurate method is to form a comprehensive dataset regardless of the prediction method used to include as many scenarios as possible.

4.3 Machine Learning and Deep Learning-Based Consequence Prediction Models

With the development of machine learning and deep learning algorithms, many studies have attempted to develop consequence prediction models using machine learning and deep learning algorithms. However, most of them are focused on consequence prediction such as gas dispersion and source term estimation due to its greater data availability and relatively simple mechanism (Jiao et al. 2020a). A summary of machine learning and deep learning-based consequence prediction models is shown in Table 4.1.

One of the disadvantages of the conventional consequence prediction model is that the source condition is required for model setup, which is usually unknown during the actual incident. Wang et al. (2015) used an artificial neural network to correlate the detected gas

Table 4.1 Summary of machine learning-based consequence analysis.

Name	Year	Algorithm	Dataset	Application
Cho et al.	2018	ANN/RF	CFD simulated data	Leak source tracking
Jiao et al.	2020b	GB	PHAST simulated data	Dispersion prediction
Jiao et al.	2021	GB/RF/DNN	PHAST simulated data	Dispersion prediction
Kim et al.	2018	ANN	CFD simulated data	Leak source tracking
Ma and Zhang	2016	ANN/SVM	Project Prairie Grass field test data	Dispersion prediction and leak source tracking
Ni et al.	2020	DBN/CNN	Project Prairie Grass field test data	Dispersion prediction
Qian et al.	2019	ANN/SVM	Project Prairie Grass field test data	Dispersion prediction
Qiu et al.	2018	ANN	Indianapolis field test data	Dispersion prediction
Shi et al.	2019	ANN	FLACS simulated data	Dispersion prediction
Shi et al.	2020	ANN	FLACS simulated data	Explosion risk analysis
Sun et al.	2019	ANN	PHAST simulated data	Fire consequence prediction
Wang et al.	2015	ANN	PHAST simulated data	Dispersion prediction
Wang et al.	2018	ANN/SVM	Project Prairie Grass and Indianapolis field test data	Dispersion prediction

concentration from gas detectors with the gas concentration in target locations to develop a novel consequence prediction model. This is used in lieu of the source condition parameters that are difficult to obtain, and the developed model is validated against the PHAST simulation results of chlorine release. Qiu et al. (2018) further utilized an artificial neural network to develop a dispersion source estimation model, and particle swarm optimization and expectation maximization algorithms were also used to accelerate the convergence. The model was validated against Indianapolis field test data for accuracy. These studies all demonstrated the ability of machine learning and deep learning to accomplish tasks that cannot be completed using conventional consequence prediction models.

During machine learning and deep learning model training processes, the unstable selection of input parameters may increase difficulty in training the model and consequently reduce the prediction accuracy. To solve this problem, Wang et al. (2018) used the integrated Gaussian parameters rather than the original monitoring parameters for artificial neural networks and support vector machine (SVM) models using Prairie Grass field test data, which shows the performance improvements over models using only original monitoring parameters. Ma and Zhang (2016) combined the classical Gaussian dispersion model with the SVM algorithm to develop a point source dispersion prediction model for contaminant dispersion, which generated satisfactory results. However, the Gaussian model has proven to be less accurate compared to the CFD and integral methods, which illustrates that the dataset developed for machine learning algorithm implementation is less accurate.

With larger datasets, deep learning algorithms are preferable to machine learning algorithms due to their ability to train large amounts of data with large input parameter sets with much higher accuracy. Qian et al. (2019) proposed a novel long short-term memory (LSTM) network to develop a toxic gas dispersion model based on the same dataset with the dropout technique implemented to reduce overfitting and improve generalization. Ni et al. (2020) further used deep belief networks (DBN) and convolution neural networks (CNN) to compare the performance of empirical, CFD, and machine learning models for Prairie Grass field experiment data prediction. This showed that the CNN had greater performance over the CFD and empirical methods. However, both studies possess limited experimental data, which makes the model less universally applicable for varied leak scenarios.

Sun et al. (2019) and Jiao et al. (2020b, 2021) used PHAST simulated data to construct consequence databases for fire radiation distance, flammable dispersion, and toxic dispersion. The data were then used to train the model to develop quantitative property–consequence relationship (QPCR) models that are more universally applicable for rapid consequence prediction. The authors proposed a novel method for further consequence prediction model development, and the methodology was shown to be expandable for other consequence modeling and combined application with quantitative structure–property relationship (QSPR). The detailed mechanism and QPCR procedure will be introduced in the next section.

The other major implementation of machine learning and deep learning in dispersion analysis is for leak source tracking. Cho et al. (2018) and Kim et al. (2018) used random forest and deep neural networks to develop a chemical leak source-tracking model, and the developed system is validated against CFD simulated data of 640 leak scenarios, which has been shown to have very high prediction accuracy.

Implementation of machine learning and deep learning algorithms to explosion consequences is still limited. However, progress has been made in this field despite the associated explosion mechanisms being fairly complex. Shi et al. (2019, 2020) used FLACS simulation results to train a modified artificial neural network for explosion risk analysis and flammable dispersion modeling, and the developed network can provide reliable results to further improve the field of study.

4.4 Quantitative Property–Consequence Relationship Models

Using machine learning algorithms to train large consequence databases for comprehensive consequence prediction was first implemented by Sun et al. Jiao et al. later formalized the procedure and implemented it in flammable dispersion in 2020, naming it QPCR analysis. The QPCR method was originally inspired by the QSPR analysis method that uses structural attributes as descriptors to build mathematical relationships between the properties of interest and structures at a quantum chemistry level (Jiao et al. 2019a). The procedural diagram of QSPR model development is shown in Figure 4.2. The QSPR method is a well-developed methodology for property prediction that has been widely implemented for hazardous property prediction. QPCR, on the other hand, uses the property descriptors as

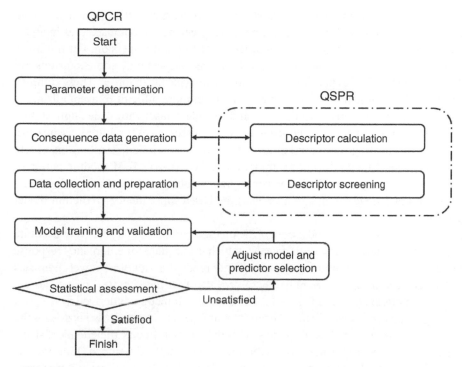

Figure 4.2 QPCR development procedure diagram.

independent variables and quantifies consequence values as dependent variables for consequence prediction. Therefore, as shown in Figure 4.2, the descriptor calculation and screening steps are substituted for the consequence data generation and data collection steps. The QPCR method can serve to bridge the gap between microscale scenarios and chemical properties with macroscale consequences, which serves a promising role in the development of more reliable and widely applicable predictions (Jiao et al. 2020b).

As can be seen from the summary of machine learning and deep learning-based consequence prediction models, all of them have limited applicability. Therefore, much work remains to be completed before implementation occurs. The greatest bottleneck is the training dataset that limited the development of an efficient and effective consequence prediction model. First, the data availability is limited due to the difficulty of conducting field tests. This differs from implementation for chemical property/activity prediction, for which the experimental data can be obtained from laboratory experiments. The consequence data such as flammability or toxic dispersion information is much more difficult to obtain due to the extremely costly field experiment setup and hazardous nature of the chemicals.

Furthermore, the currently available field test data possess limited consequence scenarios. However, the actual consequences will possess a wide range of influencing factors including chemical properties, weather conditions, storage conditions, leak conditions, etc. Therefore, it would be impossible to construct a comprehensive and experimentally based database that includes all possible scenarios.

Third, the field of chemistry is continuously changing, with countless numbers of new chemicals being synthesized. Thus, it is not possible to generate a database that includes all of their properties and subsequent consequences of fire and dispersion. On the other hand, for process plant and storage facilities where fires, dispersions, and explosions are more likely to occur, multiple chemicals and mixtures are typically involved. These complex compounds result in consequence prediction being even more difficult to accomplish.

Therefore, in order to construct an effective and reliable dataset with wide applicability, the database must be significantly enlarged to include consequences for different scenarios. One method for mitigating these difficulties is to include simulation results from widely validated conventional methods. For example, the GNV PHAST UDM module is one of the most prevalent integral methods for consequence data generation, and it has been widely used for machine learning and deep learning-based consequence prediction model development.

Compared to CFD-based methods, the QPCR method can provide a relatively accurate prediction at a much lower computational cost, which is adequate for emergency response planning. It is also capable of conducting consequence predictions for other novel chemicals that are not included in the software database provided that the chemical properties are known. Furthermore, as shown in Figure 4.3, since the QPCR method is derived from QSPR and can bridge the gap between chemical properties and consequences, it also possesses the potential to be combined with QSPR models. This will allow it to link the chemical structures to the consequences to further expand the applicability of the developed QSPR models and allow the QPCR model to predict consequences involving all chemicals and mixtures.

The QPCR procedure is shown in Figure 4.2, with there being five main steps for QPCR model development. The consequence parameters are determined first to include as many scenarios as possible within the targeted range and the consequence data are then obtained to construct the database. The following step is to determine the suitable property descriptors for specific consequence types, and the model is then constructed and trained using the consequence database. After this, the model will be further tuned and optimized to achieve the greatest performance.

4.4.1 Consequence Database

The consequence database is the first and the most critical step for QPCR model development, with experimental data being the most preferable data source. However, due to the limitations mentioned in the previous section, simulation results from reliable conventional models can be used to construct the consequence database. CFD modeling possesses the highest prediction accuracy, which can include obstacles and real-life geometry into the simulations. However, for a widely applicable database that may include hundreds or even

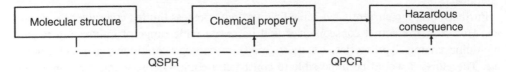

Figure 4.3 Linkage between QSPR and QPCR.

Table 4.2 Summary of consequence database for QPCR studies.

Study	Consequence type	Chemical	Database size
Sun et al. (2019)	Fire thermal radiation distance	35	15 399
Jiao et al. (2020b)	Flammable dispersion	41	19 579
Jiao et al. (2021)	Toxic dispersion	19	30 022

thousands of scenarios, the integral method is a more suitable choice since it requires fewer computational resources and has also been widely validated against field test data. Among integral models, the PHAST UDM of DNV is the most prevalently used software, as it is capable of predicting various consequences such as fire thermal radiation distances, flammable and toxic gas dispersion, and toxic dispersion casualties. The prediction accuracy has also been broadly validated against field test data. All three currently published studies used PHAST UDM as the database construction tool for QPCR model development. An overview of the consequence database constructed from these studies is shown in Table 4.2.

It can be seen from the table that by using this data gathering process, the data acquiring capabilities are greatly improved. The aforementioned consequences are simulated based on 450 different scenarios, which involve common process operating ranges. Therefore, the constructed prediction model will have much greater applicability and reliability.

4.4.2 Property Descriptors

Similar to QSPR, descriptors are essential for QPCR model development. For consequence modeling, different descriptors should be used to represent each case more accurately. Property descriptors can be broadly categorized into three groups. These are source properties, which are the consequence source conditions such as chemical temperature, pressure, and leak size; physical properties, which can define the chemical involved in the scenario; and criteria properties, which are concentrations that are contoured by the prediction results. The source properties tend to be well defined by the given scenario and typically do not require descriptor screening and selection. However, the property descriptors must be chosen carefully to eliminate unnecessary ones, especially when machine learning-based algorithms are used. This is because the overuse of variables will compromise the reliability of the constructed models. The summary of property descriptors used for QPCR studies is shown in Table 4.3.

4.4.3 Machine Learning and Deep Learning Algorithms

For algorithm selection, both machine learning and deep learning-based algorithms have been found to be suitable for QPCR model development based on published research. A summary is shown in Table 4.4. A comparison has also been conducted for different machine learning and deep learning-based algorithms, with the results showing that the deep neural network provides significantly better performance compared with single- or double-layer artificial neural networks. A more advanced machine learning-based model also provided satisfactory results for dispersion prediction. More advanced deep learning

Table 4.3 Summary of property descriptors for QPCR studies.

Study	Consequence type	Descriptor used
Sun et al. (2020)	Fire thermal radiation distance	Storage temperature, operation pressure, leak size, material volume, negative heat of combustion, auto-ignition temperature, flashpoint, flammability rating, lower flammability limit, upper flammability limit, burning rate
Jiao et al. (2020b)	Flammable dispersion	Storage temperature, operating pressure, leak size, material volume, lower flammability limit, vapor density
Jiao et al. (2021)	Toxic dispersion	Storage temperature, operating pressure, leak size, material volume, molecular weight, vapor density, specific gravity

Table 4.4 Summary of algorithms used for QPCR studies.

Study	Consequence type	Algorithm
Sun et al. (2019)	Fire thermal radiation distance	Artificial neural network
Jiao et al. (2020b)	Flammable dispersion	XGBoost
Jiao et al. (2021)	Toxic dispersion	Random forest
		XGBoost
		Deep neural network

frameworks have not been implemented since only a limited number of variables are included for model development. Therefore, a continued increase of network complexity may not significantly affect the prediction results.

4.5 Challenges and Future Directions

The implementation of machine learning and deep learning for consequence prediction has occurred only within the past 10 years. Although significant progress has been made, its development is still in an early stage and possesses many limitations.

The first challenge is the lack of reliable data. Although the simulation pathway can circumvent this issue temporarily with satisfactory prediction, the complex nature of fires, explosions, and dispersions still require validation with experimental data. Particularly for the QPCR model, even though the majority of the data can be simulated using proven methods, it still must be validated against field test data for its final proof of viability. There is also a need for a comprehensive database that can summarize and streamline all field test and simulation data for model development, which can facilitate horizontal comparisons and model accuracy improvements.

For consequences that have still not been comprehensively studied due to their complex mechanisms, machine learning and deep learning-based methods would be useful for correlation studies since they can bypass the complex internal relationships and directly correlate the input variables and target consequences. However, data availability is a more prominent issue in this case since the simulation data would not be reliable, making the experimental data a necessity.

Another limitation for machine learning and deep learning-based consequence prediction models is the interpretability. Since machine learning and deep learning methods are "black box" methods that do not provide a relationship that can be expressed using equations, many researchers still doubt their reliability. However, the accuracies of the machine learning and deep learning-based models have been validated, which show their superiority in performance compared to conventional methods. As more data is included with the implementation of advanced algorithms, the machine learning and deep learning-based consequence prediction models will continue to improve, resulting in greater performance.

Furthermore, deployment of the developed models is also very challenging. For practical implementation, a deployed model must possess an intuitive and easy-to-use interface to be truly beneficial, which is particularly challenging since it requires the researchers to be highly skilled in programming.

References

Broughton, E. (2005). The Bhopal disaster and its aftermath: a review. *Environmental Health* 4 (1): 1–6. https://doi.org/10.1186/1476-069x-4-6.

Cho, J., Kim, H., Gebreselassie, A.L., and Shin, D. (2018). Deep neural network and random forest classifier for source tracking of chemical leaks using fence monitoring data. *Journal of Loss Prevention in the Process Industries* 56: 548–558. https://doi.org/10.1016/j.jlp.2018.01.011.

Crowl, D.A. and Louvar, J.F. (2019). *Chemical Process Safety: Fundamentals with Applications*. Hoboken, NJ: Prentice Hall.

Derudi, M., Bovolenta, D., Busini, V., and Rota, R. (2014). Heavy gas dispersion in presence of large obstacles: selection of modeling tools. *Industrial & Engineering Chemistry Research* 53 (22): 9303–9310. https://doi.org/10.1021/ie4034895.

Jiao, Z., Escobar-Hernandez, H.U., Parker, T., and Wang, Q. (2019a). Review of recent developments of quantitative structure-property relationship models on fire and explosion-related properties. *Process Safety and Environmental Protection* 129: 280–290. https://doi.org/10.1016/j.psep.2019.06.027.

Jiao, Z., Yuan, S., Ji, C. et al. (2019b). Optimization of dilution ventilation layout design in confined environments using computational fluid dynamics (CFD). *Journal of Loss Prevention in the Process Industries* 60: 195–202. https://doi.org/10.1016/j.jlp.2019.05.002.

Jiao, Z., Hu, P., Xu, H., and Wang, Q. (2020a). Machine learning and deep learning in chemical health and safety: a systematic review of techniques and applications. *ACS Chemical Health & Safety* 27 (6): 316–334. https://doi.org/10.1021/acs.chas.0c00075.

Jiao, Z., Sun, Y., Hong, Y. et al. (2020b). Development of flammable dispersion quantitative property–consequence relationship models using extreme gradient boosting. *Industrial & Engineering Chemistry Research* 59 (33): 15109–15118. https://doi.org/10.1021/acs.iecr.0c02822.

Jiao, Z., Ji, C., Sun, Y. et al. (2021). Deep learning based quantitative property-consequence relationship (QPCR) models for toxic dispersion prediction. *Process Safety and Environmental Protection* 152: 352–2260. https://doi.org/10.1016/j.psep.2021.06.019.

Kim, H., Yoon, E.S., and Shin, D. (2018). Deep neural networks for source tracking of chemical leaks and improved chemical process safety. *Computer Aided Chemical Engineering* 44: 2359–2364. https://doi.org/10.1016/B978-0-444-64241-7.50388-8.

Ma, D. and Zhang, Z. (2016). Contaminant dispersion prediction and source estimation with integrated Gaussian-machine learning network model for point source emission in atmosphere. *Journal of Hazardous Materials* 311: 237–245. https://doi.org/10.1016/j.jhazmat.2016.03.022.

Mi, H., Liu, Y., Jiao, Z. et al. (2020). A numerical study on the optimization of ventilation mode during emergency of cable fire in utility tunnel. *Tunnelling and Underground Space Technology* 100: 103403. https://doi.org/10.1016/j.tust.2020.103403.

Ni, J., Yang, H., Yao, J. et al. (2020). Toxic gas dispersion prediction for point source emission using deep learning method. *Human and Ecological Risk Assessment: An International Journal* 26 (2): 557–570. https://doi.org/10.1080/10807039.2018.1526632.

Paté-Cornell, M.E. (1993). Learning from the piper alpha accident: A postmortem analysis of technical and organizational factors. *Risk Analysis* 13 (2): 215–232. https://doi.org/10.1111/j.1539-6924.1993.tb01071.x.

Qian, F., Chen, L., Li, J. et al. (2019). Direct prediction of the toxic gas diffusion rule in a real environment based on LSTM. *International Journal of Environmental Research and Public Health* 16: 2133. https://doi.org/10.3390/ijerph16122133.

Qiu, S., Chen, B., Wang, R. et al. (2018). Atmospheric dispersion prediction and source estimation of hazardous gas using artificial neural network, particle swarm optimization and expectation maximization. *Atmospheric Environment* 178: 158–163. https://doi.org/10.1016/j.atmosenv.2018.01.056.

Shen, R., Jiao, Z., Parker, T. et al. (2020). Recent application of computational fluid dynamics (CFD) in process safety and loss prevention: a review. *Journal of Loss Prevention in the Process Industries* 67: 104252. https://doi.org/10.1016/j.jlp.2020.104252.

Shi, J., Li, X., Khan, F. et al. (2019). Artificial bee colony based bayesian regularization artificial neural network approach to model transient flammable cloud dispersion in congested area. *Process Safety and Environmental Protection* 128: 121–127. https://doi.org/10.1016/j.psep.2019.05.046.

Shi, J., Chang, B., Khan, F. et al. (2020). Stochastic explosion risk analysis of hydrogen production facilities. *International Journal of Hydrogen Energy* 45: 13535–13550. https://doi.org/10.1016/j.ijhydene.2020.03.040.

Sun, Y., Wang, J., Zhu, W. et al. (2019). Development of consequent models for three categories of fire through artificial neural networks. *Industrial & Engineering Chemistry Research* 59 (1): 464–474. https://doi.org/10.1021/acs.iecr.9b05032.

Wang, B., Chen, B., and Zhao, J. (2015). The real-time estimation of hazardous gas dispersion by the integration of gas detectors, neural network and gas dispersion models. *Journal of Hazardous Materials* 300: 433–442. https://doi.org/10.1016/j.jhazmat.2015.07.028.

Wang, R., Chen, B., Qiu, S. et al. (2018). Comparison of machine learning models for hazardous gas dispersion prediction in field cases. *International Journal of Environmental Research and Public Health* 15 (7): 1450. https://doi.org/10.3390/ijerph15071450.

5

Machine Learning in Process Safety and Asset Integrity Management

Ming Yang[1,2], Hao Sun[1,3], and Rustam Abubarkirov[4]

[1] Safety and Security Science Section, Department of Values, Technology, and Innovation, Faculty of Technology, Policy, and Management, Delft University of Technology, The Netherlands
[2] Australia Maritime College, University of Tasmania, Launceston, Tasmania, Australia
[3] College of Mechanical and Electronic Engineering, China University of Petroleum (East China), Qingdao, China
[4] Department of Civil, Chemical, Environmental, and Materials Engineering, University of Bologna, Bologna, Italy

5.1 Opportunities and Threats

Artificial Intelligence (AI) is a scientific subject investigating and developing theories, methods, technologies, and application systems to simulate, extend, and expand human intelligence (Russell and Norvig 2016). Research in AI includes robotics, language recognition, image recognition, natural language processing, expert systems, etc. As a comprehensive frontier technology, **Machine Learning** (ML), an essential part of AI, has drawn widespread attention. ML has been utilized in many fields, including process systems (Mao et al. 2019; Li et al. 2021a, b).

Ensuring process safety is essential due to storing and processing enormous hazardous substances in the process industries. Process accidents are characterized by low frequency and severe consequences, leading to casualties, loss of assets, environmental damages, and damage to the reputation of a company and/or government. To prevent accidents and ensure process system safety, practitioners and researchers have been working on approaches for risk assessment and management (Abbassi et al. 2015; Adedigba et al. 2018; Casciano et al. 2019; Sun et al. 2020; Yang 2020). In general, risk assessment (RA) comprises three categories:

1) *Qualitative methods* include checklists, preliminary hazard analysis (PHA), hazard and operability analysis (HAZOP), and failure mode and effects analysis (FMEA), etc. Qualitative methods mainly use textual descriptions or descriptive numerical ranges to express relative levels. As an essential part of RA, qualitative methods can effectively identify potential hazards in the system. Still, they cannot estimate the correlation between hazards or quantify the occurrence probability of hazards and the possible consequences.

Machine Learning in Chemical Safety and Health: Fundamentals with Applications, First Edition.
Edited by Qingsheng Wang and Changjie Cai.
© 2023 John Wiley & Sons Ltd. Published 2023 by John Wiley & Sons Ltd.

2) *Semi-quantitative methods* include fault tree analysis (FTA) and event tree analysis (ETA).
3) *Quantitative methods* measure the failure probability of a system and the severity of its consequence.

Nevertheless, after decades of process system development, the interdependency and interaction among components (including human, technological, and organizational factors) are becoming more complex and automated than before. Recurring accidents prove that conventional RA methods are insufficient to ensure process safety (Sun et al. 2021). These changes bring challenges to equipment reliability assessment, failure mechanism identification, determination of the relationship between process parameters and system states, etc. For instance, the deterioration and corrosion of equipment (e.g. valves and pipelines) is a stochastic process influenced by process parameters and "non-equipment factors" such as attended time, environmental factors, and climates. It is difficult to use the aforementioned methods to evaluate the system risk under those conditions. Automated and digital systems generate a large amount of data, leading to new opportunities for process safety and asset integrity assessment. ML-based approaches are widely investigated and utilized to solve those challenges. ML can deal with massive and multidimensional data to establish a more comprehensive evaluation system and improve production efficiency and system safety by continuously optimizing deficiency and problematic links (Sattari et al. 2021).

ML has been applied to solve many process safety problems. Since the scale of process systems is large and may consist of many facilities, it is challenging to locate loss of containment (LOC). Image recognition methods can track and locate the leakage source (Abdulla and Herzallah 2015; Cho et al. 2018). ML can also be used with numerical simulation methods for RA. When applying computational fluid dynamics (CFD) techniques to assess the explosion risk of process systems, we may encounter a problem of too many scenarios to simulate (Shen et al. 2020). Thus, it is impossible to evaluate every scenario. CFD can be used as a virtual experiment to generate data. ML-based approaches can then be applied to investigate the relationship between input parameters and results. A trained network can be obtained to predict the consequences of explosions with known input parameters (Hemmatian et al. 2020). While ML brings new opportunities, it also comes with problems. The application of ML technology makes process systems more automated, thereby increasing the complexity of systems, the attractiveness of cyberattack, the probability of cascading fault, etc. For instance, ML is employed as a function of image recognition to discover the LOC in process systems. When the LOC is found by ML, the relevant alarms and valves will work to prevent escalation accidents. To some extent, the ML contributes to the assurance of system safety. However, an appropriate interlocking setting may increase the complexity of the system. At the same time, the environment of the plant is complicated, which may lead to wrong recognition. The inaccurate recognition will cause unnecessary shutdown or even fail to detect the LOC, resulting in severe consequences. In other words, erroneous input data and noise information will lead to incorrect RA and decision-making, leading to undesired consequences.

This chapter aims to discuss the application of ML in process safety and asset integrity management (AIM). The remaining parts of this paper are organized as follows. Section 5.2 gives a brief literature review of the state-of-the-art of AI in process safety and AIM. The use of ML approaches in probabilistic RA is discussed in Section 5.3. Finally, Section 5.4 presents a conceptual model for big-data-driven AIM.

5.2 State-of-the-Art Reviews

As process systems become more complex, traditional methods cannot handle the increasing dynamic system interactions and establish the logical models for risk analysis. AI techniques, e.g. ML approaches, can derive the mathematical relationship between input and output data without knowing their underlying mechanism or relationship. This attracts growing attention in safety assessment. Hinton et al. (2006) developed the technology of deep learning and achieved success in language recognition. After decades of development, AI technology, especially ML, has become mature. Researchers proposed many novel methods that focus on different problems in process safety assessment (Sarbayev et al. 2019; Jiao et al. 2020; Li et al. 2021).

Each technology of AI has its advantage and application field. In process safety and AIM, the most used AI technology is ML. AI represented by ML technology, especially ANNs, has drawn a wide range of investigations and applications in the past two decades. ML comprises two main parts, namely, supervised learning and unsupervised learning. Supervised learning contains classification algorithms (e.g. support vector machine, random forest, logistic regression) and regression algorithm (e.g. linear regression, least square regression, neural network). Unsupervised learning consists of two sections: clustering algorithm (e.g. K-means clustering and hierarchical clustering) and dimension-reduction algorithm (e.g. principal component analysis).

Some of the aforementioned methods have been applied in two main aspects to ensure production safety and continuity in a process system. According to process systems' characters, ML technology is used to measure and quantify system and equipment risk, including the failure probability and the corresponding consequence, to assure process safety. Second, ML is utilized to detect corrosion and equipment faults to develop a reasonable proactive maintenance strategy and enhance AIM. ML, as a novel data-driven technology, is capable of dealing with those problems. Especially it is effective in situations where it is impossible to define rules or procedures to solve the problem. The two most common methods of ML and one method of evolutionary computation, applied to process safety and AIM, are discussed next. The specific information can be seen in Table 5.1.

Table 5.1 The summary of the ML-related literatures in the process system.

Keywords	Research fields	References	Methods
Artificial neural networks (ANN)	Risk assessment	Sarbayev et al. (2019)	ANN + FT
		Adedigba et al. (2017)	MLP
		Ayhan and Tokdemir (2019)	LCCA + ANN + CBR
		Wen et al. (2019)	Hybrid ANN
		Li et al. (2021)	CFD + GRNN
	Detection of fault modes, corrosion, and degradation rate	Benkouider et al. (2012)	EKF + PNNC

(Continued)

Table 5.1 (Continued)

Keywords	Research fields	References	Methods
		Tan et al. (2012)	MLPNN
		Chen et al. (2021)	CNN
		Ossai (2020)	FFSCNN + PSO
		Li et al. (2021)	BPMLP + SVM
		Arunthavanathan et al. (2021)	CNN + LSTM
		Sattari et al. (2021)	BN + ANN
Principal component analysis (PCA)	Data compression, fault diagnosis	Harrou et al. (2013)	PCA + GLR
		Kaced et al. (2021)	PCA
		Ji et al. (2021)	PCA
		Xie et al. (2019)	PCA + PLSR
		Chen et al. (2019)	PCA
Genetic algorithm GA	Determination of leakage location	Jia et al. (2019)	GA + SVR
	Balance of safe design and cost	Piri et al. (2021)	GA + ANN
	Assessment of safety improvement schedule	Zhang and Tan (2018)	GA + CREAM + ANFIS
	Accident prediction	Sarkar et al. (2019)	GA + PSO
	Prediction of repair and maintenance-related accidents	Zaranezhad et al. 2019	GA + ANN
	Structural reliability assessment	Zhao et al. (2019)	Hybrid GA

5.2.1 Artificial Neural Networks (ANNs)

The main structure of the ANN is composed of three parts: neurons, layers, and networks. A typical ANN has three layers: the input, hidden layer, and output. Between the input and hidden layers, and hidden layer and output, the basic neurons are connected by weights. In other words, the connection only exists between layers, and there is no connection within a layer since the information flows in one direction. ANN belongs to an adaptive method. It is capable of optimizing its structure according to new data and information. Besides, ANN can learn from experience to enhance its performance.

Quantitative risk assessment (QRA) plays a pivotal role in ensuring process safety. Nevertheless, the fast development of process systems made a large quantity of data, which makes process safety analysis a challenge. Besides, those massive data will also include incomplete information and noisy data, posing a threat to preprocess and classification. Those challenges encourage the development of new approaches for RA.

Sarbayev et al. (2019) proposed a practical algorithm to convert fault tree (FT) into ANN and combined FT with ANN to overcome the limitations of FT (e.g. static structure and reliance on expert knowledge) to assess the risk of the process system. Adedigba et al. (2017)

utilized multilayer perceptron (MLP) and probability analysis to evaluate the safety of the process system. The MLP is used to determine nonlinear relationships among process variables. The results show that the proposed approach can measure the dynamic risk of the process system over time. Ayhan and Tokdemir (2019) introduced a hybrid approach, which uses latent class clustering analysis (LCCA) to reduce the complexity of data and utilizes ANN and case-based reasoning (CBR) to evaluate the consequence of the accident to determine the potential scenarios and corresponding measures. Wen et al. (2019) developed an optimal ANN methodology to quantify the reliability of pipelines in process systems. The proposed method is compared with Monte Carlo (MC) and non-optimal methods. The results show that the proposed hybrid approach provides an opportunity to assess the reliability of units and systems. Li et al. (2021) combined CFD with a general regression neural network (GRNN) to assess rescue risk when an explosion accident happens.

The faults, degradation, and corrosion of equipment pose a significant threat to process systems. Wood et al. (2013) pointed out that corrosion has been the primary cause of 20% of major accidents in European Union refineries since 1984. Moreover, the faults and degradation will reduce equipment functionality, making the system state fluctuate and decrease. With this regard, early identification and detection of fault modes, corrosion and degradation rate, and development of proactive maintenance strategy play a crucial role in ensuring asset integrity. Rectors are vital devices in process systems with characteristics of transient operation conditions, unstable states, nonlinear state changes, etc., which present a significant challenge to detect and diagnose faults. Benkouider et al. (2012) proposed a novel comprehensive approach for fault detection and diagnosis of rectors. According to the Extended Kalman Filter (EKF), the fault detection method is determined, while the fault diagnosis approach is achieved using a probabilistic neural network classifier (PNNC). The experimental tests are conducted to confirm the results of different simulation studies. Tan et al. (2012) introduced a multilayer perceptron neural network (MLPNN), comprising a double hidden layer, to detect various faults of reactors. Chen et al. (2021) presented a novel approach based on a one-dimensional convolution neural network (CNN) to detect and classify fault for process systems. The proposed method overcame the shortcoming of traditional methods in large images input data, enhancing the speed for application and promoting the explicit definition of spatial features. Ossai (2020) proposed a new data-driven ML-based approach based on feed-forward subspace clustered neural network (FFSCNN) and particle swarm optimization (PSO) to assess corrosion defect depth of pipelines for process systems. The results show that this technology has broad application prospects in the integrity management of corroded and aged pipelines. It can facilitate the development of a proactive maintenance strategy. Li et al. (2021) developed a backpropagation (BP) MLP, which is a 5-8-1 type ANN, to process massive data to determine the corrosion rate of the carbon steel. The same data is also input to the support vector machine (SVM) model to calculate the corrosion rate, and the results show that the proposed ANN model is better than SVM in measuring corrosion rate and can be used to improve the strategy of AIM. Arunthavanathan et al. (2021) pointed out a novel method to detect and diagnose the faults in the process system by evaluating various variables in the complex process system. This methodology mainly used the CNN-long short-term memory (LSTM) method to predict the system parameters and used an unsupervised one-class-SVM to detect fault symptoms. The results show that the proposed approach can detect potential fault conditions effectively in complex process systems by determining the fault symptoms in advance. Sattari et al. (2021) integrated two types of Bayesian network (BN) (e.g. Tabu and hill-climbing (HC)) and ANN to label and classify massive data reports from process system incidents to better understand threats and risks and enhance the strategy of AIM.

5.2.2 Principal Component Analysis (PCA)

PCA is the core of current fault diagnosis technology based on multivariate statistical process control. It can reduce the dimensionality of the original data by constructing a new set of latent variables, then extracts the primary change information from the new mapping space, and extracts statistical features, thereby forming an understanding of the spatial characteristics of the original data. In other words, PCA can extract the main feature components of data and is often used for dimensionality reduction of high-dimensional data. Therefore, researchers widely use it in different fields, e.g. data compression, image analysis, recognition, fault detection, etc. Harrou et al. (2013) proposed a PCA-based generalized likelihood ratio (GLR) fault detection method to detect faults in different process variables without a process model. Kaced et al. (2021) used the PCA to solve the problem of false alarms in a chemical process. Ji et al. (2021) developed a comprehensive approach based on two unsupervised learning to design a new combustion risk index. The PCA is used to visualize high-dimensional data. Xie et al. (2019) utilized PCA and partial least squares regression (PLSR) to determine essential factors and forecast the failure rates. Diao et al. (2020) proposed a hybrid approach, containing PCA and guided-wave mode, to identify the leakage location of flanged pipes. In the proposed method, PCA is used to detect leakage occurrence. Chen et al. (2019) used PCA to deal with the original data, and then the processed data is utilized to develop a classifier model to realize the online anomaly detection.

5.2.3 Genetic Algorithm (GA)

GA is a computational model that simulates the natural selection of Darwin's biological evolution theory and the biological evolution process of genetic mechanism. The algorithm uses mathematics and computer simulation operations to transform the problem-solving process into a process similar to the crossover and mutation of chromosomal genes in biological evolution. It is a method to search for the optimal solution by simulating the natural evolution process. It has been widely used in combinatorial optimization, ML, signal processing, adaptive control, etc. In the field of process systems, GA is used with ML algorithms. Jia et al. (2019) used GA to process the monitoring data and then integrated it with support vector regression (SVR) to determine leakage location on the long-distance pipeline. Piri et al. (2021) proposed a comprehensive approach, combining ANN with GA, to balance safe design and cost. The GA is used to train the ANN, and the results show the proposed method can forecast accurately. Zhang and Tan (2018) developed a comprehensive method, which integrated the fuzzy cognitive reliability and error analysis method (CREAM) with GA and adaptive neuro-fuzzy inference system (ANFIS) to assess safety improvement schedule to support decision-making. Sarkar et al. (2019) employed GA and PSO to optimize and enhance the accuracy of the classifier's parameters. Then, the processed data combined with a decision tree is used to predict a slip-trip-fall accident. Zar-anezhad et al. (2019) proposed a comprehensive approach, comprising ANNs, fuzzy systems, GA, and colony optimization algorithms, to predict repair and maintenance-related accidents of process systems. Those four methods are all based on the ability of the human brain or biological groups to learn things and react. However, each method has its limitations. To overcome this, numerous works have been conducted to combine all or part of these approaches to solve the problems that single methods cannot solve. This study indicates that the neural-GA is an efficient model to assess the accident in a process system. Zhao et al. (2019) presented a hybrid methodology consisting of radial basis function and GA to assess structural reliability. The GA is utilized to address the problem of constrained optimization.

5.3 Case Study of Asset Integrity Assessment

Material failures and their degradation or corrosion present a significant challenge to the present-day process industries. Wood et al. (2013) note that corrosion-related phenomena were the primary contributing factor of 20% of major accidents across the EU refineries since 1984. Early identification of credible damage mechanisms, their characterization, and the development of mitigation strategy is the critical task for maintaining asset integrity.

Failure mode and effect analysis (FMEA) is used for damage mode identification and cause and effect characterization. Applied in AIM, FMEA deals with the characterization of all potential damage mechanisms throughout the component's lifecycle, during normal and transient operations where the equipment is subjected to the interaction with the aggressive environment/conditions. FMEA depends on the subject matter experts' (SME) opinion in estimations of its target metrics as risk priority numbers. Quantitative methodologies like API RP 581 (2016) might be applied for these tasks as well. In API 581, the probability of failure (PoF) assessment is based on the concept of the statistically derived generic failure frequency for different types of components and damage factors (DFs) quantifying the severity of material degradation. API RP581 provides extensive calculation protocols covering a wide range of damage mechanisms inherent to the petrochemical industries. However, the massive data inputs and their preprocessing time is often not entirely available for quantitative assessment; thus, making qualitative and semi-quantitative approaches more commonly used for the RAs (Bai and Bai 2014). This explains active SME involvement in the analysis. The involvement of SMEs is not something undesirable in RA. Nevertheless, it can be subjective and dependent on the surveyor's experience and the information available for the assessment.

The main challenge for operators to perform a valid QRA is their limited access to some types of data prescribed in the standard frameworks. Moreover, some critical metrics in the assessments, such as the measured thickness and actual corrosion rates, are scarce due to their relatively high costs. It is rarely prioritized to obtain these data during the component's early operational history. To assess the component's integrity state quantitatively, we may adopt ML approaches.

In identifying critical damage mechanisms and failure modes, we may encounter some sample components with possibly missing input data on the design, operational, or other inspection parameters. In these cases, these input data need to be preprocessed with all missing values imputed. This step will help to prepare the sample ready for a standard RA. The multiple imputation algorithms like expectation-maximization, hot deck, or stochastic regression can be used for that purpose. Alternatively, models like neural networks or decision tree-based algorithms with built-in support of missing values can be powerful as well (Lopes and Bernardete 2012). The latter ones require less time to implement but at the cost of higher required computational power. However, the successful implementation of both approaches is dependent on the quality of the training datasets used. An ideal training dataset is believed to cover a full range of credible damage mechanisms inherent to the potentially assessed components (i.e. being representative); meanwhile, they should be consistent in the data entries over the relationships they carry. An example list of the damage mechanisms is presented in Table 5.2.

Regression algorithms (e.g. random forest regressor, deep neural network) can be used for predicting the target variables (DFs and CoF-s) based on the feature set provided for each sample component. Generally, the choice of the algorithm is primarily dependent on the data to which it is applied, its type, quality, and available computational power. Therefore, no universal algorithm can be applicable and valid for a wide range of cases. In some studies, such as Paltrinieri et al. (2019), they have discussed the application of ML to process risk estimation. Furthermore, the differences between ML and traditional RA method (e.g. risk-based inspection) in applying to process systems are compared. According to the results of these studies, ML can enhance the quality and precision of the assessment output and overcome RA challenges. However, the main accounted advantage of the

Table 5.2 Examples of damage mechanisms.

Number	Governing damage mechanism	Individual damage mechanism	Nomenclature
1	Internal thinning	General internal thinning	Int Thin
2	External thinning	General external thinning	Ext Thin
3		Corrosion under insulation	
4	Stress corrosion cracking	Amine cracking	SCC – AC
5		Caustic cracking	SCC – CC
6		Chloride stress corrosion cracking	SCC – ClSCC
7		Hydrogen induced cracking	SCC – HIC
8		Sulfide stress cracking	SCC – SSC
9	High-temperature hydrogen attack	High-temperature hydrogen attack	HTHA
10	Brittle fracture	Brittle fracture	BF
11		885° embrittlement	885F
12	Mechanical fatigue	Mechanical fatigue	MF

chosen methods was their ability to handle missing inputs. The standard algorithms and especially the networks cannot handle the missing values independently; it generally requires extensive modifications. Nevertheless, given the successful model performance on the complete dataset, retraining a model on the dataset with reduced dimensionality, which fits a specific problem, makes it possible to deal with missing values. This is believed to significantly save the overall computational time compared with imputation techniques with the prediction efficiency retained.

Deep neural networks (DNN) use multiple layers to progressively extract the features in increasing complexity from the input information (Kelleher et al. 2015). Within its structure, the first layer will represent the input set of unnormalized features, the last layer defined as the output with the layer in between being a set of model-derived layers. The overall length of the model chain there defines its deepness, referring to the term "deep" in its name.

Random forest regressor (RFR) is an ensemble algorithm that comprises a set of decision trees built independently and with a different structure. Making several decision trees and taking their weighted vote, rather than relying on a single decision tree, generally prevents overfitting, which is inherent in decision tree-based algorithms (Pal 2005). Overfitting is an undesired condition when the obtained model fits extensively to the data at which it was trained, including noise in data. Random forest algorithm can also be used to calculate feature importance distribution based on Gini impurity reduction. For the practical application, to some extent, it allows tracking the structure of the model and its decision criteria, which is challenging otherwise with DNN.

There are several metrics used for the model performance assessment in regression problems. The list includes but is not limited to the mean absolute error (MAE), mean squared error (MSE), mean absolute percentage error (MAPE), with each having its application specifications dependent on the type of model, data structure, the fraction of outliers, etc. For illustration purposes, MAPE was used due to its intuitive simplicity and independence from the scale of the assessed parameters. MAPE is given by:

$$\text{MAPE} = \frac{1}{n} \sum_{t=1}^{n} \left| \frac{A_t - P_t}{A_t} \right| \tag{5.1}$$

where A_t is the actual value; P_t is the predicted value of the target variable.

The following example is presented to provide a concrete discussion on the use of ML approaches in AIM. For illustrative purpose, Figure 5.1 presents the probabilities of the failure caused by the damage mechanisms. They are predicted by the DNN-based model based on different data availability scenarios. Figure 5.2 gives an example of a risk value chart for each damage mechanism, based on which the prioritization and identification of critical failure modes can be conducted. The data used to plot the figure, including all the DF, are taken from the same predicted outcomes as those in Figure 5.2 are constructed; thus, not accounting for the natural randomness in the model. The consequence of failure (CoF) for each case was predicted by the same DNN-based model, input data for which is based on Level 1 CoF Analysis as per API RP 581.

As indicated by Figures 5.1 and 5.2, the central question of interest is concerned with the quality of the obtained predictions and the overall reliability of the proposed approach as a

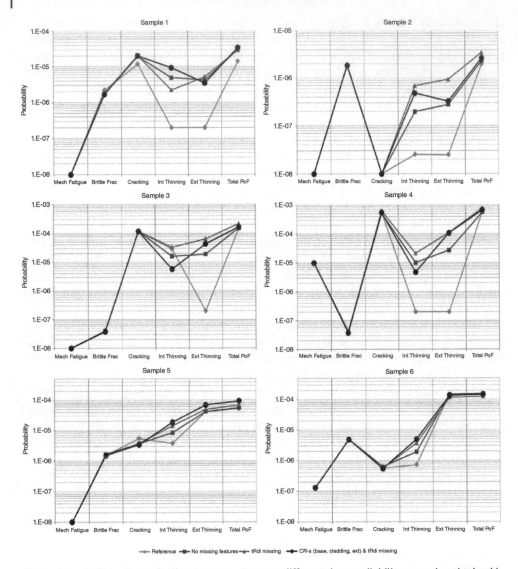

Figure 5.1 PoFs by the major damage type groups at different data availability scenarios obtained by DNN model (dark grey lines). Note: Light grey line indicates the reference predictions as per API RP 581.

whole. The problem statement required the identification of critical failure modes and their risk ranking. As Figure 5.2 illustrates, the hierarchical order of the failure mode risks was successfully predicted even in the cases with limited input data. However, the significant discrepancies were inherent to the magnitudes of the obtained predictions. In some cases, as for Samples 2, 3, and 4 from Figure 5.1, the difference between the reference values and baseline predictions with no missing values for PoF values accounted for two orders of magnitude, i.e. between 2E-7 and 2E-5. There is ample space for discussion on these two numbers. In the domain of statistics, various assessment metrics, e.g. MAE, MSE, or MAPE, being applied would eventually indicate tremendous discrepancies.

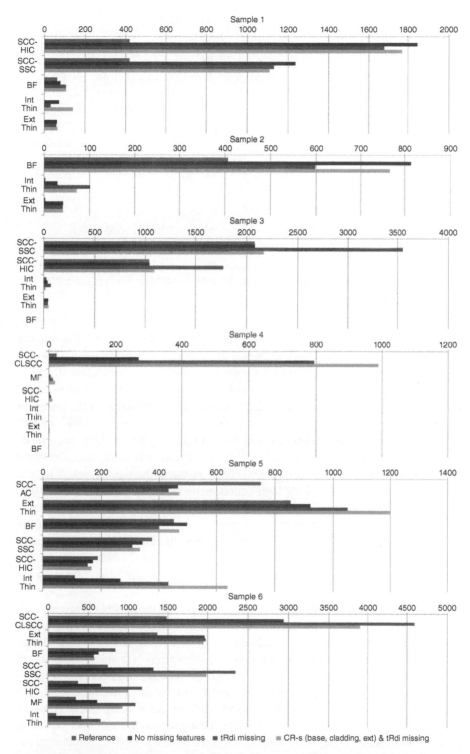

Figure 5.2 Estimated risk values ($/year) for each individual failure mode inherent to the corresponding process component. Note: Predictions produced by DNN model at different data availability scenarios are marked by light grey, reference calculations as per API RP 581 by dark grey.

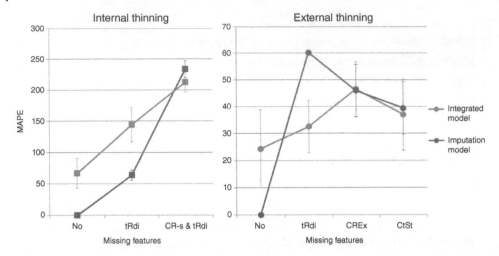

Figure 5.3 Stochastic regression-based versus integrated model performance on both Internal and External DF predictions. Note: the error bars on the chart indicate standard deviation of the results with dot indicating their mean.

The primary assumption of the ML method lies in the premise that the distribution of the test samples would be identical to the distribution of training. Once either sample sets are not comparable to each other and do not represent the overall actual population, the condition is called bias. The phenomenon can be classified into several types depending on the quality of the samples or prejudice of their measurements. The distribution of the results in Figures 5.2 and 5.3 shows the signs of sample bias in the training data. For example, according to the review of the training samples, the absolute majority (93%) of them has a moderate or severe level of internal thinning (DF > 10); meanwhile, only 2% of the entries had the DF of around 1; thus, explaining the general conservatism of the predictions. A similar trend is inherent to the CoF data, where high-consequence entries were dominant as well. Despite the seeming simplicity of the phenomenon, managing bias in the data may not be easy: eventually, the properties distribution is dependent on the user's expectations of the test samples. Ideally, the training dataset represents evenly distributed data with the complete set of relevant features. However, even with the ideal data model, precision is critically dependent on the used ML model itself. One of the critical properties is the bias-variance trade-off.

Underfitting and overfitting are the two opposing states in the model's bias-variance balance. Overfitting occurs when the model includes excessive needed features and fits the training data replicating its structure. Overfitting is undesired in ML as it complicates the model, wasting resources, and leads to incorrect predictions (Hawkins 2004). From Figure 5.3, the significant difference in the accuracy of predictions was observed between Samples 1–4 and Samples 5–6, i.e. between the component sample groups being slightly different in the property's distribution. The model performance was significantly better in the predictions on simulated samples, which were firmly related to the model's training samples. This indicates overfitting. Considering the conservative values of predictions, the DNN model, in analogy with the discussed bias phenomenon, possibly prioritized the widely represented biased high corrosion rates related features over the actual thickness measurements when available in the reference model. Later, when the thickness and corrosion rates were not available, the extended fluid property and material specifications features were used to support the conservative trend. The general ways to combat the overfitting are improving procedures of the model cross-validation by providing more test data to it and tightening the feature selection process.

Stochastic regression is another proposed option for performing the RA with limited data concerning the imputation of the missing features. Unlike the integrated models, which predicted the required risk metrics directly, the approach implies a prediction of the missing metrics first to be used in the standard RA calculations next. Theoretically, prediction of simpler metrics, e.g. corrosion rates or thickness instead of the DF, allows better control over the input set of features; thus, reducing the chances of overfitting or bias. Similar to the integrated models, two DNN models have been trained to predict Internal and External Thinning DFs at different data availability scenarios. To account for the natural randomness of the models, the predictions have repeatedly been made multiple times. This allows obtaining the whole distribution of the predicted metrics. Models' assessment has been carried on the test dataset to diminish the effect of the data bias, with MAPE chosen as the assessment metrics. Figure 5.3 presents the obtained results in comparison with the integrated models set to predict the DFs directly.

Despite the seemed competitiveness of the imputation approach, the results of the simulations did not reveal any significant advantages. In cases with the measured thickness, corrosion rates, or coating integrity state missing, the approaches' performance was approximately equal, with imputation results being generally less variable. Meanwhile, there was no practical meaning to assess imputation with no missing features. In general, the obtained results might indicate that the main complexity associated with predictions of the safety parameters with limited data lies in recovering thickness, corrosion rate information, or any other metrics missing. Thus, regardless of the applied approach of the chosen, the models' performance is critically dependent on the training data quality and the set of model features chosen to predict the optimal parameters.

5.4 Data-Driven Model of Asset Integrity Assessment

Khan et al. (2016a) have proposed a conceptual framework for real-time monitoring and maintenance and claimed that it would be the next generation of the IM approach. In this approach, dynamic risk assessment (DRA) is conducted to support IM planning based on real-time condition monitoring (CM) data. This approach is a way forward with the advancement of process digitalization. However, a bigger step forward could be achieved if other advanced technologies such as the Internet of Things (IoT), big data, cloud computing, smart equipment, and cyber-physical systems (CPS) are applied and integrated into the AIM framework. With the progression of Industry 4.0, one of the significant trends in production system, the CM and DRA-based AIM could be further enhanced to a higher level to ensure process safety and improve production efficiency.

Gobbo Junior et al. (2018) have conducted a study on identifying the links and non-links between process safety and Industry 4.0. As one of the conclusions of their study, new research directions could emerge from these non-links, such as cloud computing, data mining, information technology, the IoT, radio-frequency identification (RFID), and CPS. This

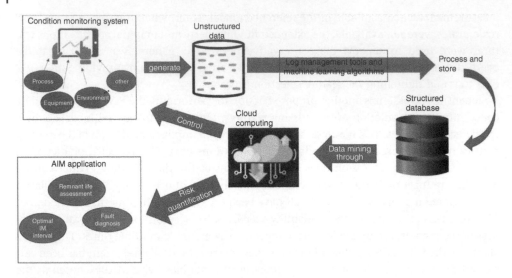

Figure 5.4 The conceptual framework.

paper aims to propose a conceptual framework for a new generation of AIM approach that incorporates the aforementioned technologies under the umbrella of IoT to enhance safety and production efficiency during the process of Industry 4.0 of process industries.

Figure 5.4 presents the proposed conceptual framework. Each of the main elements of the proposed model will be discussed in the following sections.

5.4.1 Condition Monitoring Data Collection

Data pertaining to progressive defects in equipment, deviations in process operation, variations of environmental factors, and others need to be collected through multiple sensors. The data related to process operations have the following features (Wu et al. 2019):

- Spatial dispersion: different types of process equipment are installed in large underground spaces, such as roadways, chambers, and longwall faces.
- Diversity: various types of data coming from the monitoring of underground working environment (e.g. temperature, humidity, smoke, wind speed, stress, groundwater activity, geological structure) and auxiliary data.
- Dynamic variation: CM data changes in real time. Using representative data within a time slot is impossible due to the existence of multiple influencing factors.

5.4.2 Data Processing and Storage

Yuan (2017) has discussed the pros and cons of different data acquisition, transmission, and storage tools in the context of process safety. The literature review presented in the paper concludes that the IoT can overcome the weakness of the existing approaches. IoT can automatically identify, locate, track, and monitor process systems in real time. This extensive network comprises a large number of sensors, signal frequency labels, cameras, and signal recognizers (Wu et al. 2019). The big data collection technologies (e.g. online data filing) enable multiple data collection and integration of heterogeneous data. In contrast, the big data storage technologies use a big data architecture (e.g. Spark, Yarn) to enable the distributed storage of structured data with the functions of full data lifecycle management, data backup, and recovery.

5.4.3 Data Mining for Risk Quantification and Monitoring Control

Multiple anomalies exist in the data collected and stored from the previous step. This step intends to extract the abnormal data from the normal operation data. Statistical classification approaches are applied for this purpose. ML-based approaches can be used to perform the risk-based inspection screen assessment to identify the safety-critical equipment. For the identified equipment, a QRA will be conducted. More research work is needed to investigate how ML-based approaches can be applied to generate the possible incident scenarios (e.g. cracking with pressure, temperature, vibration, and flowrate; corrosion with pressure, temperature, vibration, and flowrate) and estimate their joint probabilities. The PoF data could be generated through MC simulation. Figure 5.5 gives an example of how the joint PoF can be estimated.

The CoF can be estimated by historical data. Loss function can also be used to estimate the real-time loss due to process deviations and asset defect condition (Khan et al. 2016b). The parameters of these loss functions can be learned from historical data. Risk is estimated by the product of PoF and CoF.

5.4.4 AIM Application

The estimated risk can be applied to support the development of IM strategy and quantify the remnant life of each equipment. For instance, the inspection interval can be optimized

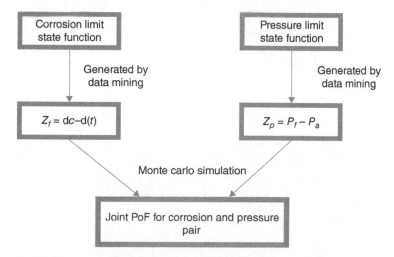

Figure 5.5 An example of steps for joint probability of failure estimation. *Source:* Khan et al. (2016a)/ with permission of Elsevier.

based on identifying the time where the IM cost is minimized subject to an acceptable risk level.

5.4.5 The Application of the Framework

The application of the proposed framework needs to be implemented through different layers (Figure 5.6) under the defined scope of AIM, which includes:

- Hardware and software layer: hardware may consist of the host server, data and its backup clusters, and computer room. Software systems comprise of data management, data storage, and operating systems. It provides technical support for data collection and transmission.
- Data sensing layer: the function of this layer is to collect CM data for underground coal mines in a real-time manner. It needs the devices such as sensors, positioning equipment, communication equipment, and RFID tags. The sensing devices are targeted for equipment, process, environment, and personnel.
- Data transmission layer: aims to transmit the data to the process and storage units. Considering the harsh environment of underground coal mines, it is recommended to use wireless sensor networks (WSNs) because the wired network would be challenging to design its layout, install, and maintain good performance (Wu et al. 2019).
- Data storage and process layer: the received data are cleaned, reduced, integrated, transformed, and stored for standardization purpose. It constructs the big data database and knowledge base.
- Data extraction and analysis layer: statistical classification techniques and other advanced ML-based approaches are used to classify abnormal data, generate incident

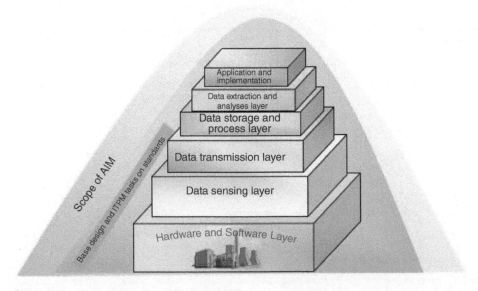

Figure 5.6 Technical layers of CM-based AIM.

scenarios, and quantify operational risks. BN-based risk models are recommended to deal with inter-dependent influencing factors such as PoF and CoF calculations.

- Application and implementation layer: apply the quantified risk to support any decision-making in AIM. The achieved IM strategies and tasks need to comply with applicable standards on the base design.

5.5 Conclusion

With technology advancement, smart equipment and systems are developed and applied to monitor and control process systems to ensure safety. Process systems become more complex and automated, while they also generate enormous data that can be used for safety analysis. ML-based approaches provide promising ways to utilize these data and formulate a new generation of process safety and asset integrity assessment methods. A high-level conceptual model of data driven is proposed in this chapter to serve as a generic framework for developing ML-based AIM. Similarly, this framework can be adapted for process safety management. Researchers in the safety domain endeavor to build up the chemistry between safety and ML algorithms.

References

Abbassi, R., Khan, F., Garaniya, V. et al. (2015). An integrated method for human error probability assessment during the maintenance of offshore facilities. *Process. Saf. Environ. Prot.* 94: 172–179.

Abdulla, M.B. and Herzallah, R. (2015). Probabilistic multiple model neural network based leak detection system: experimental study. *J. Loss Prev. Process Ind.* 36: 30–38.

Adedigba, S.A., Khan, F., and Yang, M. (2017). Dynamic failure analysis of process systems using neural networks. *Process. Saf. Environ. Prot.* 111: 529–543.

Adedigba, S.A., Khan, K., and Yang, M. (2018). An integrated approach for dynamic economic risk assessment of process systems. *Process. Saf. Environ. Prot.* 116: 312–323.

Arunthavanathan, R., Khan, F., Ahmed, S. et al. (2021). A deep learning model for process fault prognosis. *Process. Saf. Environ. Prot.* 154: 467–479.

Ayhan, B.U. and Tokdemir, O.B. (2019). Safety assessment in megaprojects using artificial intelligence. *Saf. Sci.* 118: 273–287.

Bai, Y. and Bai, Q. (2014). *Subsea Pipeline Integrity and Risk Management*. New York: Gulf Professional Publishing.

Benkouider, A.M., Kessas, R., Yahiaoui, A. et al. (2012). A hybrid approach to faults detection and diagnosis in batch and semi-batch reactors by using EKF and neural network classifier. *J. Loss Prev. Process Ind.* 25: 694–702.

Casciano, M., Khakzad, N., Reniers, G. et al. (2019). Ranking chemical industrial clusters with respect to safety and security using analytic network process. *Process. Saf. Environ. Prot.* 132: 200–213.

Chen, W.Z., Liu, T.J., Tang, Y. et al. (2019). Multi-level adaptive coupled method for industrial control networks safety based on machine learning. *Saf. Sci.* 120: 268–275.

Chen, S.M., Yu, J.B., and Wang, S.J. (2021). One-dimensional convolutional neural network-based active feature extraction for fault detection and diagnosis of industrial processes and its understanding via visualization. *ISA Trans.* 122: 424–443. https://doi.org/10.1016/j.isatra.2021.04.042.

Cho, J., Kim, H., Gebreselassi, A.L. et al. (2018). Deep neural network and random forest classifier for source tracking of chemical leaks using fence monitoring data. *J. Loss Prev. Process Ind.* 56: 548–558.

Diao, X., Chi, Z.Z., Jiang, J.C. et al. (2020). Leak detection and location of flanged pipes: an integrated approach of principle component analysis and guided wave mode. *Saf. Sci.* 129: 104809.

Harrou, F., Nounou, M.N., Nounou, H.N. et al. (2013). Statistical fault detection using PCA-based GLR hypothesis testing. *J. Loss Prev. Process Ind.* 26: 129–139.

Hawkins, D.M. (2004). The problem of overfitting. *J. Chem. Inf. Comput. Sci.* 44: 1–12.

Hemmatian, B., Casal, J., Planas, E. et al. (2020). Prediction of BLEVE mechanical energy by implementation of artificial neural network. *J. Loss Prev. Process Ind.* 63: 104021.

Hinton, G.E., Osindero, S., and The, Y.W. (2006). A fast learning algorithm for deep belief nets. *Neural Comput.* 18: 1527–1554.

Ji, C.X., Jiao, Z.R., Yuan, S. et al. (2021). Development of novel combustion risk index for flammable liquids based on unsupervised clustering algorithms. *J. Loss Prev. Process Ind.* 70: 104422.

Jia, Z.G., Ho, S.C., and Li, Y. (2019). Multipoint hoop strain measurement-based pipeline leakage localization with an optimized support vector regression approach. *J. Loss Prev. Process Ind.* 62: 103926.

Jiao, Z., Hu, P., Xu, H., and Wang, Q. (2020). Machine learning and deep learning in chemical health and safety: a systematic review of techniques and applications. *ACS Chem. Health Safety* 27 (6): 316–334.

Junior, J., Busso, C., Gobbo, S., and Carreao, H. (2018). Making the links among environmental protection, process safety, and industry 4.0. *Process. Saf. Environ. Prot.* 117: 373–382.

Kaced, R., Kouadri, A., Baiche, K. et al. (2021). Multivariate nuisance alarm management in chemical processes. *J. Loss Prev. Process Ind.* 72: 104578.

Kelleher, J.D., Mac Namee, B., and D'Arcy, A. (2015). *Fundamentals of Machine Learning for Predictive Data Analytics: Algorithms, Worked Examples, and Case Studies*. Cambridge: The MIT Press.

Khan, F., Thodi, P., Imtiaz, S. et al. (2016a). Real-time monitoring and management of offshore process system integrity. *Curr. Opin. Chem. Eng.* 14: 61–71.

Khan, F., Wang, H.Z., and Yang, M. (2016b). Application of loss functions in process economic risk assessment. *Chem. Eng. Res. Des.* 111: 371–386.

Li, Q., Wang, D.X., Zhao, M.Y. et al. (2021a). Modeling the corrosion rate of carbon steel in carbonated mixtures of MDEA-based solutions using artificial neural network. *Process. Saf. Environ. Prot.* 147: 300–310.

Li, Q.Z., Zhou, S.N., and Wang, Z.Q. (2021b). Quantitative risk assessment of explosion rescue by integrating CFD modeling with GRNN. *Process. Saf. Environ. Prot.* 154: 291–305.

Lopes, N. and Bernardete, R. (2012). Handling missing values via a neural selective input model. *Neural Netw. World* 22 (4): 357–370.

Mao, S., Wang, B., Tang, Y. et al. (2019). Opportunities and challenges of artificial intelligence for green manufacturing in the process industry. *Engineering* 5: 955–1002.

Ossai, C.I. (2020). Corrosion defect modelling of aged pipelines with a feed-forward multi-layer neural network for leak and burst failure estimation. *Eng. Fail. Anal.* 110: 104397.

Pal, M. (2005). Random forest classifier for remote sensing classification. *Int. J. Remote Sens.* 26 (1): 217–222.

Paltrinieri, N., Comfot, L., and Reniers, G. (2019). Learning about risk: machine learning for risk assessment. *Saf. Sci.* 118: 475–486.

Piri, J., Pirzadeh, B., Keshtegar, B. et al. (2021). Reliability analysis of pumping station for sewage network using hybrid neural networks – genetic algorithm and method of moment. *Process. Saf. Environ. Prot.* 145: 39–51.

Russell, S.J. and Norvig, P. (2016). *Artificial Intelligence: A Modern Approach*. Kuala Lumpur: Pearson Education Limited.

Sarbayev, M., Yang, M., and Wang, H.Q. (2019). Risk assessment of process systems by mapping fault tree into artificial neural network. *J. Loss Prev. Process Ind.* 60: 203–212.

Sarkar, S., Raj, R., Vinay, S. et al. (2019). An optimization-based decision tree approach for predicting slip-trip-fall accidents at work. *Saf. Sci.* 118: 57–69.

Sattari, F., Macciotta, R., Kurian, D. et al. (2021). Application of Bayesian network and artificial intelligence to reduce accident/incident rates in oil & gas companies. *Saf. Sci.* 133: 104981.

Shen, R., Jiao, Z., Parker, T. et al. (2020). Recent application of computational fluid dynamics (CFD) in process safety and loss prevention: a review. *J. Loss Prev. Process Ind.* 67: 104252.

Sun, H., Wang, H.Q., Yang, M. et al. (2020). On the application of the window of opportunity and complex network to risk analysis of process plants operations during a pandemic. *J. Loss Prev. Process Ind.* 68: 104322.

Sun, H., Wang, H.Q., Yang, M. et al. (2021). Resilience-based approach to safety barrier performance assessment in process systems. *J. Loss Prev. Process Ind.* 73: 104599.

Tan, W.L., Nor, N.M., Bakar, M.Z.A. et al. (2012). Optimum parameters for fault detection and diagnosis system of batch reaction using multiple neural networks. *J. Loss Prev. Process Ind.* 25: 138–141.

Wen, K., He, L., Liu, J. et al. (2019). An optimization of artificial neural network modeling methodology for the reliability assessment of corroding natural gas pipelines. *J. Loss Prev. Process Ind.* 60: 1–8.

Wood, M.H., Arellano, A.L.V., and Wijk, L.V. (2013). *Corrosion-Related Accidents in Petroleum Refineries*. Luxembourg: European Commission, Institute for the Protection and Security of the Citizen.

Wu, Y.Q., Chen, M.M., Wang, K. et al. (2019). A dynamic information platform for underground coal mine safety based on internet of things. *Saf. Sci.* 113: 9–18.

Xie, L., Habrekke, S., Liu, Y.L. et al. (2019). Operational data-driven prediction for failure rates of equipment in safety instrumented systems: a case study from the oil and gas industry. *J. Loss Prev. Process Ind.* 60: 96–105.

Yang, M. (2020). Safety system assessment using safety entropy. *J. Loss Prev. Process Ind.* 66: 104174.

Yuan, L. (2017). Framework and key technologies of internet of things for precision coal mining. *Indus. Minc. Automat* 43 (10): 1–7.

Zaranezhad, A., Mahabadi, H.A., and Dehghani, M.R. (2019). Development of prediction models for repair and maintenance related accidents at oil refineries using artificial neural network, fuzzy system, genetic algorithm, and ant colony optimization algorithm. *Process. Saf. Environ. Prot.* 131: 331–348.

Zhang, R.Y. and Tan, H. (2018). An integrated human reliability-based decision pool generating and decision-making method for power supply system in LNG terminal. *Saf. Sci.* 101: 86–97.

Zhao, J., Chen, J.Q., and Li, X. (2019). RBF-GA: an adaptive radial basis function metamodeling with genetic algorithm for structural reliability analysis. *Reliab. Eng. Syst. Saf.* 189: 42–57.

6

Machine Learning for Process Fault Detection and Diagnosis

Rajeevan Arunthavanathan[1], Salim Ahmed[1], Faisal Khan[2], and Syed Imtiaz[1]

[1] Centre for Risk, Integrity, and Safety Engineering (C-RISE), Faculty of Engineering and Applied Science,
Memorial University of Newfoundland, St. John's, NL, Canada
[2] Mary Kay O'Connor Process Safety Center, Artie McFerrin Department of Chemical Engineering,
Texas A&M University, College Station, TX, USA

6.1 Background

Fault detection and diagnosis (FDD) are crucial for safe operation of process systems. Continuous operation, and thus economic and production objectives, of a plant can be achieved by accurately detecting fault conditions and their proper diagnoses before the faults leading to failures. Many fault detection and diagnosis approaches have been developed over the years and these methods are generally classified as analytical model-based, knowledge-based, and data-driven approaches (Chiang et al. 2001; Zhang and Jiang 2008a; Alzghoul et al. 2014). Analytical model-based approaches use first principle models of the system components in deriving a mathematical description of the system (Gertler 1997; Venkatasubramanian et al. 2003b; Isermann 2011). For a complex process plant, it is difficult to develop first principle models that can capture all aspects of the system (Zhang and Jiang 2008b). Therefore, for such plants, the analytical model-based approach has significant limitations. In the absence of a proper analytical model with the required inputs, outputs, and states, knowledge-based approaches can provide a better solution for FDD (Fan and Lu 2008). However, to develop the knowledge-based algorithms, human knowledge regarding the relevant process system is essential (Frank 1996; Alzghoul et al. 2014). Without sufficient knowledge about the process plant, the knowledge-based method may lead to inaccurate outcomes. Data-driven approaches are becoming more popular in recent years because these methods can translate processed data into useful knowledge without using much human expertise about the process plant (Venkatasubramanian et al. 2003a; Yin et al. 2014).

The data-driven approach is further classified into traditional statistical approach and machine learning (ML) approach. The traditional statistical models use a sample population and a hypothesis to infer connections between variables in order to develop a model. ML models, on the other hand, are built on the basis of accurate predictions using supervised or unsupervised learning.

Machine Learning in Chemical Safety and Health: Fundamentals with Applications, First Edition.
Edited by Qingsheng Wang and Changjie Cai.
© 2023 John Wiley & Sons Ltd. Published 2023 by John Wiley & Sons Ltd.

The traditional statistical procedures are divided into two categories: univariate and multivariate. In the univariate approach, each variable is monitored separately in order to detect its associated faults using the variable's threshold limit (Kano et al. 2003). To investigate the univariate characteristics of process systems, statistical process control (SPC) charts such as the Shewhart control chart, cumulative control charts (CUSUM), and exponentially weighted moving average (EWMA) control charts are widely used (MacGregor and Kourti 1995; Alauddin et al. 2018; Arunthavanathan et al. 2021a). These approaches use estimations to create upper and lower control boundaries in order to distinguish between normal and abnormal operations. For uncorrelated processed data, the univariate approaches are useful. However, for almost all process systems, variables are highly correlated; this necessitates the use of multivariate statistical analysis.

Multivariate approaches have been widely employed in process FDD due to the highly correlated nature of data resulting from the complexity of process systems. The principal component analysis (PCA) and the partial least squares (PLS) are among the frequently used multivariate methods; the ability to handle highly correlated data without much preprocessing is a key benefit of these methods (Qin 2012). These two approaches are applied to minimize feature dimension and are integrated with various statistical hypothesis testing algorithms such as T^2, ANOVA, or MANOVA test to detect a fault condition (Ge and Song 2013; Ge 2017).

In industry 4.0, ML techniques are influencing engineering and scientific breakthroughs in important ways. Feature generation, fault classification, and decision making are the basic and traditional ML tasks relevant to FDD (R. Liu et al. 2018; Arunthavanathan et al. 2021a). In the recent development, intelligent FDD is a term that refers to the application of ML models, such as artificial NNs (ANNs), support vector machines (SVMs), and deep NNs (DNNs) (Lei et al. 2020), to process plant fault diagnosis.

6.2 Machine Learning Approaches in Fault Detection and Diagnosis

ML refers to a nexus of computer science and statistics in which algorithms are utilized to do a task without being explicitly programmed for the same; instead, they recognize patterns in the data and generate predictions as a new data point arrives (Nor et al. 2020). In general, these algorithms' learning processes can be either supervised or unsupervised, depending on the data used to feed them.

Supervised learning requires a set of erroneous and normal data. All fault condition data must be collected in order to create the necessary models using supervised learning (Dai and Gao 2013; Lo et al. 2019). NNs and SVMs are two of the most popular supervised ML algorithms for categorizing fault scenarios. However, because of the scarcity of faulty data, it is difficult to implement this strategy for a newly constructed system. Also for an existing plant, generating or having such labeled data for all faulty conditions is difficult for many process units. As a result, the unsupervised learning strategy has gained popularity in recent years in ML FDD.

Unsupervised ML models are trained on historical data that lacks categorization labels for fault conditions (Dike et al. 2018). However, these models employ an algorithm to pick classification tags by identifying hidden patterns in the data. Clustering and anomaly detection are two unsupervised techniques that are frequently employed in FDD. Clustering is a technique for classifying a collection of data into smaller groups that can be used for multiclass classification. Anomaly detection is a two-step classification process that identifies the outliers.

6.3 Supervised Methods for Fault Detection and Diagnosis

In this section, the two most commonly used supervised ML methods for FDD, namely the NN and the SVMs, are discussed.

6.3.1 Neural Network

The use of NN for fault detection in process systems began in the late 1980s. (Hoskins and Himmelblau 1988) and (Venkatasubramanian and Chan 1989) investigated and used NN-based techniques for FDD in process systems. The FDD methods based on NNs are applicable for both linear and nonlinear dynamic systems. As a result, there was significant research interest in the early 1990s; however, due to the need of demanding computational power and data collection for normal and abnormal situations, constructing models based on NN faced growing challenges.

6.3.1.1 Neural Network Theory and Algorithm

The NN is a supervised learning method and its outputs classify each fault category based on the fault label. A typical NN consists of three elements: the structure, the learning algorithm, and the activation function. The structure includes the input signals, the hidden layers, and the outputs. Figure 6.1 illustrates the simplest NN model structure with a single hidden layer. In the case of FDD using NN, the input nodes are the features/variables involved in the fault condition and the outputs are the faults. For an example, if a process system data contained pressure, temperature, and flow rate information to detect the two different system fault conditions, then the NN model structure will have three input layers to feed the information to the network and two output layers to classify the two different fault conditions. As shown in Eq. (6.1), network inputs nodes (x_i) are subjected to each synaptic weight in the hidden layers, respectively $(x_i \cdot \theta_i^j)$. The number of hidden layers and the number of nodes in each hidden layer depend on the application. After the processing of summing with the bias (b^j) and the activation function (g), the classification output y is obtained for each data samples. The purpose of activation function in the NN is to introduce the nonlinearity into the output neuron. There are several activation functions as summarized in Table 6.1. Mathematically an activation function is described as

$$a_i^j = g\left(x_i \cdot \theta_i^j + b^j\right) \tag{6.1}$$

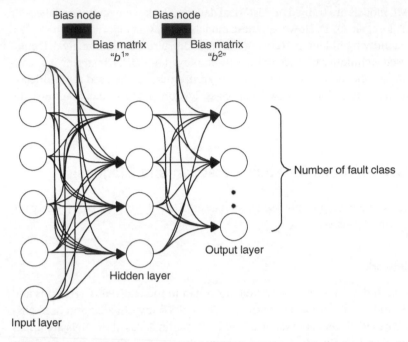

Figure 6.1 Neural network model.

Table 6.1 Neural network activation functions.

Function	Output Range
Sigmoid	0,1
Tanh	−1.1
ReLu	0,+∞
LeakyRelu	−∞, + ∞
Softmax	0,1

where a_i^j is an activation of node i in layer j, $\theta_i^j i$ is the weight matrix from layer j to $j+1$, b^j is the bias matrix in the jth layer. Initial data samples are defined as (x_i) in the input layer, and for hidden and output layers $(j-1)$, node value is defined as (x_i).

To determine the classified output using an NN model, training data must include the classification information for each sample. For an example, if a process system contained the non-faulty and faulty samples in the data, those should be exemplified by a fault index in the training data; for example, non-faulty may be given an index "0," fault data may be given "1."

6.3.1.2 Neural Network Learning for Fault Classification

The backpropagation algorithm is the key in the NN model to classify a fault condition. Backpropagation can be adopted to obtain the error through the feedforward network transmission as illustrated in Figure 6.1. The backpropagation algorithm operates in the feedforward stage and the backpropagation phase. In the feedforward stage training, input data are fed into the input layer and forward propagate to the output layer through hidden layers to generate the output. In this stage, the NN weights (θ) in each node and each layer are determined randomly. Then the mean squared error (MSE) between the measured output ($\hat{y}$) and target output (y) is calculated. The MSE is defined in Eq. (6.2).

$$MSE = \frac{1}{N}\sum\nolimits_{i=1}^{N}(y-\hat{y})^2 \tag{6.2}$$

In the backpropagation phase, with the calculated MSE, the NN information backpropagates from the output layer to the input layer and the weights and bias in the corresponding nodes are updated using the delta rule.

$$\theta^+ = -\mathrm{lr}\cdot\frac{\partial MSE}{\partial\theta} + \theta \tag{6.3}$$

In Eq. (6.3), θ^+ denotes updated weight, θ denotes current weight, and lr denotes the learning rate. Similarly, NN bias will get updated using the learning rate and a prior bias value.

However, when using backpropagation to train the feedforward NN, it is required to randomly select hyperparameters such as the network topology, a learning rate value, the activation function, and a value for the gain in the activation function. Depending on the selection of these parameters, there may be slow convergence or even network paralysis. Another problem is that the bias of the gradient descent technique, which is used in the training process, can easily get stuck at the local minima. Therefore, improving the application of backpropagation remains an important research topic.

To overcome these issues, several approaches have been proposed by a number of research studies. For efficient learning in feedforward NN, momentum acceleration was proposed (Perantonis and Karras 1995). However, this method requires more training time and higher computational cost. The universal acceleration techniques were introduced to backpropagation based on the extrapolation of each node's weight (Kamarthi and Pittner 1999). However, to perform extrapolation at the end of each iteration, the convergence behavior of each network weight in the algorithm needs to be individually examined. The use of the Levenberg–Marquardt algorithm for feedforward NN training was proposed by (Lera and Pinzolas 2002). In this approach, the computational time depends on the neighborhood size.

In recent years, heuristic algorithms and search algorithms such as genetic algorithm, particle swarm optimization, and harmonic search algorithm have become popular in NN parameter optimizations. Unlike backpropagation, these algorithms generate a pool of initial solutions in which the selection of the near-optimal solution can be made for a choice. However, these algorithms require more computational time to conduct a search process.

6.3.1.3 Algorithm for Fault Classification Using Neural Network

This section illustrates the algorithmic steps to implement the supervised NN approach for fault classification.

Algorithm: Neural Network Offline Training and Online Classification

Offline train model

1a. **Input** Non-faulty preprocessed data with a non-faulty label (one-hot encoding)

2a. **Output** NN model.

3a. Randomly initialize the weights.

4a. Implement the forward propagation and determine the cost function.

5a.   **for** (1: length_data):

6a.   Perform forward propagation and backpropagation to optimize the w and b in Eq. (6.1) using Eq. (6.3).

7a. Verify the backpropagation by using MSE and determine the appropriate learning rate for the optimization algorithm.

8a. NN_Model is developed for fault classification.

Online/Real-time classification

1b. **Input** Preprocessed data is sampled using the "n" window.

2b. **Output** Fault classification for a known (supervised) fault condition.

3b. Developed/updated NN model (input sample)

4b.   **for** (1: length_input sample):

5b.   Predict the one-hot encoding classification for each sample.

6b.   Convert the predicted one-hot encoding label to a classified label.

7b. Using the classified label, define the test accuracy.

6.3.2 Support Vector Machine

In ML, SVMs are supervised learning methods with associated learning algorithms that analyze data and recognize patterns. SVMs are non-probabilistic binary linear classifiers that divide a dataset into two classes based on the hyperplane (a decision boundary) that has the greatest distance between support vectors in each class.

6.3.2.1 Support Vector Machine Theory and Algorithm

Consider a training dataset for SVM containing the fault and no-fault classes as $\mathbb{X}^{NxF_e}$, in which N represent the number of observed samples, while F_e, denote the variables/features in the processed data samples. Similarly, corresponding $\mathbb{Y}$ serves as a class label containing two entries -1 and 1. $y_i = 1$ denotes positive class (non-fault class), and $y_i = -1$ denotes negative class (fault class). Therefore, the supervised dataset denotes:

$$\mathbb{D} = \left\{ (x^1, y^1), \ldots, (x^N, y^N) \right\}, \text{where } x \in \mathbb{R}, y \in \{-1, 1\} \tag{6.4}$$

If the dataset is linearly separable, the SVM will try to separate by a linear hyperplane:

$$y_i f(x_i) = y_i(\langle w, x_i \rangle + b) \geq 1, i = 1, \ldots, N \tag{6.5}$$

However, not every dataset is linearly separable. Constraints must be relaxed to allow for the processing of linearly inseparable cases using the linearly separating hyperplane. This is done by introducing the slack variables

$$\xi_i \geq 0, i = 1, \ldots, N \tag{6.6}$$

where ξ_i represents a measure of distance from hyperplane to misclassified points.

Therefore, the relaxed constraints are given by

$$y_i(\langle w, x_i \rangle + b) \geq 1 - \xi_i, i = 1, \ldots, N \tag{6.7}$$

The primal optimizations then become

$$\min_{w, b, \xi} \frac{1}{2} \|w\|^2 + C \sum_{i=1}^{N} \xi_i \tag{6.8}$$

Subject to: $y_i(\langle w, x_i \rangle + b) \geq 1 - \xi_i,$

$\xi_i \geq 0, i = 1, \ldots, N.$

where the parameter C is a constant corresponding to the selection of $\|w\|^2$.

The aforementioned optimization problem can be presented using the Lagrangian optimization method

$$L(w, b, \xi, \alpha, \gamma) = \frac{1}{2} \|w\|^2 + C \sum_{i=1}^{N} \xi_i - \sum_{i=1}^{N} \alpha_i [y_i(\langle x, w \rangle + b - 1 + \xi_i)] - \sum_{i=1}^{N} \gamma_i \xi_i \tag{6.9}$$

where α and γ are the Lagrangian multipliers.

This optimization problem can be solved by first resolving the quadratic optimization problem described here.

$$\max_{\alpha} W(\alpha) = \max_{\alpha} \left\{ \sum_{i=1}^{N} \alpha_i - \frac{1}{2} \sum_{i,j=1}^{N} y_i y_j \alpha_i \alpha_j \langle x_i, x_j \rangle \right\} \tag{6.10}$$

Subject to $\sum_{i=1}^{N} y_i \alpha_i = 0, \alpha_i \in [0, C], i = 1, \ldots, N$

To transform the input space into a higher-dimensional feature space suitable for classification, a nonlinear kernel function, K can be used. The frequently used kernel functions and optimization parameters are described in Table 6.2.

$$K(x_i, x_j) = \langle \phi(x_i), \phi(x_j) \rangle \tag{6.11}$$

Table 6.2 Different Kernel functions for SVM.

Kernel function	Formula	Optimization parameter
Liner or Dot product	$K(x_i, x_j) = x_i \cdot x_j + C$	C (regularization)
Polynomial	$K(x_i, x_j) = (x_i, x_j + C)^d$	d (degree of the polynomial)
Sigmoid	$K(x_i, x_j) = \tanh\left(\alpha(x_i^T \cdot x_j) + r\right)$	α (slope), r (intercept constant)
Radial basis function (RBF)	$K(x_i, x_j) = \exp\left(-\dfrac{\|x_i - x_j\|^2}{2 \cdot \sigma^2}\right)$	σ (variance)

where ϕ represents a mapping from the input space to the feature space. Therefore, the classifier implementing the optimal separating hyperplane comes out in the following form:

$$f_{(x)} = \text{sgn}\left(\sum_{i,j=1}^{N} \alpha_i y_i \langle x_i, x_j \rangle + b\right) \tag{6.12}$$

6.3.3 Support Vector Machine Model Selection and Algorithm

The SVM algorithm is a powerful data classification approach and constructing a defect detection model based on the SVM algorithm is not difficult. In the process, FDD training and testing datasets are used to classify the normal and abnormal behavior. The basic procedure for classifying data with the SVM algorithm is as follows: first, construct a classifier using the training set, and then use it to predict the target value of the data in the testing set where just the attributes are known. To construct the classifier model and perform FDD, the following algorithm can be used.

Algorithm: SVM Classification

```
Offline Training
    Input: Normal and abnormal mixed supervised data samples
```
$\left(\mathbb{X}^{N \times F_e}, y^{(i, c)}\right)\big|_{i=0}^{N}$, length of the data samples (N), features in the data sample (F_e), and c denotes the number of a fault condition.
```
    Output: SVM model for classification.
 1. Data preprocessing.
 2. Find the best Kernel by experimenting with a variety of them
    and then looking for the factors that make it work best.
 3. Use the optimal parameters and appropriate Kernel function
    (as shown in Table 6.2) to build the classifier.
```

4. Do a test on the created classifier with the testing data. As a result, the incorrect data will be discovered, and the problematic station will be identified.

Online testing

5. **Input: Online processed data sample** ($\mathbb{X}^{N \times F_e}$), N can be varied based on the data sampled window and feature in the data sample F_e cannot vary.
6. **Output:** Fault and non-fault class.
7. **Online data preprocessing**
8. Fed, $\mathbb{X}^{N \times F_e}$ dataset into the trained SVM model to classify fault and non-fault using margin.

6.3.4 Support Vector Machine Multiclass Classification

The SVM is fundamentally a two-class classifier. Therefore, when applying these methods in FDD, it can classify only the fault and non-fault conditions. However, in process system monitoring, it is important to diagnose a fault condition by classifying the different kinds of faults. In order to generate a multiclass classifier, various approaches have been proposed to combine multiple two-class SVMs.

In the one against-the-rest approach (Liu and Zheng 2005), assume the dataset is to be classified into M distinct categories. As a result, M binary SVM classifiers may be developed, each trained to discriminate between one class and the remaining $M-1$ classes. For instance, a binary classifier for class one is designed to discriminate between class one data vectors and data vectors from other classes. Other SVM classifiers are similarly created. Classification of data vectors occurs during the testing or application phase; it is done by determining the margin from the linear separation hyperplane. The final result is the class that corresponds to the largest margin SVM. However, in this approach, individual classifier judgments can generate inconsistent outcomes when an input is assigned to numerous classes concurrently.

A variant of one-against-the-rest approach was proposed by (Lee et al. 2004) where the model target values are modified as +1 for positive classes and the negative classes has a target of $-1/(K-1)$.

Instead of creating many binary classifiers to determine the class labels, multiclass objective function attempts to directly solve a multiclass problem (Weston and Watkins 1998). This is accomplished by altering the binary class objective function for each class and adding a constraint to it. However, this can significantly slow the training process because, instead of solving K separate optimization problems each over N samples with an overall cost, a single optimization problem of size $(K-1)/N$ must be solved giving an overall cost.

Error-Correcting Output Coding (ECOC) multi-class approach is another way to handle multi-class classification problems using binary (two-class) classifiers (Dietterich and Bakiri 1994). This technique works by transforming multiclass classification issues into a large number of two-class classification problems. (Dietterich and Bakiri 1994) proposed to use codes with maximum Hamming distance between each other. (Allwein et al. 2000) proposed another scheme using a margin-based binary learning algorithm to replace the Hamming distance-based decoding.

6.4 Unsupervised Learning Models for Fault Detection and Diagnosis

Most of the FDD procedures for process system monitoring are developed using supervised methods. However, processed data for all fault conditions are not always available. Furthermore, in recent years, with the advancement of digital technologies, the processing system FDD is intended to develop in an autonomous environment. When developing an autonomous FDD method, it is necessary to monitor the process condition in real time using unlabeled data.

Based on the above two viewpoints, developing an unsupervised learning model for process monitoring has become a topic of significant interest in recent research. In ML, clustering is a commonly used unsupervised learning tool used to detect anomalies. Anomaly detection is the process of identifying abnormal events in a dataset. In a process system, the signals generated during a faulty condition have different patterns that differ from the signals during normal operation; the patterns indicate a change in the process behavior. Using a method that indicates changes in the current state of the equipment does not require a labeled historical dataset for training.

Anomaly detection techniques are capable of encapsulating the state of a multivariate system using a single quantitative indicator, which is commonly referred to as the anomaly score. While many approaches provide guidelines on how to define outliers based on the anomaly scores, the quantitative nature of the anomaly score allows for the implementation of several strategies that facilitate a trade-off between false positives and false negatives depending on the application. The mathematical formulation and optimization details of selected unsupervised models are detailed in the following sections.

6.4.1 K-Nearest Neighbors

K-nearest neighbors (KNN) are mainly used in classification and regression. The density for a particular location may be approximated locally using the distance between the spots in the KNN. KNN may also be used in the context of anomaly detection; given a sample, the distance to its kth-nearest neighbors can be regarded as an abnormality score. The anomaly score of the KNN is formulated as

$$\text{score}_{\text{KNN}} = D^K(x)$$

where $D^K(x)$ denotes the distance of the kth-nearest neighbor from observation x. Any metric distance function may be used as the distance function. The most often used approaches for selecting a distance function are the maximum distance, which uses the distance to the kth neighbor, and the least distance, which uses the distance to the kth neighbors. As the anomaly score, the mean distance, which is the average or median of all k neighbors' distances, is used.

6.4.2 One-Class Support Vector Machine

Using a one-class support vector machine (OCSVM), unsupervised raw sensor real-time data can be used to monitor the system condition online. An OCSVM formulation is a specific instance of SVM (Alam et al. 2020). In the traditional binary SVM classification, a hyperplane separates the two classes with a large possible margin, and it is supported by the support vectors (Xiao et al. 2016). In the one-class classification, the hyperplane exhibits only positively labeled data during the training.

In OCSVM, the origin of the coordinate system is assigned to the hyperplane corresponding to the negative class (Alam et al. 2020). As a result, the goal of OCSVM is to discover the hyperplane that is further away from the origin and where positively labeled data occur in the positive half-space of the hyperplane. Therefore, the OCSVM model output provides the +1 value for data within the region and the −1 value for the region outlier, as shown in Eq. (6.13).

$$d_n = \begin{cases} +1 \text{ is data point } x_i \text{ is in the} \\ -1 \text{ if data point } x_i \text{ is not in the region} \end{cases} \tag{6.13}$$

where x_i is a data point available in the dataset, $x_i \subseteq$ (training or testing data). Let x be a training input sample; then, to separate the data points from the origin, OCSVM initially solves the quadratic programming equation shown in Eq. (6.14) and optimizes the parameters.

$$\min_{w,\varepsilon,b} \frac{1}{2}\|w\|^2 + \frac{1}{nr}\sum_{i=1}^{N}\varepsilon_i - b \tag{6.14}$$

Subject to $\langle w, \Phi(x_i)\rangle \geq b - \varepsilon_i, \varepsilon_i \geq 0$

where ε_i is a slack variable corresponding to the ith training sample that allows it to lie on the other side of the decision boundary. Φ is a nonlinear mapping function that maps x_i to kernel space. 'b' is a bias term, 'r' is the regularization parameter, and n is the size of the dataset. When the optimization is complete, the condition may be used to infer query sample testing data using Eq. (6.15).

$$d_n = \text{sgn}\left(\langle w, \Phi(x_i)\rangle - b\right) \tag{6.15}$$

The best value of parameters in the problem described in Eq. (6.14) can be obtained using the kernel function K, without explicitly specifying the mapping function Φ using the dual form of the problem, as shown in Eq. (6.14). Therefore, the decision function for any test data (x_{test}) can also be expressed in terms of the kernel function using the dual variables and vectorized training samples as follows:

$$d_n = \text{sgn}\left(\sum_{i}^{n}\alpha_i K(x_i, x_{\text{test}}) - b\right) \tag{6.16}$$

The one-class SVM algorithm is summarized as follows:

Algorithm: OCSVM

Offline Training of Model/Online Update

1. **Input**   Non-faulty preprocessed data with non-faulty label/detected faulty data with the updated label.
2. **Output**   Optimized OCSVM/updated OCSVM decision function.
3.   **for** (1: length_data):
4.   Optimize the *w* and *b* in the SVM decision function defined in Eq. (6.7).
5.   **end**
6.   Run the validation non-faulty preprocessed dataset to validate the decision function.
7.   Set the number of anomaly margins using validation data.

Online/Real-Time Test

8. **Input**   Preprocessed data sampled using "*n*" window.
9. **Output**   Set of decision score/anomaly count for each windowed sample.
10.   **for** (1: length_data):
11.   Compute the decision score for each sample: *g*(*n*)
12.   if (*f*(*n*) == -1) then
13.   Anomaly_point
14.   else
15.   Normal operation condition.
16.   **end**

6.4.3 One-Class Neural Network

One-class neural network (OCNN) is similar to the OCSVM that separates all the data points from the origin. OCNN can exploit and improve features obtained from unsupervised learning specifically for anomaly detection. In this approach, the model is not trained with predicted output. Unlike supervised learning, this model is trained and tested using processed data. OCNN evaluates data samples based on decision scores to differentiate anomalies in complex datasets where the decision boundary between normal and anomalies is highly nonlinear.

Chalapathy et al. proposed a shallow NN with one output model to detect anomalies (Chalapathy et al. 2018). The suggested model is shown in Figure 6.2.

In the above NN, "γ" is the scalar matrix from the hidden to the output layer, "α" is the weight matrix from the input to the hidden layer, F is the feature space, "β" is the bias matrix, g is the NN activation function.

The OCNN objective function is formulated as

$$\min_{\gamma,\alpha,\beta} \frac{1}{2}\|\gamma\|_2^2 + \frac{1}{2}\|\alpha\|_F^2 + \frac{1}{\alpha}\cdot\frac{1}{N}\left[\sum_{n=1}^{N} \max\left(0, \beta - \langle\gamma, g(\alpha X_{n:})\rangle\right) - \beta\right] \tag{6.17}$$

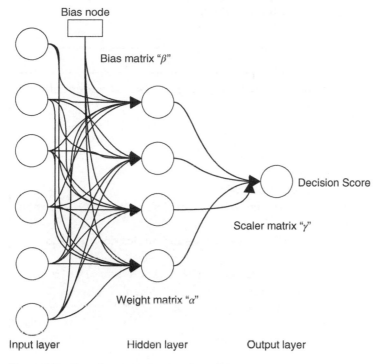

Figure 6.2 One-class neural network model.

In Eq. (6.17), "X_n, F, and g" denotes the training samples, feature space, and activation function, respectively. From the equation, bias "β," weight matrix "α," and the output matrix "γ" can be optimized. Chalapathy et al. described the optimization algorithm for OCNN by optimizing "α" and "γ" separately while updating the "β" based on optimized "α" and "γ" (Chalapathy et al. 2018).

$$\min_{\gamma, \alpha} \frac{1}{2} \|\gamma\|_2^2 + \frac{1}{2} \|\alpha\|_F^2 + \frac{1}{\alpha} \cdot \frac{1}{N} \left[\sum_{n=1}^{N} \max\left(0, (\beta - [\gamma, g(\alpha X_{n:})])\right) \right] \tag{6.18}$$

Similarly, the optimization for β

$$\underset{\beta}{\text{argmin}} \left[\frac{1}{N\alpha} \cdot \sum_{n=1}^{N} \max\left(0, (\beta - [< \gamma, g(\alpha X_{n:}) >]) - \beta\right) \right]. \tag{6.19}$$

In the OCNN algorithm, the input layer feeds the set of multivariate samples and calculates the data sample decision score by a single output for a single data sample (Chalapathy et al. 2018). Equation (6.18) shows the optimization of "α" and "γ" using the "β" value. Similarly, Eq. (6.19) shows the optimization of "β" by using the optimized "α" and "γ" values. Initial decision scores are managed to compute for each data sample based on the optimized γ, α, and β values. By using the decision score, normal and abnormal data samples are classified. Finally, with the optimized model, positive decision score data will be identified as a normal condition, and a negative decision score will be defined as an abnormal condition.

The One-Class NN algorithm is described as follows:

Algorithm: OCNN

```
1  Input     Non-faulty data, "n" sample window time-dependent
             data. X_{n:}, n : 1, ..., N
2  Output    Set of decision scores (for one data frame)/update
             OCNN
3  while     (1: "n" sample window):
4            Define "β" value
5            for  (1: No. of epochs)
6                  Update ("α" and "γ"); optimize the NN weights
                   and output layer
7                  Optimize the r-value using updated "α" and "γ"
8            end
9            Compute the decision score for each sample: d(n)
10           if   decision score (d(n) ≥ 0), then
11                Normal data point
12           else
13                Anomaly_data point
14 end
```

6.4.4 Comparison Between Deep Learning with Machine Learning in Fault Detection and Diagnosis

FDD methodologies have generally been based on SVMs, k-nearest neighbor (KNN), decision tree (DT), and ANN models. These ML approaches are similar to the traditional statistical models. To perform well using these models, a massive number of features should be carefully chosen to minimize the computational complexity of the ML model while maintaining classification accuracy. The feature selection procedure is used in the ML model to eliminate unnecessary and duplicated features.

Additionally, ML models have a limitation in terms of training size. In comparison, when substantial training data are used, deep learning models are highly successful; as the size of the training dataset increases, the diagnostic performance of a deep learning model improves.

Traditional fault detection and diagnostic systems using statistical and ML involve five critical phases, data acquisition, data processing, feature extraction, feature selection/dimension reduction, and feature classification. Developing appropriate methods for each phase is critical and needs a trial-and-error approach. Each phase of the standard FDD approach employs a unique set of techniques. Deep learning models eliminate the need to combine approaches for each step of the traditional FDD process; instead, the process is accomplished through the use of several hidden layers of deep learning architecture.

However, deep learning models may also require a data preprocessing step under certain scenarios, such as when the data is too noisy.

6.5 Intelligent FDD Using Machine Learning

One of the main objectives of applying ML algorithms in process monitoring is to develop autonomous intelligent models to detect and diagnose faults. To develop an intelligent model using data, it is required to detect the fault condition using unsupervised data and classify the fault condition using supervised data.

As illustrated in Figure 6.3, to cluster the unknown data samples and update the models online, an unsupervised learning model is used. For classifying the known cluster, a supervised learning model is integrated with the unsupervised model.

6.5.1 Model Development

To illustrate the step-by-step model development procedure, this section uses the Tennessee Eastman (TE) simulation process model testing as described in (Arunthavanathan et al. 2020, 2022). The TE processed data are generated using a simulated model developed by (Downs and Vogel 1993). As shown in Figure 6.4, the TE process consists of a reactor, a condenser, a compressor, a separator, and a stripper. By feeding the A, C, D, and E to

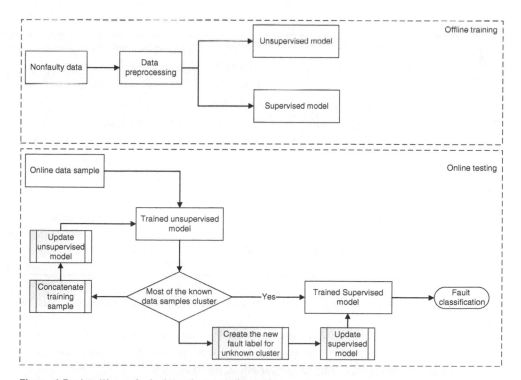

Figure 6.3 Intelligent fault detection and diagnosis.

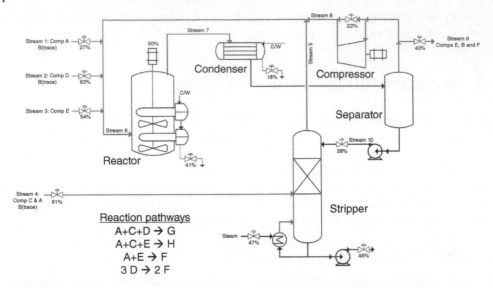

Figure 6.4 Tennessee Eastman process flow diagram. *Source:* Udugama et al. (2020)/With permission of Elsevier.

Table 6.3 TE process continuous process variables.

Index	Description	Index	Description
XMEAS1	A feed (stream 1)	XMEAS12	Separator level
XMEAS2	D feed (stream 2)	XMEAS13	Separator pressure
XMEAS3	E feed (stream 3)	XMEAS14	Separator underflow (stream 10)
XMEAS4	Total feed (stream 4)	XMEAS15	Stripper level
XMEAS5	Recycle flow (stream 8)	XMEAS16	Stripper pressure
XMEAS6	Reactor feed rate (stream 6)	XMEAS17	Stripper underflow (stream 11)
XMEAS7	Reactor pressure	XMEAS18	Stripper temperature
XMEAS8	Reactor level	XMEAS19	Stripper stream flow
XMEAS9	Reactor temperature	XMEAS20	Compressor work
XMEAS10	Purge rate (stream 9)	XMEAS21	Reactor cooling water outlet temperature
XMEAS11	Separator temperature	XMEAS22	Condenser cooling water outlet temperature D

the reactor liquid, products G and H are produced. The species F is a by-product of the process.

In the TE process simulation, 22 process variables, 19 composite measurement variables, and 12 manipulated variables can be simulated. To demonstrate the performance of the intelligent FDD methodology, 22 process variables, listed in Table 6.3, are considered in the algorithm development.

Table 6.4 Selected TE process fault condition for testing.

Fault ID	Description variable	Type
IDV1	A/C feed ratio, B composition constant (Stream 4)	Step
IDV4	Reactor cooling water inlet temperature	Step
IDV11	Reactor cooling water inlet temperature	Random variation

In the TE simulation process, 21 different process faults/disturbances can be generated by changing the independent disturbance variables (IDV). The faults are named accordingly; e.g. fault IDV 1 refers to a change in the variable IDV 1. Out of the 21 faults, IDV 1–8 are related to step changes in the related variables. IDV 9–12 are related to random variation of the variables. Slow drift in a reaction kinetics parameter fault is demonstrated in IDV 13. IDV 14, 15, and 21 are related to sticky valves. The remaining IDV 16–20 are faults with unknown causes. Table 6.4 lists the faults that are tested using the proposed models.

6.5.2 Data Collection

As shown in Table 6.5, TE process fault condition data are concatenated to provide unique as well as repeated fault conditions over the different operating points to test the intelligent FDD method.

6.5.2.1 Model Development Steps

In an offline condition, 300 non-faulty raw sensor data samples are standardized, as illustrated in the following equation:

$$\hat{x} = \frac{x - \mu_x}{\sigma_x} \tag{6.20}$$

In Eq. (6.20), x is the value of input data, and $\hat{x}$ is the standardized data. μ_x and σ_x are the mean and standard deviation of the input data, respectively.

Table 6.5 TE fault condition for automated test.

Data sample	Fault condition
1–1000	No fault
1001–1500	Fault 1
1501–2000	No fault
2001–2500	Fault 4
2501–3000	Fault 11
3001–3400	Fault 4
3401–3700	Fault 1

Preprocessed non-faulty data samples are then used to train unsupervised and supervised models. In this experiment, one-class SVM is used as unsupervised anomaly detection and the NN model is used as a supervised model. All of the 22 process variables from the TE processed data are used in the one-class SVM. To optimize w and ρ parameters in the one-class SVM model, the radial basis function kernel with a 1e-3 learning rate is used in the model architecture.

The supervised three-layer NN model is developed using 22 neurons input layer (to get 22 process variables input), 68 neurons hidden layer and initially 1 neuron in the output layer. To optimize the θ and b parameters, the Adam optimizer learning rate is tested over a range of learning rates. By properly selecting the NN model hyperparameters, namely the learning rate and the number of epochs, accuracy can be maintained at over 90%. Figure 6.5 shows the impacts of the learning rate and the number of epochs on the accuracy.

For the initial training, 0.1 learning rate with 150 epochs was found to give a better performance for this system and models.

In online, preprocessed multivariate sensor data (with multiple sensors) are windowed with 100 samples and fed into the trained OCSVM to determine any process deviation. If the number of anomalies in the data sampling window exceeds the marginal level, the data window is defined as a possible fault condition and the OCSVM and dynamic output NN are updated. In the tested data sample window, if the number of anomalies are within the margin as illustrated in Eq. (6.21), the data window is fed into the dynamic output NN to classify the fault condition.

$$\text{Margin} = \max\left(\text{anomaly}_{\text{count}}\right) + \delta * \max\left(\text{anomaly}_{\text{count}}\right) \tag{6.21}$$

In Eq. (6.21), $\max(\text{anomaly}_{\text{count}})$ defines the maximum number of anomalies in the non-faulty training dataset, and δ defines the noise margin.

To test the NN and OCSM model to detect the fault condition autonomously, initially non-faulty data were used to train the models offline. As shown in Figure 6.6a, window samples 1–9 and window samples 14–20 were detected as a normal fault condition, and window samples 10–15 were detected as an abnormal condition. Once the abnormal condition is detected at the 10th sampling window, the OCSVM and NN model self-learn using normal and detected abnormal data. As shown in Figure 6.6b, window samples 15–20 and window samples 33–40 are identified as faulty. After the fault 4 detection at the 20th sample window, the NN model self-updates using no-fault, fault 1, and fault 4 data. As shown in the result, 20–25 window samples and 30–34 window samples are identified as faulty.

In this way, whenever a new fault is detected, OCSVM is updated to further detect the new fault condition, and NN is updated with a new label to classify the same fault condition. Also, Figure 6.6 illustrates that once the fault is detected and updated by the OCSVM, the same fault will not be considered further as a new fault condition.

After testing the anomaly condition using OCSVM, and if the data samples are under the anomaly margin, samples are fed into the NN to classify the fault condition. Table 6.6 summarizes the model update, detection, and classification test for this concatenated TE processed data.

6.5.2.2 Result Comparison

Performances of the unsupervised learning anomaly detection methods, namely one-class SVM and one-class NN models, are compared with statistical models, namely PCA and

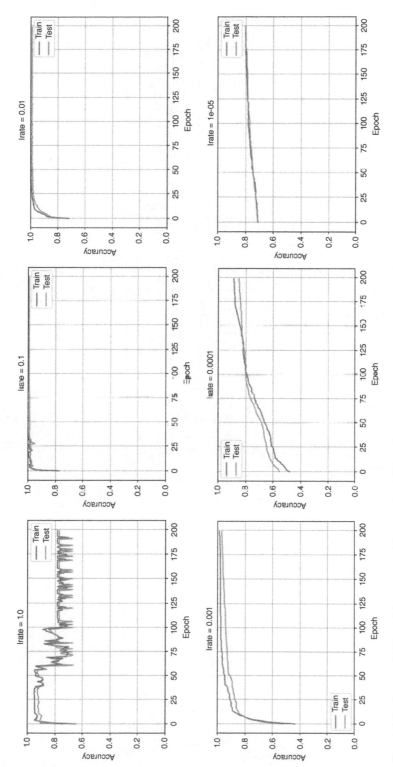

Figure 6.5 Proposed NN model accuracy with a number of iterations.

(a)

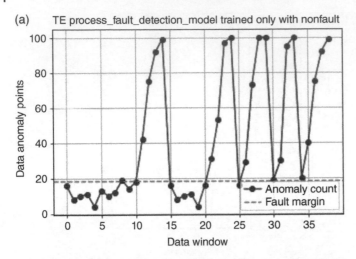

(b)

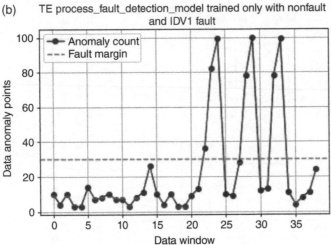

(c)

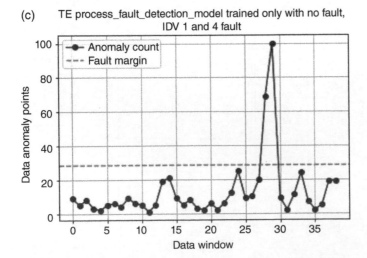

Figure 6.6 Autonomous and model self-update test (a) model trained with non-faulty data (b) model trained with detected fault 1 (c) model trained with detected fault 1 and 4. *Source:* Republished with permission of American Chemical Society, from Arunthavanathan et al. (2022).

Table 6.6 Data samples and obtained results using OCSVM and NN.

Data sample	Model update and results
1–600	Offline trained one-class SVM and NN
601–1000	No-fault classified (100%)
1001–1300	Fault 1 detected and one-class SVM and NN updated
1301–1500	Fault 1 classified using NN (98%)
1501–2000	No fault classified using NN (97%)
2001–2300	Fault 4 was detected and Onelass-SVM and NN were updated
2301–2500	Fault 4 classified (97%)
2501–2800	Fault 11 detected and model update
2801–3000	Fault 11 classified (96%)
3001–3400	Fault 4 classified (94%)
3401–3700	Fault 1 classified (96%)

Source: Arunthavanathan et al. (2022).

Table 6.7 Fault detection result comparison.

TE fault	Mahadevan and Shah (2009)				Arunthavanathan et al. (2020, 2021b)	
	PCA-T^2	PCA-Q	DPCA-T^2	DPCA-Q	Oneclass SVM	Oneclass NN
IDV 1	99.2	99.8	99.4	99.5	99.9	100
IDV 4	4.4	96.2	6.1	100	100	100
IDV 5	22.5	25.4	24.2	25.2	100	100
IDV 6	98.9	100	98.7	100	100	100
IDV 11	20.6	64.4	19.9	80.7	100	100

dynamic PCA integrated with T^2 and square prediction error (Q), using the TE processed data for different fault conditions. Table 6.7 shows a summary of the results.

Performance of the supervised fault classification models described in Section 6.3, namely, the Neural Network and the Support Vector Machines, are compared with that of PCA for fault classification; Table 6.8 shows the results.

The preceding results show a better performance of the ML approach compared with the traditional statistical analysis. However, when applying ML approaches, fault detection and classification accuracy may differ based on various factors. Apart from model parameters and hyperparameters described in Sections 6.3 and 6.4, the number of samples used in the training and testing window, quality of data, nature of the process dynamics, and performance of the computational tools may also impact the accuracy.

Table 6.8 Comparison of fault classification results.

TE fault	Jing et al. (2014)		Arunthavanathan et al. (2020, 2021b)
	PCA + statistical	SVM	Dynamic output NN
IDV 1	97	93	98
IDV 4	92	65	97
IDV 5	98	73	100
IDV 6	99	77	99
IDV 11	83	30	96

6.6 Concluding Remarks

The recent literatures on ML-based FDD demonstrate widespread applicability and potential benefits of the approach for process systems. In this chapter, an SVM- and NN-based ML approach for process FDD is presented; performance of the method is also demonstrated. The results are promising in terms of accuracy when compared with traditional statistical analysis. ML methods are becoming ubiquitous; like other fields of applications, process safety in general, and FDD in particular, are benefiting from the use of the ML tools. However, the black box nature of the tools remains a concern for FDD applications. In this regard, research efforts on how explainability and interpretability of the ML tools can be enhanced are required. Another issue that needs attention is the inclusion of available process knowledge in the analysis. The data-based approaches, like ML, flourished due to limited availability of process knowledge. However, these tools do not offer proper mechanisms to include available process knowledge that can further enhance fault diagnosis and root cause analysis. Efforts are needed to transform useful operational experience and contextual information into useful data suitable for use in ML and thus make the tools more inclusive of process knowledge.

References

Alam, S., Sonbhadra, S.K., Agarwal, S., and Nagabhushan, P. (2020). One-class support vector classifiers: a survey. *Knowledge-Based Systems* 196: 105754. https://doi.org/10.1016/j.knosys.2020.105754.

Alauddin, M., Khan, F., Imtiaz, S., and Ahmed, S. (2018). A bibliometric review and analysis of data-driven fault detection and diagnosis methods for process systems [review-article]. *Industrial and Engineering Chemistry Research* 57 (32): 10719–10735. https://doi.org/10.1021/acs.iecr.8b00936.

Allwein, E.L., Schapire, R.E., and Singer, Y. (2000). Reducing multiclass to binary: a unifying approach for margin classifiers. *Journal of Machine Learning Research* 1 (2000): 113–141.

Alzghoul, A., Backe, B., Löfstrand, M. et al. (2014). Comparing a knowledge-based and a data-driven method in querying data streams for system fault detection: a hydraulic drive system application. *Computers in Industry* 65 (8): 1126–1135. https://doi.org/10.1016/J. COMPIND.2014.06.003.

Arunthavanathan, R., Khan, F., Ahmed, S. et al. (2020). Fault detection and diagnosis in process system using artificial intelligence-based cognitive technique. *Computers & Chemical Engineering* 134: 106697. https://doi.org/10.1016/J.COMPCHEMENG.2019.106697.

Arunthavanathan, R., Khan, F., Ahmed, S., and Imtiaz, S. (2021a). An analysis of process fault diagnosis methods from safety perspectives. *Computers and Chemical Engineering* 145: 107197. Elsevier Ltd. https://doi.org/10.1016/j.compchemeng.2020.107197.

Arunthavanathan, R., Khan, F., Ahmed, S., and Imtiaz, S. (2021b). A deep learning model for process fault prognosis. *Process Safety and Environmental Protection* 154: 467–479. https://doi. org/10.1016/J.PSEP.2021.08.022.

Arunthavanathan, R., Khan, F., Ahmed, S., and Imtiaz, S. (2022). Autonomous fault diagnosis and root cause analysis for the processing system using one-class SVM and NN permutation algorithm. *Industrial & Engineering Chemistry Research* 61 (3): 1408–1422. acs.iecr.1c02731. https://doi.org/10.1021/ACS.IECR.1C02731.

Chalapathy, R., Menon, A. K., and Chawla, S. (2018). Anomaly detection using one-class neural networks. (accessed 19–23 August 2022). http://arxiv.org/abs/1802.06360

Chiang, L.H., Russell, E.L., and Braatz, R.D. (2001). Fault Detection and Diagnosis in Industrial Systems. *Springer International Publishing* https://doi.org/10.1007/978-1-4471-0347-9.

Dai, X. and Gao, Z. (2013). From model, signal to knowledge: a data-driven perspective of fault detection and diagnosis. *IEEE Transactions on Industrial Informatics* 9 (4): 2226–2238. https:// doi.org/10.1109/TII.2013.2243743.

Dietterich, T.G. and Bakiri, G. (1994). Solving multiclass learning problems via error-correcting output codes. *Journal of Artificial Intelligence Research* 2: 263–286. https://doi.org/10.1613/ JAIR.105.

Dike, H. U., Zhou, Y., Deveerasetty, K. K., and Wu, Q. (2018). Unsupervised learning based on artificial neural network: a review. *2018 International Conference on Cyborg and Bionic Systems, IEEE (CBS 2018)*, (25–27 October 2018). Shenzhen, China. 322–327. https://doi.org/ 10.1109/CBS.2018.8612259

Downs, J.J. and Vogel, E.F. (1993). A plant-wide industrial problem process. *Computers & Chemical Engineering* 17 (3): 245–255. https://doi.org/10.1016/0098-1354(93)80018-I.

Fan, C.M. and Lu, Y.P. (2008). A bayesian framework to integrate knowledge-based and data-driven inference tools for reliable yield diagnoses. *Proceedings - Winter Simulation Conference* 2323–2329. https://doi.org/10.1109/WSC.2008.4736337.

Frank, P.M. (1996). Analytical and qualitative Model-based fault diagnosis – a survey and some new results. *European Journal of Control* 2 (1): 6–28. https://doi.org/10.1016/S0947-3580(96) 70024-9.

Ge, Z. (2017). Review on data-driven modeling and monitoring for plant-wide industrial processes. *Chemometrics and Intelligent Laboratory Systems* 171: 16–25. https://doi.org/ 10.1016/J.CHEMOLAB.2017.09.021.

Ge, Z. and Song, Z. (2013). Multivariate Statistical Process Control. London: Springer https://doi. org/10.1007/978-1-4471-4513-4.

Gertler, J. (1997). Fault detection and isolation using parity relations. *Control Engineering Practice* 5 (5): 653–661. https://doi.org/10.1016/S0967-0661(97)00047-6.

Hoskins, J.C. and Himmelblau, D.M. (1988). Artificial neural network models of knowledge representation in chemical engineering. *Computers and Chemical Engineering* 12 (9–10): 881–890. https://doi.org/10.1016/0098-1354(88)87015-7.

Isermann, R. (2011). *Fault Diagnosis Applications.* Springer Berlin Heidelbe https://doi.org/10.1007/978-3-642-12767-0_9.

Jing, C., Gao, X., and Zhu, X. (2014). Fault classification on tennessee eastman process: PCA and SVM. *2014 International Conference on Mechatronics and Control (ICMC), IEEE*, (3–5 July 2014). Jinzhou, China, 2194–2197. https://doi.org/10.1109/ICMC.2014.7231958

Kamarthi, S.V. and Pittner, S. (1999). Accelerating neural network training using weight extrapolations. *Neural Networks* 12 (9): 1285–1299. https://doi.org/10.1016/S0893-6080(99)00072-6.

Kano, M., Tanaka, S., Hasebe, S. et al. (2003). Monitoring independent components for fault detection. *AICHE Journal* 49 (4): 969–976. https://doi.org/10.1002/aic.690490414.

Lee, Y., Lin, Y., and Wahba, G. (2004). Multicategory support vector machines: theory and application to the classii cation of microarray data and satellite radiance data. *Journal of the American Statistical Association* 99 (465): 67–81. https://doi.org/10.1198/016214504000000098.

Lei, Y., Yang, B., Jiang, X. et al. (2020). Applications of machine learning to machine fault diagnosis: a review and roadmap. *Mechanical Systems and Signal Processing* 138: 106587. https://doi.org/10.1016/J.YMSSP.2019.106587.

Lera, G. and Pinzolas, M. (2002). Neighborhood based Levenberg-Marquardt algorithm for neural network training. *IEEE Transactions on Neural Networks* 13 (5): 1200–1203. https://doi.org/10.1109/TNN.2002.1031951.

Liu, Y. and Zheng, Y.F. (2005). One-against-all multi-class SVM classification using reliability measures. *Proceedings of the International Joint Conference on Neural Networks* 2: 849–854. https://doi.org/10.1109/IJCNN.2005.1555963.

Liu, R., Yang, B., Zio, E., and Chen, X. (2018). Artificial intelligence for fault diagnosis of rotating machinery: a review. *Mechanical Systems and Signal Processing* 108: 33–47. https://doi.org/10.1016/J.YMSSP.2018.02.016.

Lo, N. G., Flaus, J. M., and Adrot, O. (2019). Review of machine learning approaches in fault diagnosis applied to IoT systems. *2019 International Conference on Control, Automation and Diagnosis, IEEE (ICCAD 2019)* (2–4 July 2019). Grenoble, France, 1–6. https://doi.org/10.1109/ICCAD46983.2019.9037949

MacGregor, J.F. and Kourti, T. (1995). Statistical process control of multivariate processes. *Control Engineering Practice* 3 (3): 403–414. https://doi.org/10.1016/0967-0661(95)00014-.

Mahadevan, S. and Shah, S.L. (2009). Fault detection and diagnosis in process data using one-class support vector machines. *Journal of Process Control* 19 (10): 1627–1639. https://doi.org/10.1016/J.JPROCONT.2009.07.011.

Nor, N.M., Hassan, C.R.C., and Hussain, M.A. (2020). A review of data-driven fault detection and diagnosis methods: applications in chemical process systems. *Reviews in Chemical Engineering* 36 (4): 513–553. https://doi.org/10.1515/REVCE-2017-0069/MACHINEREADABLECITATION/RIS.

Perantonis, S.J. and Karras, D.A. (1995). An efficient constrained learning algorithm with momentum acceleration. *Neural Networks* 8 (2): 237–249. https://doi.org/10.1016/0893-6080 (94)00067-V.

Qin, S.J. (2012). Survey on data-driven industrial process monitoring and diagnosis. *Annual Reviews in Control* 36 (2): 220–234. https://doi.org/10.1016/j.arcontrol.2012.09.004.

Udugama, I.A., Gernaey, K.V., Taube, M.A., and Bayer, C. (2020). A novel use for an old problem: the tennessee eastman challenge process as an activating teaching tool. *Education for Chemical Engineers* 30: 20–31. https://doi.org/10.1016/j.ece.2019.09.002.

Venkatasubramanian, V. and Chan, K. (1989). A neural network methodology for process fault diagnosis. *AICHE Journal* 35 (12): 1993–2002. https://doi.org/10.1002/aic.690351210.

Venkatasubramanian, V., Rengaswamy, R., and Kavuri, S.N. (2003a). A review of process fault detection and diagnosis: Part II: qualitative models and search strategies. *Computers & Chemical Engineering* 27 (3): 313–326. https://doi.org/10.1016/S0098-1354(02)00161-8.

Venkatasubramanian, V., Rengaswamy, R., Yin, K., and Kavuri, S.N. (2003b). A review of process fault detection and diagnosis: Part I: quantitative model-based methods. *Computers & Chemical Engineering* 27 (3): 293–311. https://doi.org/10.1016/S0098-1354(02)00160-6.

Weston, J. and Watkins, C. (1998). *Multi-class Support Vector Machines*. England: Royal Holloway.

Xiao, Y., Wang, H., Xu, W., and Zhou, J. (2016). Robust one-class SVM for fault detection. *Chemometrics and Intelligent Laboratory Systems* 151: 15–25. https://doi.org/10.1016/j. chemolab.2015.11.010.

Yin, S., Ding, S.X., Xie, X., and Luo, H. (2014). A review on basic data-driven approaches for industrial process monitoring. *IEEE Transactions on Industrial Electronics* 61 (11): 6418–6428. https://doi.org/10.1109/TIE.2014.2301773.

Zhang, Y. and Jiang, J. (2008a). Bibliographical review on reconfigurable fault-tolerant control systems. *Annual Reviews in Control* 32 (2): 229–252. https://doi.org/10.1016/J. ARCONTROL.2008.03.008.

Zhang, Y. and Jiang, J. (2008b). Bibliographical review on reconfigurable fault-tolerant control systems. *Annual Reviews in Control* 32 (2): 229–252. https://doi.org/10.1016/J. ARCONTROL.2008.03.008.

7

Intelligent Method for Chemical Emission Source Identification

Denglong Ma

School of Mechanical Engineering, Xi'an Jiaotong University, Xi'an, Shannxi, China

7.1 Introduction

7.1.1 Development of Detecting Gas Emission

Metal oxide semiconductor (MOS) sensor is one kind of most widely used methods to detect gases or volatile organic components (VOCs) due to its low cost and high sensitivity. For different gases, the different sensors with different materials are varied, such as SnO_2 is always used to detect combustible gases and WO_3 has great response to H_2 and NO. In addition, some gases can absorb the light with certain wavelength, thus many spectrum methods have been proposed based on the optical properties of gas molecules, including differential optical absorption spectroscopy (DOAS), tunable diode laser absorption spectroscopy (TDLAS), laser radar, and nondispersive infrared detector (NDIR). These optical detection methods have high detection sensitivity, selectivity, and stability than non-optical method, but complex structure and operation are required as well as huge size and high cost. Furthermore, electrochemical sensor has advantages in good selection, high precision, low cost, compact structure, and portability, but there are still limitations in weak stability, short work life, and potential electrolyte leakage. Besides, there are some methods to detect gases based on acoustics principle, such as surface acoustic wave (SAW) and photoacoustic spectrum method. However, it is still hard to balance the selectivity and sensitivity of the gas detection with acoustic method. Gas chromatography (GC) is based on the separation principle of different substances, and almost all VOCs can be detected accurately when GC is combined with other detection methods like mass spectrum (MS). GC is always equipped to analyze the gases or VOCs in the laboratory due to its high cost, huge size, and nonprobability.

Animals and humans can recognize the trace smell with olfactory system, where the olfactory organ cells feel the gases or VOCs in the smell environment and then the brain neuron system can react to the certain smell and analyze the information about it according to the experiences. Inspired by natural animal and human's olfactory, researchers proposed

Machine Learning in Chemical Safety and Health: Fundamentals with Applications, First Edition.
Edited by Qingsheng Wang and Changjie Cai.
© 2023 John Wiley & Sons Ltd. Published 2023 by John Wiley & Sons Ltd.

the caption of "electronic nose (EN)" or "artificial olfactory system (AOS)." The features of the whole reaction process of AOS indicate the species and concentrations of the gases, and the multidimensional response matrix data can be viewed as the image form. Therefore, the common support vector machine (SVM) algorithms and deep learning models (DLMs) can be used to classify the gases.

7.1.2 Development of Source Term Identification

Estimating the sources of contaminant or hazard emissions is important for pollution control and safety management. Source parameters, including source location, strength (mass emission rate), and emission characters etc., are required to be estimated as soon as possible when some dangerous gases release to the atmosphere. There have been many methods to be applied for source identification. Besides the direct ways by portable instruments or widely distributed sensors, indirect method by computation algorithms coupled with measurement results is another useful tool to determine gas emissions source parameters. In fact, it is a typical inverse problem to identify the gas emissions source based on measurement data while the gas emissions gas dispersion from a known source is viewed as a forward problem, as shown in Figure 7.1.

Optimization method has been proved to be a feasible way to identify the gas emissions source parameters (Haupt 2005, Addepalli et al. 2009, Khlaifi et al. 2009, Long et al. 2010, Ma et al. 2013). Different optimization methods are genetic algorithms (GAs) (Haupt et al. 2008), simulated annealing (SA) (Thomson et al. 2007), quasi-random sampling, and regularized gradient optimization (Addepalli et al. 2009). Ma et al. (2018, 2019) compared the intelligent group algorithms including particle swarm optimization (PSO), ant colony optimization (ACO), and firefly algorithm (FA), to identify the source term. An improved

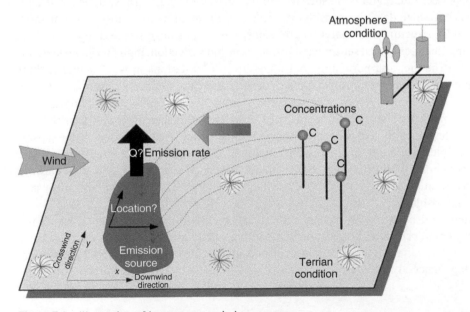

Figure 7.1 Illustration of inverse gas emission process.

active FA was also proposed by introducing active particle generation strategy to enhance the estimation accuracy and efficiency of the source parameters. However, the reasonability of the estimation results is doubtful by using the optimization method due to the random variabilities existed in the real world. Yet, stochastic approximation method (Yin 1988, Heemink and Segers 2002) is a suitable method based on inferences theory, and a more reasonable probability distribution result under some confidence levels can be obtained.

Stochastic approximation is a method based on the Bayesian inferences theory, from which the probability distribution results under some confidence levels may be more reasonable because the errors of instruments and the dispersion model really exist. Chow et al. (2008) reconstructed the source character by a combination of Bayesian inference and Markov chain Monte Carlo (MCMC) methods. Keats et al. (2006, 2007), Rao (2007), and Yee (2008) have addressed this problem by using the adjoint advection–diffusion equation in conjunction with MCMC methods to perform the computations efficiently. The minimum relative entropy (MRE) has been further proposed to estimate source parameters with probability outputs (Ma et al. 2014). However, the estimated results based on the MRE greatly depend on prior information. Therefore, Tikhonov parameter regularization has been employed to improve the identification of source parameters with lesser dependence on prior information (Ma et al. 2017, 2018).

In this chapter, we will discuss some intelligent methods to detect the hazardous gases including atmospheric CO_2 leakage in geosequestration project and unknown gases and VOCs in Section 7.2. Further, the intelligent method to identify the emission source will be presented in Section 7.3.

7.2 Intelligent Methods for Recognizing Gas Emission

7.2.1 Leakage Recognition of Sequestrated CO_2 in the Atmosphere

Carbon capture, utilization, and storage (CCUS) network is an effective way to reduce stationary CO_2 emissions and their adverse environmental impacts with low costs. However, sequestrated CO_2 gas has the risk of escaping from the storage site or leaking from the transportation pipe. Once the leakage occurs, there will be a negative impact on the underground water, geochemical environment, and even the microbial ecosystem and growth of plants and animals. Additionally, CO_2 leakage will reduce the climate benefit from long-term CCUS project. Therefore, it is necessary to monitor and verify the CO_2 leakage during CCUS process.

One of the most challenging problems is to recognize and determine CO_2 leakage signal in the complex atmosphere background due to its complex emission sources and variations in the atmosphere. The signals of CO_2 leakage from storage may be hidden in the natural variation. In order to solve this problem, some technologies have been developed to monitor CO_2 leakage in the atmosphere, such as infrared gas analyzer (IRGA), eddy covariance (EC) network, gas chamber, isotopic composition, gas tracers (e.g. SF6), laser open-path laser, and others.

However, most of the methods still have problems in complex operation, high instrument cost, or incapability of real-time monitoring. In order to find a simple and low-cost method to recognize CO_2 leakage in the atmosphere, Leeuwen et al. designed a measurement system to recognize CO_2 leakage by calculating the exchange ratio of atmospheric CO_2 and O_2 before and after injection. Ma et al. proposed a series of parameters based on correlation analysis between atmospheric CO_2 and O_2 to identify CO_2 leakage in real time. The test results with simulation and experiments proved that it is a feasible method to recognize CO_2 leakage in real time. Although compounded atmospheric CO_2 and O_2 measurement is an easy and low-cost tool to recognize CO_2 leakage, a deficiency for this method is that it is difficult to obtain an accurate O_2 measurement due to its much higher atmospheric concentration than CO_2, and it is hard to recognize very small O_2 variation in the atmosphere. Hence, a method to identify CO_2 leakage just by monitoring CO_2 is required, especially for small leakage cases.

7.2.1.1 Gas Leakage Recognition for CO_2 Geological Sequestration

For daily variation of atmospheric CO_2 at certain area, the sink and emission source of CO_2 is relatively stable, thus the natural variation of CO_2 in one day can be viewed as a periodic change. If CO_2 concentration variation tendency can be predicted with a mathematical model, the abnormal leak CO_2 will make change for variation law of atmospheric CO_2. Hence, the abnormal signal can be identified by comparing the results of prediction and measurement. Based on this idea, how to predict daily atmospheric CO_2 variation becomes a critical issue. Daily atmospheric CO_2 is a non-stationary time series. Therefore, the models to predict time series can be considered. More details were published in the *Chinese Journal of Chemical Engineering* (Ma et al. 2020).

There have been many models proposed to predict non-stationary time series. Data-driven method is one of the widely used prediction method in carbon capture and sequestration (CCS) technology, such as prediction of CO_2 absorption property and parameters in sequestration process. However, time series model is specific method to predict periodic data. Autoregression (AR) model is a simple model for linear time series analysis based on the correlation relationship between past and future data. Different from AR, moving average (MA) model predicts the time series by adding the white noise time series with certain weights. Additionally, autoregressive and moving average model (ARMA) was proposed by combing the AR and MA. ARMA is a very general class of linear models used for forecasting purposes, and it is always used to model the linear stationary process. It utilizes the advantages of MA and AR, where AR analyzes the relationship between past and future data series while MA solves the problem of the noises. For nonlinear stationary process, ARMA framework can also be modeled by differencing the ARMA model to transform the data to a new series with the properties of stationary and autocovariance, which is autoregressive integrate moving average model (ARIMA). ARMA framework is a structural time series model, which is specified in terms of components such as trend, seasonality, and noise.

Further, there exists large classes of time series, such as those with nonlinear MA components. They are not well modeled by feed-forward networks or linear models, but can be modeled by recurrent network, which is a type of nonlinear autoregressive moving average (NARMA) model. Recurrent networks have advantages over feed-forward neural networks similar to that ARMA models have over autoregressive models for some types of time series. Recurrent networks are well suited for time series that possess MA components, are state dependent, or have trends. Among different NARMA models, the nonlinear autoregressive moving average with exogenous input (NARMAX) models are the most popular representations in discrete-time domain. Because prediction errors are always hard to be known for common NARMAX model, a special case of NARMAX model, the nonlinear autoregressive with exogenous input (NARX) model, is proposed. It is also called recurrent NARX network. NARX has been evaluated to predict time series variation in both long and short time periods.

Because daily atmospheric CO_2 is a non-stationary time series, the models for nonlinear dynamic system identification can be used in this case. The nonlinear autoregressive moving average with exogenous input (NARMAX) models are the most popular representations in discrete-time domain, which was first introduced in 1981 and developed in subsequent publication. The NARMAX model is defined as

$$y(k) = f\big(y(k-1), ..., y(k-n_y), x(k-d), ..., x(k-d-n_x), e(k-1), ..., e(k-n_e)\big) + e(k)$$
(7.1)

where $y(k)$, $x(k)$, and $e(k)$ are the system output, input, and noise sequences, respectively; n_y, n_x, and n_e are the maximum lags for the system output, input, and noise; $f(\cdot)$ is some nonlinear function; and d is the time delay. The noise terms $e(k)$ is normally defined as the prediction errors $e(k) = y(k) - \hat{y}(k|k-1)$. Many different types of model structures have been proposed to approximate the unknown mapping $f(k)$, such as power-form polynomial models, rational models, neural network, fuzzy logic-based models, wavelet expansions, and many more.

Because prediction errors are always hard to be known, the NARX as a special case of NARMAX model is proposed so that it does not include any noise-dependent model terms and can be formulated as:

$$y(k) = f\big(y(k-1), ..., y(k-n_y), x(k-d), ..., x(k-d-n_x)\big) + e(k)$$
(7.2)

It is also called recurrent NARX network sometimes. For atmospheric emission gas prediction, there are many models such as scalar concentration fluctuation models based on Lagrangian particles and computational fluid dynamics (CFD) models based on solving dispersion equations. Due to the complex variation of atmospheric CO_2, neural network is considered to approximate the model of atmospheric CO_2 variation. Therefore, a NARX network is designed, as shown with Figure 7.2. The output can be formulated mathematically with Eq. (7.3).

$$y(k) = f[\mathbf{u}(k)] = w_0 + \sum_{i=1}^{m}(w_i \phi_i(\mathbf{u}(k)) + b_i) + e(k)$$
(7.3)

where $\mathbf{u}(k) = [\mathbf{y}(k-1), ..., \mathbf{y}(k-n_y), \mathbf{x}(k-1), ..., \mathbf{x}(k-n_x)]^T$, $\phi_i(\cdot)$ $(i = 1, 2, ..., m)$ are predefined. w_i $(i = 0, 1, 2, ..., m)$ are weight parameters of network. b_i $(i = 1, 2, ..., m)$ are residual parameters.

In this case, future values of a time series $y(t)$ can be forecasted from past values of that time series and past values of a second time series $x(t)$ with NARX network. The prediction process of NARX approach consists of following steps: structure detection, parameter estimation, model validation, prediction, analysis.

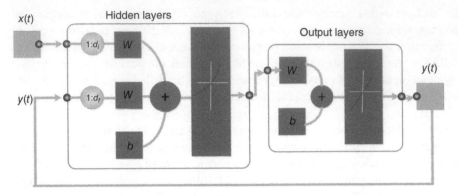

Figure 7.2 A recurrent NARX network model.

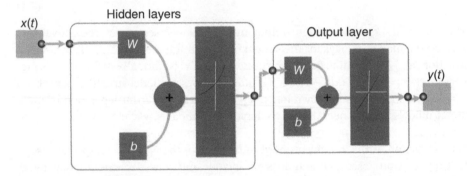

Figure 7.3 A fitting network model.

For a certain place, the daily variation of atmospheric CO_2 will retain a similar periodic tendency in recent few days. Therefore, it is possible to forecast the variation of CO_2 with recent monitoring data near this area by time series prediction method, and then comparison results of measured and predicted concentration can be used to recognize the leakage signal. The prediction process could be updated with newest real-time data. In order to compare the results obtained by recurrent NARX network, a conventional fitting neural network was first used to approximate the variation of atmospheric CO_2.

The structure of convention fitting neural network is manifested in Figure 7.3. In this case, the outputs are not fed back for prediction as well as the time delay.

7.2.1.2 Case Studies for CO_2 Recognition

The CO_2 concentration during the monitoring periods without extra CO_2 releases are illustrated in Figure 7.4. Note that the concentrations of CO_2 in atmosphere vary periodically due to the production and consumption of CO_2 in ecosystem. The results of the release experiments are shown in Figure 7.5, where the gas emission stage is marked in rectangle.

The average release rates for the cases of Figure 7.5a–c were 15, 20, and 25 $m^3 h^{-1}$, respectively (release case 1–3). Obviously, the leakage signal can be easily captured for large

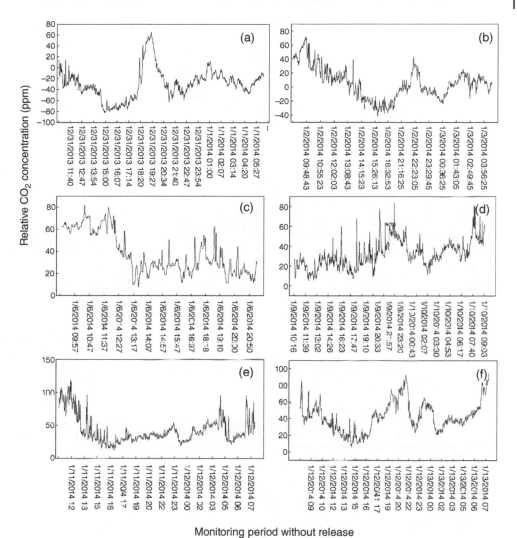

Figure 7.4 Monitoring results of CO_2 variation in the atmosphere without CO_2 release. (a)–(f) are the results in different monitoring periods.

release rate, e.g. Figure 7.5b and c. But small leakage signal may be hidden in the variation of nature CO_2 fluctuation, e.g. Figure 7.5a. In this case, it is hard to identify the leakage signal from the monitoring data directly.

The approximation error monitoring results of release case 1 with recurrent NARX network demonstrate that the leakage signal is observed clearly even in such small release case, as illustrated in Figure 7.6b.

The predicted results with recurrent NARX network in release period 1 are manifested in Figure 7.6. It is obvious that the prediction of recurrent NARX network can fit the CO_2 atmospheric variation very well due to making use of time-delayed feedback of past response. Although the abnormal variations have impact on the prediction and make

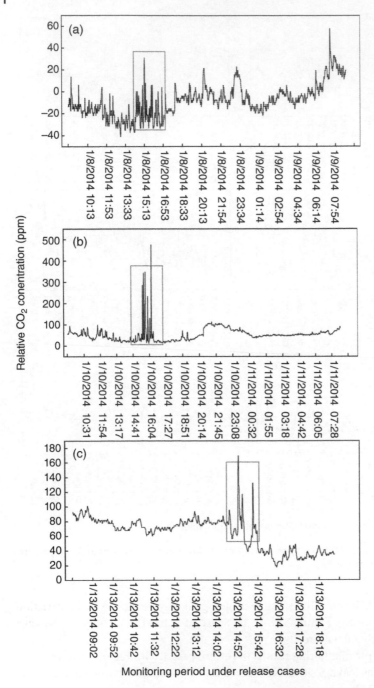

Figure 7.5 Monitoring results of CO_2 variation in atmosphere with CO_2 release. (a)–(c) are the monitoring periods 1–3 with different release rates of 15, 20, and 25 m^3 h^{-1}, respectively. CO_2 emissions occur in the time stage of rectangle.

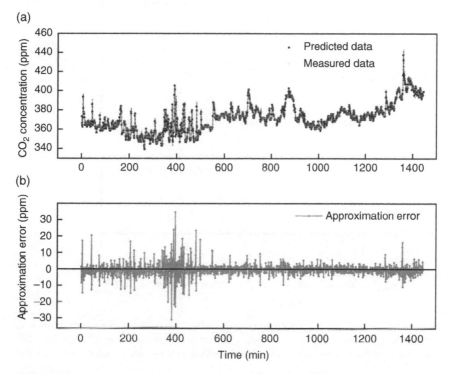

Figure 7.6 The prediction and residual results of recurrent NARX network for release case 1. (a) Measured and predicted data is the results of measured and predicted data and (b) approximately error between measurement and prediction is the approximation error results between measured and predicted results.

the predicted concentrations overestimated during the release time as shown in Figure 7.6a, the baseline of the residual errors are clear, as shown in Figure 7.6b. The correlation coefficient between predicted and measured results in this period is 0.9644, and the mean square between the results is 144.71 ppm The approximation error monitoring results with recurrent NARX network demonstrate that the leakage signal is observed clearly even in such small release case, as illustrated in Figure 7.6b.

The approximation results with recurrent NARX network for release period 2 are manifested in Figure 7.7. The correlation efficient between the prediction and measurement is 0.7108, and the mean square error (MSE) between the results is 820.34 ppm in this case. The baseline of the approximation errors with recurrent NARX is stabilized near zero line, which is useful to determine the leakage signal. Figure 7.8 presents the results with recurrent NARX network for release monitoring period 3. The similar conclusion as discussed from Figure 7.7 can be gained from Figure 7.8. The correlation coefficient between prediction and measurement is 0.9711 while the MSE is 134.04 ppm in this case.

The peak fit method with Gaussian form was used to capture the abnormal leakage signals for different release monitoring periods with approximation error of recurrent NARX network, as shown in Figure 7.9. It is noted that all abnormal leakage signals can be captured easily and accurately with peak fit model even in the little emission case.

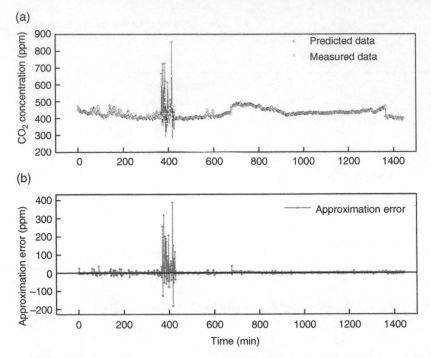

Figure 7.7 The prediction and residual results of recurrent NARX network for release monitoring 2. (a) Measured and predicted data is the result of measured and predicted data and (b) approximately error between measurement and prediction is the approximation error results between measured and predicted results.

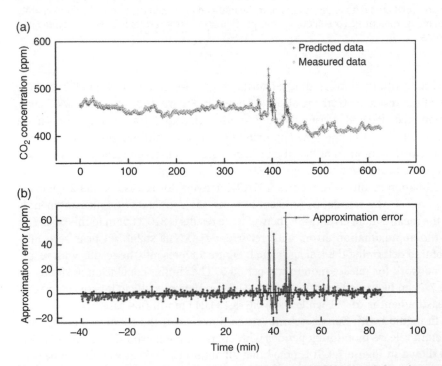

Figure 7.8 The prediction and residual results of recurrent NARX network for release monitoring 3. (a) Measured and predicted data is the result of measured and predicted data and (b) approximately error between measurement and prediction is the approximation error results between measured and predicted results.

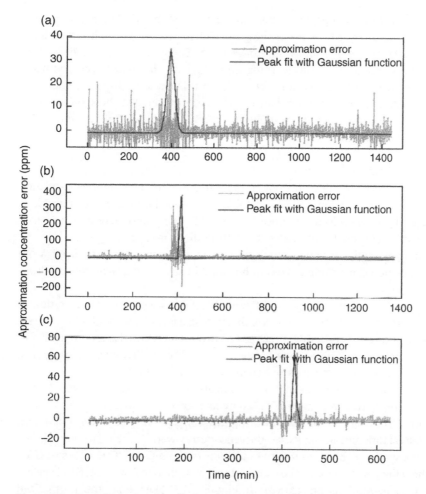

Figure 7.9 Peak fit results of approximation errors with recurrent NARX net using Gaussian model for release period 3. (a) Release monitoring period 1 is for the results of release monitoring period 1; (b) release monitoring period 2 is for release monitoring period 2; and (c) release monitoring period 3 is for release monitoring 3.

7.2.2 Emission Gas Identification with Artificial Olfactory

For many cases, it is hard to know the components of the emission source; therefore, it is an important issue to detect and identify the gases or VOCs in the emission event.

MOS sensor is one kind of most widely used methods to detect gases or VOCs in chemical engineering field. But one MOS sensor can be sensitive to various kinds of gases, thus cross-response to different gases makes MOS sensor difficult to identify the component accurately, which is a natural defect for such conventional detection module. AOS uses only the information provided by the sensor array for the gases. The features of response matrix of sensor array will change following with either the variation of the kind of gas or the concentration for certain type of gas. Then the pattern recognition algorithms were utilized to identify the status of the object gas. The data source used in current AOS is usually the stable response values or parts of dynamic response signals. Hence, some useful information may be wasted. The response process of the sensors to the gases, such as MOS and polymer sensor, is the chemical reaction process. Then, the features of the whole reaction process indicate the species and concentrations of the gases. Furthermore, the multidimensional response matrix data can be viewed as the image form. Therefore, the method used for image identification such as machine learning and DLM can be used to deal with the dynamic response data of AOS in the form of image in this research. In this section, the dynamic signals captured from the sensor array will be transformed to the standard form and then the image recognition methods including common SVM algorithms and DLMs will be applied to discern different gases.

7.2.2.1 Features of Responses in AOS

In this section, the samples of different VOCs, including ethanol, methanol, isopropanol, n-pentane, n-hexane, furan, ethyl acetate, acetaldehyde, and the mixtures of two kinds of them, captured by PEN 3 electronic nose (AIRSENSE, Germany) were used to verify the feasibility of different intelligent models to identify the component.

A gas recognition method based on machine learning and gas sensors is introduced. More details about this section can be found in our publication in *Sensors and Actuators B: Chemical* (Ma et al. 2021a).

The response results of sensor array for different substances with the concentration of 10 ppm are demonstrated in Figure 7.10. Here, the response is indicated with G/G_0, where G_0 is the response of certain sensor to the zero gas environment while G is that in measurement. It is noted from Figure 7.10 that one sensor may respond to different substances. The response curves of sensor array have varied forms for different VOCs.

The contour maps of different VOCs with 1000 ppm are shown in Figure 7.11, which indicates that the features of response maps of different substances are variable. After normalizing the measurement data of each sensor during the whole period in the range of 0–255, the response maps of different substances to the sensor array were formed. The vertical axis indicates the time while the horizontal is the indicator of the sensors. The values of G/G_0 were used to show the response map. The standard response map of furan and ethyl acetate under different concentrations are shown in Figure 7.12. Note that, the normalized response maps of certain substances with different concentrations have similar patterns especially under larger concentration (>100 ppm). Although the response maps under lower concentrations (<100 ppm) are different from that under larger concentrations, which may be caused by the disturbance of background, the response maps still have some relationship with each other. Therefore, the patterns of response maps are specific for certain substance, which can be used to identify the VOCs or gases.

7.2.2.2 Support Vector Machine Models for Gas Identification

SVM is a good binary classification; therefore, the error-correcting output codes (ECOC) model using SVM binary learners could be used to identify different substances. An ECOC model reduces the problem of classification with three or more classes to a set of binary classification problems. Therefore, a coding design is required.

Here, one-versus-one coding design is used to exhaust all combinations of class pair assignments, where one class is positive, another is negative, and the software ignores the rest. The learners are SVMs, the number of which is $K(K\text{-}1)/2$, where K is the number of the class labels. The kernel function for SVM is a linear form. The decoding scheme uses loss-weighted decoding.

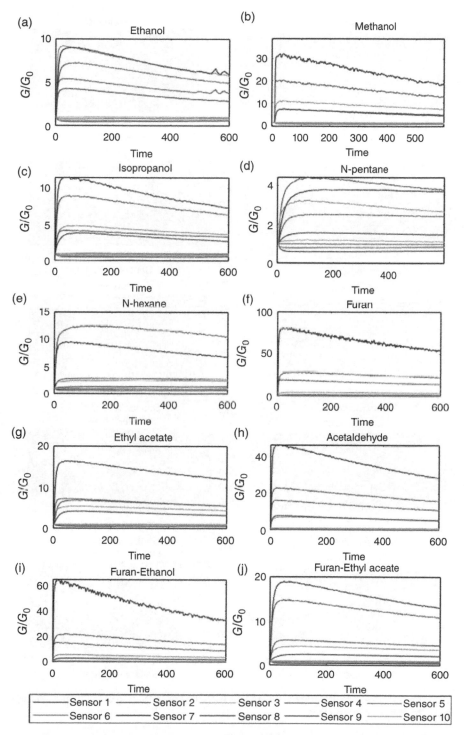

Figure 7.10 The responses of sensor array of PEN 3 for different substances with the concentration of 10 ppm; (a)–(j) are the results for different substances.

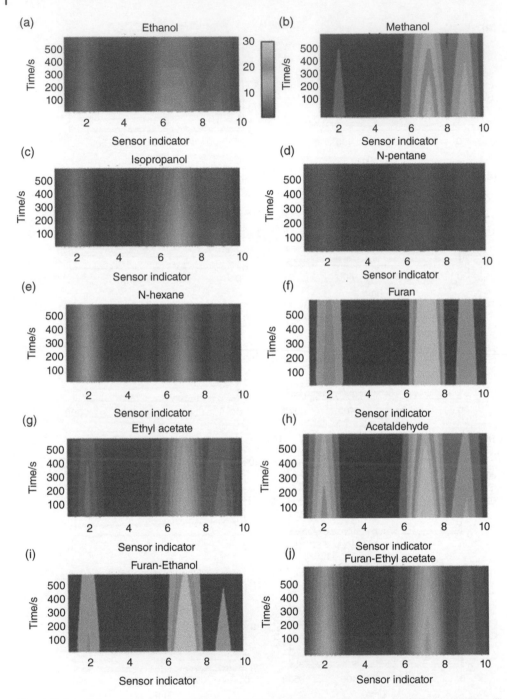

Figure 7.11 The response map of different substances with the concentration of 1000 ppm. (a) to (j) are the results for different substances.

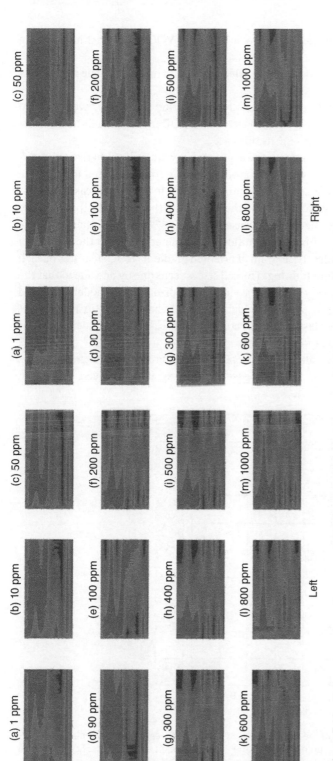

Figure 7.12 The response map of sensor array for furan (left) and ethyl acetate (right) under different concentrations. The vertical and horizontal axis are sensor indicator and time, respectively.

Example:

First, the data from sensor 7 were selected to predict different VOCs with machine learning method.

In this case, the values at 80 s of sensor 7 were selected to build the model. The accuracy of all predicted labels is 29% in this case.

Further, the responses during the whole measurement period of sensor 7 were used to train and test the model. The same set was also tested by ECOC model with SVM learners. The total accuracy of classification is 42% in this case.

Then, the response values of all sensors in the sensor array were utilized to recognize different gases. One response point for each sensor of the sensor array was used to build the model first. The accuracy of classification was improved greatly to 81% in this case. Further, all response points of 10 sensors were used to be the features of the model. In this case, the classification accuracy with the same method as used earlier is 87%. On the other hand, the feature extraction method used for image recognition is considered to extract the features of response maps for different samples. Histogram of oriented gradient (HOG) features of each response map were extracted before training. The cell size is set as one by one due to that the scale of our dataset is small and every data can be viewed as a pixel. ECOC model with SVM learners was adopted to train and test the model. The confusion matrix in this case is demonstrated in Figure 7.13. The classification accuracy in this case is 86%, which is similar with that trained with primary response values.

According to preceding results, although one gas sensor is cross-sensitive for different substances, it is hard to recognize different gases accurately with one certain sensor.

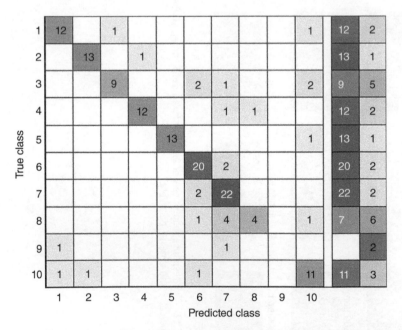

Figure 7.13 The confusion matrix of the prediction by SVM with HOG features. The labels of different numbers indicate different VOCs. 1: Furan; 2: N-pentane; 3: N-hexane; 4: ethyl acetate; 5: aldehyde; 6: ethanol; 7: methanol; 8: isopropanol; 9: furan-ethanol; 10: furan-ethyl acetate.

However, the machine learning model with responses from sensor array could improve the performance of gas recognition.

7.2.2.3 Deep Learning Models for Gas Identification

Although the machine learning model like SVM could provide accurate prediction, its classification accuracy is still not satisfied, as discussed earlier. Thus, it is required to improve the accuracy of the prediction model.

Because the conventional feature extraction method used for image processing did not improve the classification accuracy greatly, deep learning network, which has been used in image identification, human authentication, and other high-dimensional data analysis, is considered to extract more effective features of the response map to enhance the performance of gas recognition

Two types of DLM could be built to identify the gases with response map. First, the simple DLM model, where the parameters of the model can be obtained with training set. Here, a simple basic convolutional neural network (CNN) model is built. This network is a series network, and it consists of 15 layers, as shown in Figure 7.7. The input size of this network is [28,28,1]. Stochastic gradient descent with momentum (SGDM) was used to train the network. The basic parameters for SGDM solver including momentum factor, initial learning rate, minimum batch size, and maximum epochs were optimized with random grid search method.

Second, the transfer learning with pretrained deep learning network is also a good method to identify the gases with response map. The structure of DLM that has been trained with a larger number of samples could be retained, and most of convolution blocks should be frozen. Some weights of last blocks where the parameters are not frozen should be tuned slightly. Then the samples of AOS are used to train and optimize the hyperparameters of new model. In this method, only small size of samples is required to build a DLM with good generalization performance. Here, VGG-19 DLM developed by the Visual Geometry Group (VGG) at the University of Oxford for large-scale image recognition was used in this chapter. VGG-19 model is a series deep learning network. The size of output layer is tuned to the number of gas classes in this case. First three convolution blocks were frozen, and the weights of last two convolution blocks were tuned slightly. Then the new model was optimized with random grid search.

Example:
The confusion matrix with simple DLM for identifying different VOCs with the response map data discussed earlier is shown in Figure 7.14. It is noted that substances of N-pentane and aldehyde can be classified successfully with the accuracy of 100%. It is shown from curves of training process (Figure 7.15) that the classification accuracy of training set fast reaches to 100% in 100 iterations, which seems overfitting.

Then the same training and testing set were utilized to build pretrained transfer learning VGG-19 model. Compared with the process of simple DLM shown in Figure 7.15, the variations of accuracy and loss during the training process in this case are smoother, which indicates that the transfer learning model with VGG-19 is not overfitting. The final accuracy of validation is 90% in Figure 7.16. The confusion matrix shown in Figure 7.17 indicates that N-pentane, ethyl acetate, and aldehyde can be predicted successfully with the accuracy of 100% in this case.

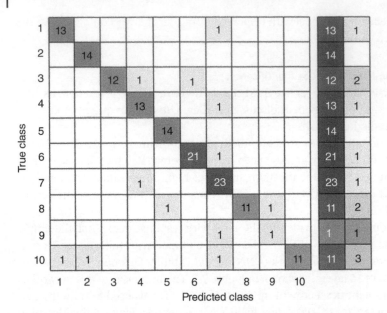

Figure 7.14 The confusion matrix of the prediction by simple deep learning network.

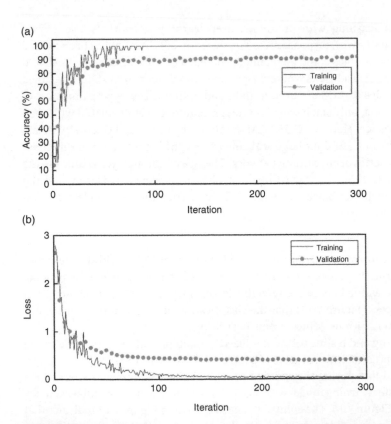

Figure 7.15 The training process curves of gas classification with simple DLM. (a) is the curve of accuracy variation during training process and (b) is the curves of loss parameter for training set and validation set.

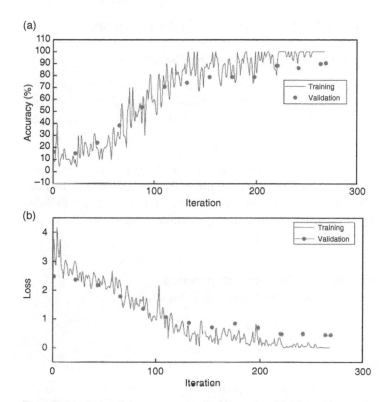

Figure 7.16 The training process curves of gas classification with transferred VGG-19 model; (a) is the curve of accuracy variation during training process and (b) is the curves of loss parameter for training set and validation set.

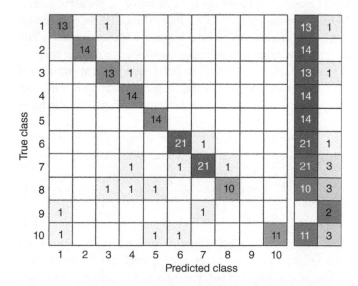

Figure 7.17 The confusion matrix of the prediction by transferred VGG-19 model.

7.3 Intelligent Methods for Identifying Emission Sources

7.3.1 Source Estimation with Intelligent Optimization Method

The problem of parameters estimation for emission source term is an inverse problem. Therefore, three kinds of inverse solution methods have been investigated, including direct inverse solution (Enting 2002; Yang et al. 2007), stochastic approximation (Yin 1988, Heemink and Segers 2002, Keats et al. 2007), and optimization method. Because the source term estimation is always an ill-pose inverse problem (Isakov and Kindermann 2000), it is not suitable to solve the atmospheric dispersion model directly to obtain the source parameters. The optimization method utilizes the measurement results and a forward dispersion model to obtain simulation results, which maximally match the measurement information. More details about the optimization method for source term estimation can be found in our previous publications (Ma et al. 2013, 2018, 2019).

7.3.1.1 Principle of Source Estimation with Optimization Method

The cost function of the optimization problem is shown in Eq. (7.4).

$$\min \ f = \sum_{i=1}^{N} [C_{\text{mea},i} - C_{\text{model},i}(Q,x,y,z)]^2 \tag{7.4}$$

$$\text{s.t.} \quad Q > 0; x > 0; -\infty < y < \infty; z > 0.$$

where $C_{\text{mea},i}$ (g·m^{-3}) is the measurement concentration at sensor i; $C_{\text{model},i}$ is the result from dispersion model, which is depend on the leakage rate Q (g·s^{-1}), downwind distance x (m), crosswind distance y (m), and height z (m) above ground and environment conditions; N is the number of sensors. Therefore, it is a constrained minimization problem.

Usually, there are three kinds of models to predict the gas dispersion in the atmosphere, which are the Gaussian model based on semianalytical solution, the CFD model based on N–S equations, and the Lagrangian stochastic (LS) model based on Markov process. The CFD and LS dispersion models are not suitable in fast source identification process because they will consume more time than the Gaussian model to finish one forward simulation process. The Gaussian model has advantages like simple implementation and low time cost with sufficient accuracy. Therefore, it has been adopted as the forward dispersion model in many applications.

The Gaussian plume model is a simple mathematical model, which is typically applied for point source emitters. The expression for a continuous point emission source is:

$$C(x,y,z) = \frac{Q_c}{2\pi u \sigma_y \sigma_z} \left\{ \exp\left(\frac{-(z-h-z_0)^2}{2\sigma_z^2}\right) + \exp\left(\frac{-(z+h+z_0)^2}{2\sigma_z^2}\right) \right\} \left\{ \exp\left(\frac{-(y)^2}{2\sigma_y^2}\right) \right\} \tag{7.5}$$

where $C(x,y,z)$ is the concentration of the emission at any position (x,y,z). u is the wind speed (m·s^{-1}). h is effective stack height (m), which is the sum of stack height, plume rise, and deposition height; z_0 is the roughness height of surface; σ_y(m) and σ_z(m) are the standard deviations of a statistically normal plume in the lateral and vertical dimensions, respectively. According to the results by Briggs (Briggs 1973), the standard deviations (σ_y and σ_z)

depend on atmospheric stability and downwind distance. Pasquill's stability categories are adopted to classify atmospheric stability (Hanna et al. 1982).

The objective of the optimization is to reduce the deviations of the prediction results with the mathematical model from measurements to a minimum. Source location and strength are two parameters to be determined. Both classic optimization methods such as pattern search (PS) and Nelder–Mead Simplex method (N–M Simplex) and heuristic intelligent search methods such as GA and SA could be used to identify the source term. Because the results from local optimization methods are usually largely dependent on the initial values while heuristic intelligent global optimization methods consume more time, the global search methods are used to obtain the initial values first and then the estimation values are searched by local optimization methods, which form the structure of the hybrid optimization method.

7.3.1.2 Case Studies of Source Estimation with Optimization Method

The experiment data from Prairie Grass emission experiment (Barad 1958) are used to verify the performance of different optimization methods for identifying the leakage source parameters. In this section, two sampling periods under stable (run 17) and unstable (run 33) atmospheric stabilities are selected to test the source determination algorithm. The MSE of simulation results of run 17 and 33 are 4759 and 3629 mg m^{-3}. The correlation coefficient (COR) between the experiment and the model for run 17 and 33 is 0.97 and 0.77

The optimization algorithm of GA, SA, and hybrid optimization methods of GA-NM (N–M simplex), GA-PS, SA-NM, and SA-PS were tested. The skill score was used to evaluate the performances of different methods for different source parameters including location and source strength. The skill score approaches zero when the estimated value is close to the real value. The less the skill score, the better is the performance of the optimization method. The results are shown in Figure 7.18. It is also known from Figure 7.18 that GA, SA, and GA-SA methods have slightly better performance.

It is also known from Figure 7.18 that GA, SA and GA-SA method have slightly better performance for determining source parameters using the experiment data from both experiments. However, the robustness of the SA method itself is not very good, as discussed in our previous research (Ma et al. 2013), GA-SA hybrid method is a better method than merely SA. It proves that the global optimization method has the advantage in finding optimization values near the global optimization point even if there are more errors between the forward model and the experiment. Moreover, it shows that the optimization method is capable of estimating the leakage source.

7.3.2 Source Estimation with MRE-PSO Method

Because the reasonability of the estimation results is doubtful by using the optimization method due to the random variabilities existed in the real world, stochastic approximation method may be more reasonable. The stochastic method often combines Bayesian inference and MCMC method, and it is a time cost process, especially for a wider computational area. Therefore, a less expensive algorithm based on probability principle for source identification is necessary. The MRE method could meet this requirement to obtain an estimation result with higher efficiency at certain probability level.

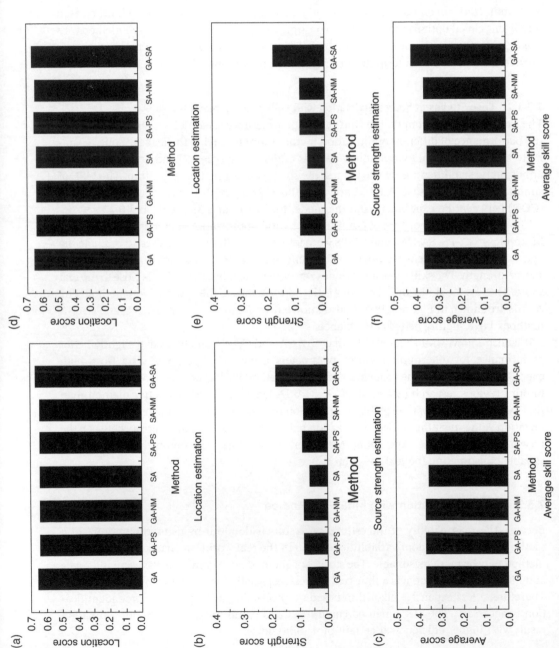

Figure 7.18 Comparison of different optimization methods for source parameters estimation for different cases. (a) to (c) are the results for run 17 and (d) to (f) are for run 33.

7.3.2.1 Principle of PSO-MRE for Source Estimation

In the MRE method, the unknown model parameters ($\mathbf{m}$) are viewed as random variables. The solution of the inverse problem is obtained from the multivariate probability density function (PDF) of $\mathbf{m}$ (Woodbury and Ulrych 1993, 1996, 1998). Let $\mathbf{x}$ be a state of a system, which has a set of possible states. $q^*(\mathbf{x})$ is its multivariate PDF. Let denote a possible PDF such that:

$$\int q^*(\mathbf{x})\mathrm{d}\mathbf{x} = 1 \tag{7.6}$$

Suppose a priori estimate of $q^*(\mathbf{x})$ is $p(\mathbf{x})$. The goal of the MRE is to obtain a reasonable estimation of $q^*(\mathbf{x})$ based on the prior information provided. The solution is to minimize the entropy of $q(\mathbf{x})$ relative to $p(\mathbf{x})$, as shown:

$$H(p,q) = \int_M q(x) \ln \left[q(x)/p(x) \right] \mathrm{d}x \tag{7.7}$$

The posterior estimate $q(\mathbf{x})$ has the form (Woodbury and Ulrych 1993, 1996, 1998) as:

$$q(x) = p(x) \exp \left[-1 - \mu - \sum_{j=1}^{N} \lambda_j f_j(x) \right] \tag{7.8}$$

where μ and λ_j are Lagrange multipliers determined from the constraints.

For a linear inverse problem, the problem can be expressed by Eq. (7.9):

$$\mathbf{d} = \mathbf{Gm} \tag{7.9}$$

where $\mathbf{d}$ is a discrete set of known data, $\mathbf{G}$ are the known transfer functions, and $\mathbf{m}$ is the vector of unknown "true" model parameters. A reasonable estimate $\hat{\mathbf{m}}$ of $\mathbf{m}$, which satisfies the preceding equation, is the final goal.

Woodbury and Ulrych (1993, 1996) derived the prior distribution for linear model in hydrogeologic applications, which has the form:

$$p(\mathbf{m}) = \prod_{i=1}^{M} \frac{\beta_i \exp \left(-\beta_i m_i \right)}{\exp \left(-\beta_i L_i \right) - \exp \left(-\beta_i U_i \right)} \tag{7.10}$$

where $[L_i, U_i]$ is the lower and upper bound of m_i and β_i is the Lagrange multiplier, which is determined by the definition of Eq. (7.11).

$$\int_m \mathbf{m} p(\mathbf{m})\mathrm{d}\mathbf{m} = \mathbf{s} \tag{7.11}$$

where $\mathbf{s}$ is the prior expected value constraints. By integrating Eq. (7.6), Eq. (7.12) is obtained to evaluate β_i (Neupauer and Borchers 2001).

$$\frac{-(\beta_i U_i + 1) \exp \left(-\beta_i U_i \right) + (\beta_i L_i + 1) \exp \left(-\beta_i L_i \right)}{\beta_i [\exp \left(-\beta_i L_i \right) - \exp \left(-\beta_i U_i \right)]} = s_i \tag{7.12}$$

In order to obtain the posterior distribution, the entropy of $q(\mathbf{m})$ relative to $p(\mathbf{m})$, as shown with Eq. (7.7), is minimized subject to the expected value constraints in Eqs. (7.6) and (7.12). Finally, the posterior estimate of the PDF $q^*(\mathbf{x})$ is obtained (Woodbury and Ulrych 1993, 1996):

$$q(\mathbf{m}) = \prod_{i=1}^{M} \frac{a_i \exp \left(-a_i m_i \right)}{\exp \left(-a_i L_i \right) - \exp \left(a_i U_i \right)} \tag{7.13}$$

where $a_i = \beta_i + \sum_{j=1}^{N} g_{ji}\lambda_j$. The posterior mean solution should fit the data within a specified tolerance, such as:

$$\|\mathbf{d} - \mathbf{G}\hat{\mathbf{m}}\|^2 \leq \xi^2 \varepsilon^2 \tag{7.14}$$

where ξ is a parameter that depends on the assumed error model and ε is the measurement error. Therefore, λ_j should satisfy the nonlinear system of Eq. (7.15) when $F(\lambda) = 0$.

$$F(\lambda) = \mathbf{d} - \mathbf{G}\hat{\mathbf{m}}(\lambda) + \frac{\xi\varepsilon}{\|\lambda\|}\lambda \tag{7.15}$$

In real operation, the Lagrange multipliers are adjusted to minimize the norm of the residuals between the measurement and modeled results.

Finally, the solution to this linear inverse problem is $\hat{\mathbf{m}}$, which is computed by $\int_{\mathbf{m}} \mathbf{m}p(\mathbf{m})d\mathbf{m} = \hat{\mathbf{m}}$, given as

$$\hat{m} = \frac{(a_iL_i + 1)\,exp\,(-a_iL_i) - (a_iU_i + 1)\,exp\,(-a_iU_i)}{a_i[\,exp\,(-a_iL_i) - exp\,(-a_iL_i)]} \tag{7.16}$$

The uncertainty in the MRE solution can be expressed with probability levels of $q(\mathbf{m})$. Confidence intervals (CIs) about the mean value $\hat{\mathbf{m}}$ is obtained by calculating the cumulative distribution function for $\mathbf{m}$, given as:

$$\int_0^{\mathbf{m}} q(\mathbf{x})d\mathbf{x} = P(\mathbf{m}) \tag{7.17}$$

Neupauer et al. (2000) defined $100(1-\alpha)\%$ probability interval with $\alpha/2$ and $1-\alpha/2$ percentile probability level. Then, the true value lies within $100(1-\alpha)\%$ interval with lower and upper bound of $\mathbf{m}_{P_{\alpha/2}}$ and $\mathbf{m}_{P_{1-\alpha/2}}$, which are the values of m when $P(\mathbf{m}) = \alpha/2$ and $1-\alpha/2$ in Eq. (7.17). For example, the 90% probability interval ($\alpha = 0.1$) lies between the estimated values at the 5th and 95th percentile probability level.

Since $a_{Mi,j} = \beta_{i,j} + \sum_{j=1}^{N} g_i\lambda_j$, a scale adjustment parameter λ_0 is considered to be added on β during the computation of a to compensate the bias from the constrains of λ. a_M thus becomes:

$$a_{Mi} = \lambda_0\beta_i + \sum_{j=1}^{N} g_{ji}\lambda_j(i = 1, ...4; j = 1, ..., N) \tag{7.18}$$

During this computation process, the determination of β with Eq. (7.12) and λ with Eq. (7.15) are two key steps. Woodbury and Ulrych (1998) and Neupauer and Borchers (2001) used the bisection method to determine the value of β. They also applied Newton–Raphson method to solve Eq. (7.15), in which Jacobian matrix is required to be calculated. But the condition number is too large, and the solution with Newton–Raphson method is impossible to be convergent when Jacobian matrix is singular. Hence, a useful intelligent optimization algorithm, PSO method, can also be used to solve Eq. (7.15). Therefore, a hybrid algorithm coupled the MRE with the PSO method with adjustment parameters that was proposed (Ma et al. 2014).

Because the linear model is required for MRE method, the nonlinear Gaussian dispersion model is not fitted for application of MRE. But a linear form can be obtained after transforming Gaussian dispersion model. Smith (1979) summarized a power law for σ_y and σ_z: $\sigma_y = ax^b$ and $\sigma_z = cx^d$. The values of the parameters a, b, c, and d are given by Martin (1979) for different atmospheric stabilities. Hence, Eq. (7.18) can be written in the form of linear and the corresponding terms are:

$$\ln\left(\pi u a c \underset{1 \times N}{\mathbf{C}}\right) = \left[1, -(b+d), -\frac{1}{2a^2}, -\frac{z^2}{2c^2}\right]$$

$$\left[\ln \underset{N \times 1}{\mathbf{Q_c}}, \ln\left(\underset{N \times 1}{\mathbf{X_0}} + x_1\right), \frac{\left(\underset{N \times 1}{\mathbf{Y_0}} + y_1\right)^2}{\left(\underset{N \times 1}{\mathbf{X_0}} + x_1\right)^{2b}}, \frac{1}{\left(\underset{N \times 1}{\mathbf{X_0}} + x_1\right)^{2d}}\right]^T$$

(7.19)

where $(\mathbf{X_0}, \mathbf{Y_0})$ is the location of sensor array in X_0OY_0 coordinate system where the origin point is the first sensor; (x_1, y_1) is the relative distance of the first sensor to gas emissions source. N is the length of monitor data. Dispersion model is built in the coordinate system of XSY with the origin point of gas emissions source. Hence, the distribution of all sensors in XS_0Y system is determined once (x_1, y_1) is estimated, which is only obtained from the first column of the parameters matrix. Therefore, only three parameters are required to be estimated and thus the problem is feasible to be solved.

7.3.2.2 Case Studies

The experiment data of release 33 from classic Prairie Grass emission experiment are used to estimate the source parameters of these two cases. The estimated results of location parameters and confidence area at different probability levels are shown in Figure 7.19. The CIs for different source parameters from 50 to 90% probability level are also calculated by Eq. (7.12), as shown in Table 7.1, where $[Q_c, x_1, y_1]$ is the estimation parameter vector. $\hat{\mathbf{m}}$ is estimated value and $\mathbf{m}$ is real value. $\mathbf{m}_{P_{0.05}}$ and $\mathbf{m}_{P_{0.95}}$ are the values at the 5th and

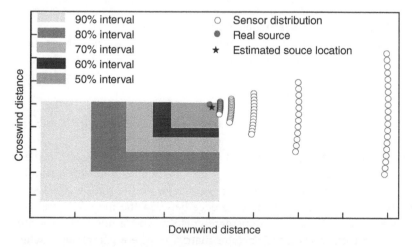

Figure 7.19 The location estimation with MRE-PSO method for run 33 case.

Table 7.1 Results of source term estimation by MRE-PSO method.

Parameter	Run 33	Run 17
$\mathbf{m}$	[94.7, 48.62, 6.09]	[56.5, 47.5, 15.5]
$\hat{\mathbf{m}}$	[90.62, 38.20, 12.05]	[41.91, 34.65, 13.63]
$\mathbf{m}_{P_{0.05}}$	[2.42, 1.48, 0.22]	[1.38, 1.37, 0.57]
$\mathbf{m}_{P_{0.95}}$	[930.12, 802.60, 219.64]	[755.16, 739.47, 151.89]

95th percentile probability levels, between which is the 90th confidence probability interval. The source parameters of run 17 were also calculated with MRE method, as shown in Table 7.1. It proves again that the MRE-PSO method can be applied for source parameters estimation of gas emissions.

7.3.3 Source Estimation with PSO-Tikhonov Regulation Method

The Bayesian-based methods like MCMC to estimate source parameters often consume more computation time than optimization methods. Additionally, a lot of prior information such as measurement errors, parameter bounds, and expected inputs should be determined previously in Bayesian estimation and MRE methods, but they are hard to be known in the real condition. Therefore, a method without prior error input and with reasonable uncertain intervals output should be considered. Regularization method is a classical inverse method based on the least squares solution. In this method, the ill-posed inverse problem is replaced with a family of similar well-posed problems through the introduction of a regularization operator and a regularization parameter. If some special method like L-curve is used to determine regularization parameter, the prior noise level is not required. For a linear inverse problem, the uncertainty at some probability levels can also be obtained by the regularization method. In this section, a method coupling Tikhonov regularization with PSO algorithm was presented to estimate the source parameters of hazardous gas emission in atmosphere.

7.3.3.1 Principle of PSO-Tikhonov Regularization Hybrid Method

Tikhonov regularization is based on the least square principle. But it is different from common least square method by introducing regularization operator and parameters. Tikhonov regularization method is often utilized to solve the linear inverse problem. The basic expression of the regularization form is shown with Eq. (7.20):

$$\text{min} \quad \phi(\mathbf{m}) = \|\mathbf{G}_{M \times N} \mathbf{m}_{N \times 1} - \mathbf{d}_{M \times 1}\|_2^2 + \lambda^2 \|\mathbf{L}_{M_L \times N} \mathbf{m}_{N \times 1}\|_2^2 \tag{7.20}$$

where M is the length of $\mathbf{d}$, and N is the number of estimation parameter; λ is the regularization parameter; $\mathbf{L}$ is an operator matrix and $\|\cdot\|$ denotes the Euclidian norm. Generally, the rank of $\mathbf{L}$ is M_L and it satisfies $M \geq N \geq M_L$. The first term on the right-hand side of Eq. (7.21) represents the square norm of the difference between the measured and the model-predicted system state. The second term represents the square norm of a specific property of the model that depends on the operator matrix, $\mathbf{L}$, where parameter λ determines how well the solution fits the data. The parameter λ has to be adjusted to make the solution fit the data in some optimal way. The error in the Tikhonov regularization solution depends both on the noise level and on the regularization parameter λ. A good regularization parameter should yield a fair balance between the data error and the regularization error in the regularized solution. The selection of the optimal regularization parameter is based on minimizing the total error. Generalized cross validation (GCV), L-curve, and quasi-optimality criterion are two most popular methods, which do not require any information about the noise level.

The L-curve is simply a logarithmic plot of residual norm $\|\mathbf{Gm} - \mathbf{d}\|_2^2$ versus the solution norm $\|\mathbf{Lm}\|_2^2$ for a set of admissible regularization parameter. In this way, the L-curve displays the compromise between the minimization of these two terms. The "corner" of the L-curve is defined as the point on the L-curve:

$$(\varsigma(\lambda), \mu(\lambda)) = (\log\|\mathbf{Gm} - \mathbf{d}\|, \log\|\mathbf{Lm}\|) \tag{7.21}$$

with the maximum curvature. Here the curvature k is given by:

$$k(\lambda) = \frac{\varsigma'\mu'' - \varsigma''\mu'}{\left(\left(\varsigma'\right)^2 + \left(\mu'\right)^2\right)^{3/2}} \tag{7.22}$$

where the differentiation is with respect to λ.

With every element in the regularization set, the damped least squares problem like Eq. (7.20) is solved. If the problem has a linear form and satisfies the structure of common Tikhonov method, singular value decomposition (SVD) can be used to solve Eq. (7.20). But gas dispersion process is a classical nonlinear problem, if multiple source parameters are required to be determined, SVD is failure in this case. PSO is a good method to solve the optimization problem, and thus it can be used to obtain the optimal estimation $\mathbf{m}$ under certain regularization parameter.

Because Eq. (7.6) is not proper to estimate multiple gas emission source parameters, Ma et al. (2017) proposed linear PSO-Tikhonov method, as shown with Eq. (7.24), after transforming the nonlinear Gaussian dispersion model to be a linear form and modified Eq. (20) to be:

$$\min \ \phi(\mathbf{m}') = \|\mathbf{G}_{1 \times N'}\mathbf{m}'_{N' \times M} - \mathbf{d}_{1 \times M}\|_2^2 + \lambda^2\|\mathbf{m}'_{N' \times M}\mathbf{L}_{M \times M_L}\|_2^2 \tag{7.23}$$

where $\mathbf{m}'$ is the transformed parameter matrix and N' is the length of $\mathbf{m}'$. In this case m' can be estimated successfully with linear PSO-Tikhonov method and then the original estimation parameters can be obtained.

Additionally, according to the presentation of Kathirgamanathan (2003), the gas emission source parameters can be determined with two steps nonlinear method to solve the problem as Eq. (7.24).

$$\mathbf{d} = \mathbf{G}(\mathbf{m}_0)\mathbf{m}_1 \tag{7.24}$$

where $\mathbf{m}_0$ is the matrix only including location parameters and $\mathbf{m}_1$ is the source strength parameter. The transfer matrix $\mathbf{G}(\mathbf{m}_0)$ is a nonlinear function about location parameters, and the source strength is linear to the concentration $\mathbf{d}$. The source location parameters are determined first with a unit emission rate by optimization method. Then, the emission rate is calculated with the estimated location parameters in the previous step. The detail of linear PSO-Tikhonov and two-step nonlinear method computation process were discussed in the paper of Ma et al. (2018).

Although linear PSO-Tikhonov method costs much less time than two-step nonlinear method as discussed in Ma et al., there are still some problems due to the assumption of linear transformation. Hence, one-step nonlinear PSO-Tikhonov method was designed to estimate multiple emission source parameters.

The source parameters including emission rate and downwind and crosswind distance to the certain sensor are required to be estimated when the emission gas concentrations at different sensors have been measured. In this case, the regularization equation becomes:

$$\min \boldsymbol{\phi}(\mathbf{m}) = \left\| \mathbf{G}(\mathbf{m})_{N \times 1} - \mathbf{d}_{M \times 1} \right\|_2^2 + \lambda^2 \left\| \mathbf{L}_{M_L \times N} \mathbf{m}_{N \times 1} \right\|_2^2 \tag{7.25}$$

where the first term is a nonlinear form including all source parameters. The computation process of one-step nonlinear PSO-Tikhonov regularization method is illustrated in Figure 7.20.

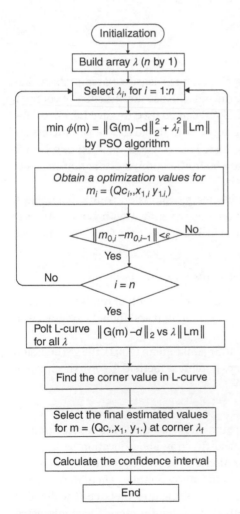

First, an array with different regularization values is set. For every regularization parameter, the best estimation for source parameters will be obtained with PSO algorithm by the cost function of Eq. (7.25). It is noted from Figure 7.20 that one-step nonlinear method proposed here only needs one PSO computation process with a certain regularization parameter. The curve of $\|\mathbf{Gm} - \mathbf{d}\|_2^2$ vs. $\|\mathbf{Lm}\|_2^2$ under all regularization values is then plotted, and the point with the maximum curvature is selected as the best estimation parameter. Finally, CI under certain probability level is calculated. The difference between one-step nonlinear and linear PSO-Tikhonov method lies in the regularization term. Hence, the CIs will be calculated with different methods compared with linear inverse model. For nonlinear method, the CIs can be determined with Eq.(7.26).

$$\hat{m}_i \pm t_{(1-\alpha)/2, f} S_{m_i} \tag{7.26}$$

which is based on t-distribution. α is the probability level and f is the degree of freedom. S_{m_i} is the square root of the diagonal elements of variance–covariance matrix $S^2(W^T W)^{-1}$. S is given by:

$$S = \frac{\sum_{i=1}^{n} |d_i - \mathbf{G}(\mathbf{m}_i)|}{\sqrt{n+1-r}} \tag{7.27}$$

Here, n is the length of data, r is the length of parameter vector. $\mathbf{m} = [Q_c, x_1, y_1]$ and $\mathbf{W}$ is a matrix of the partial derivatives of $\mathbf{G}$, which is obtained by $\mathbf{W} = [\mathbf{G} \, \mathbf{J}]$, where $\mathbf{J}$ is $n \times 2$ matrix,

Figure 7.20 Computation flow of one-step nonlinear PSO-Tikhonov nonlinear method for source identification.

$$J_{i1} = \sum_j \frac{\partial G_{i,j}}{\partial x_1} Q_c, J_{i2} = \sum_j \frac{\partial G_{i,j}}{\partial y_1} Q_c \tag{7.28}$$

The variance–covariance matrix of estimated parameter $\hat{\mathbf{m}}$ is

$$\text{var}(\hat{\mathbf{m}}) = S^2 (w^T w)^{-1}, w = \begin{bmatrix} W \\ \lambda[L \quad 0] \end{bmatrix} \tag{7.29}$$

7.3.3.2 Case Study

Run 17 and 33 in Prairie Grass emission experiment were used to test the methods. The L-curve during the test process of release 17 with one-step nonlinear PSO-Tikhonov regularization method is illustrated in Figure 7.21.

The L-curve during the test process of release 17 with one-step nonlinear PSO-Tikhonov regularization method is illustrated in Figure 7.21. The optimal regularization parameter in this case is 67.03. The estimation results are listed in Table 7.2. The results for the source estimation for run 33 were also listed in Table 7.2. The estimation results of two cases by two-step nonlinear method were also listed in Table 7.2. Compared with the results in Table 7.2, it is noticed that the CI calculated by one-step nonlinear method is narrower than that by linear and two-step nonlinear PSO-Tikhonov method and it seems more reasonable.

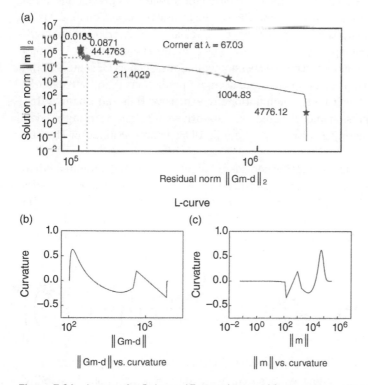

Figure 7.21 L-curve for Release 17 experiment with one-step nonlinear PSO-Tikhonov method.

Table 7.2 Results of source term parameters by two-step nonlinear and linear PSO-Tikhonov regularization method with experiment case.

Method		Optimal λ	Estimated values with uncertainty interval	Skill score $S = [S_q, S_x, S_y]$
			$\hat{m} = [\hat{Q}_c, \hat{x}_0, \hat{y}_0]$	
Two-step nonlinear PSO-Tikhonov method	Release 17	0.4385	$[35.11 \pm 36, 35.49 \pm 2690, 12.87 \pm 219]$	$[0.38, 0.25, 0.17]$
	Release 33	3.3630	$[137.12 \pm 9531, 72.86 \pm 9983, 7.80 \pm 1363]$	$[0.45, 0.47, 0.28]$
Linear PSO-Tikhonov method	Release 17	2.2381	$[35.11 \pm 66.45, 35.73 \pm 8.36, 12.87 \pm 2.85]$	$[0.36, 0.40, 0.16]$
	Release 33	0.1451	$[72.01 \pm 116.45, 30.67 \pm 36.02, 9.80 \pm 10.27]$	$[0.24, 0.38, 0.61]$

7.3.4 Source Estimation with MCMC-MLA Method

7.3.4.1 Forward Gas Dispersion Model Based on MLA

It is known that the prediction accuracies and efficiencies of forward dispersions are crucial for rapidly and precisely estimating source terms with inverse algorithm (Ma et al. 2013). However, the different dispersion models have been used for forward gas predictions in the atmosphere, such as LS, Gaussian mixture, and CFDs. The Gaussian dispersion model is an explicit model based on analysis and experiment; thus, it has a high computational efficiency. However, this type of model is not suited to some complex environments. Although the models based on solving gas advection–dispersion equations, such as CFD models and those based on the LS principle, can obtain more accurate predictions than Gaussian dispersion models, they are much more computationally expensive (Ma and Zhang 2016). Therefore, a forward dispersion model based on data-driven machine learning models were proposed to atmospheric gas dispersion (Vapnik 1998; Lauret et al. 2016; Huchuk et al. 2019).

Data-driven machine learning models are considered to offer a potentially useful method for predicting complex systems accurately and efficiently. Previously, we proposed a series of machine learning algorithm (MLA)–Gaussian models to predict the gas dispersion in atmosphere for point continuous emissions (Ma and Zhang 2016), combining MLA and Gaussian integration modules. Our results showed that the SVM–Gaussian model performed the best among different MLA models.

The Gaussian module is written as

$$C_{\text{gau}}(D_i, A_i) = \frac{1}{2\pi u \sigma_y \sigma_z} \left\{ \exp\left(\frac{-(D_z - h)^2}{2\sigma_z^2}\right) + \exp\left(\frac{-(D_z + h)^2}{2\sigma_z^2}\right) \right\} \left\{ \exp\left(\frac{-(D_y)^2}{2\sigma_y^2}\right) \right\}$$

(7.30)

The fundamental structure of the Gaussian–SVM model is illustrated in Figure 7.22. Here, 10 monitoring parameters, including wind speed, wind direction, atmospheric stability, Monin–Obukov length, effective source height, measured height, and location, were integrated into one parameter with a Gaussian module. The source strength and the parameter

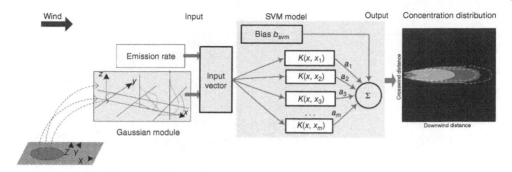

Figure 7.22 Structure of Gaussian–SVM model.

obtained with the Gaussian module were then adopted as the input vectors. Thus, only two input vectors were required for the Gaussian–SVM model. The Gaussian dispersion model, with a unit emission rate, was used to integrate the atmospheric and source term parameters.

The approximation model for the SVM used in this study is as follows (Vapnik 1998; Hsu and Lin 2002):

$$y(\theta) = \sum_{i=1}^{m} (-\alpha_i + \alpha_i^*) K(\theta_i, \theta_j) + b_{svm} \tag{7.31}$$

where α is the support vector and $K(\theta_i, \theta_j)$ is the kernel function. The radial basis function was selected as the kernel function here. For the Gaussian–SVM model, the random grid search was implemented sequentially to optimize the regularization parameter, C, spread parameter, δ, and loss function parameter, ε.

The prediction process of Gaussian–MLA model is as following: first, the input parameters are integrated with Gaussian module and then it combines with source leakage rate to form the input vector. Therefore, there are only two input parameters in the new network model, which is called Gaussian–MLA model. Finally, the concentrations at different positions are predicted with Gaussian–MLA model.

In this section the SVM model was constructed based on the data of the Prairie Grass experiment (Addepalli et al. 2009). Half of 64 release monitoring datasets (2832 data points) in the Prairie Grass experiments were utilized to train the model, while the other 3147 points were used for testing. The predicted results of test dataset and release 33 are shown in Figure 7.23.

The prediction results of run 33 in Prairie Grass emission experiment were compared in Table 7.3.

It indicated that Gaussian-SVM can predict the concentration with highest accuracy among Gaussian, LS, CFD, and Gaussian-SVM models and has the similar calculation time with Gaussian model.

7.3.4.2 Source Estimation with MCMC-MLA Method

The proposed MCMC method is based on Bayes' rule, as expressed in Eq. (7.32) (Brooks et al. 2011):

$$P(\theta \mid D) = \frac{P(D \mid \theta)P(\theta)}{P(D)} \tag{7.32}$$

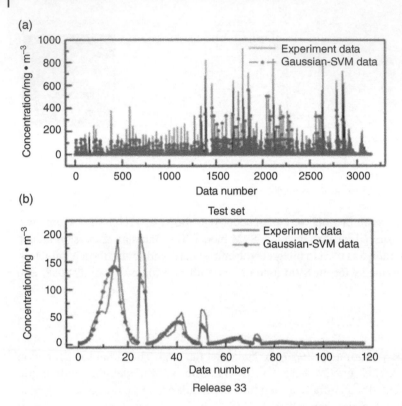

Figure 7.23 The results of predicted concentrations with SVM-MLA dispersion model. (a) is for all test cases and (b) is just for run 33.

Table 7.3 Comparison results of different common dispersion models.

Model	Mean square error	Correlation coefficient	Computation speed
Gaussian model	1676.21	0.78	1.04 s
LS model	506.17	0.89	>24 h
CFD model	1519	0.73	~6 h
Gaussian-SVM model	300.37	0.92	14.51 s

where $P(\theta)$ is the prior probability of θ, which is the strength of our certainty without considering the evidence, D, and $P(\theta \mid D)$ is the posterior, which is the strength of our certainty once the evidence D has been considered. Thus, $P(D \mid \theta)$ is the likelihood (e.g. the probability of the data D being generated by a model with parameter θ) and $P(D)$ is the evidence, which is the probability of the data as determined.

$$P(\mathbf{D}) = \int_{\Theta} P(\mathbf{D}, \theta) d\theta \tag{7.33}$$

Therefore, the evidence is an integral over all values of θ. As it is impossible to evaluate the analytically expressed integral because it is impossible to know all values of θ, a numerical method to approximate the integral results of Eq. (7.33) is necessary. Because the prior distributions could potentially have a large number of dimensions, the posterior distributions would also be high dimensional. Hence, a potentially large-dimensional integral may be evaluated numerically. To address such a high-dimensional Bayesian model, the MCMC method was adopted.

With the MCMC method, it is possible to draw samples from the posterior PDF without knowing the normalization constant; thus, a simplified version of Eq. (7.32) is used:

$$P(\theta \mid D) \propto P(D \mid \theta)P(\theta) \tag{7.34}$$

It is not necessary to calculate the evidence term, for which analytical solutions are rarer and numerical computation is expensive. After assuming that the errors between measured and model values have a mean of zero and variances of $\delta_{D,i}^2$, the likelihood term becomes:

$$P(D \mid \theta) \propto \exp\left[-\frac{1}{2}\sum_{i=1}^{N}\frac{(D_{model,i}(\theta) - D_{mea,i})^2}{\delta_{D,i}^2}\right] \tag{7.35}$$

where D_{model} is the data predicted by the model, D_{mea} is the measured data, and N is the number of observations. Thus, the closer the prediction is to the measured values, the higher the likelihood of the source parameters. Hence, $D_{model,\ i}(\theta)$ can be calculated with various values of θ, which can be sampled via the MCMC method.

The essential idea underlying MCMC is to sample from the posterior distribution by combining a Monte Carlo random search with a Markov chain, in which the sample does not depend on where the search process begins. Thus, the MCMC method is a memoryless search performed with an intelligent sample. There have been many sample algorithms proposed for the MCMC method, such as the Metropolis algorithm (Metropolis et al. 1953), Metropolis–Hastings (MH) (Hastings 1970), Gibbs sampler (Gilks et al. 1995), and Hamiltonian sampler (Neal 2011).

In this section, the concentration in Eq. (7.35) is estimated using the Gaussian–SVM model during the MCMC process, which is the proposed MCMC–MLA method (Ma et al. 2021b).

The Delayed Rejection Adaptive Metropolis (DRAM) sampling process was introduced. For regular MH sampling, the acceptance of the samples can be written as (Hastings 1970):

$$\partial_1(\theta, \varphi_1) = \min\left(1, \frac{\pi(\varphi_1)q_1(\varphi_1, \theta)}{\pi(\varphi_1)q_1(\theta, \varphi_1)}\right) = \min\left(1, \frac{N_1}{D_1}\right) \tag{7.36}$$

where θ is the current point; φ_1 is the new value drawn from proposed distribution, $q_1(\theta, \cdot)$, and π is the target distribution, which was a Gaussian normalizing distribution in this study. For the delayed rejection (DR) strategy, if φ_1 was rejected, a second candidate, φ_2, from $q_2(\theta, y, \cdot)$ was drawn, and the acceptance probability (Tierney 1994) was updated to:

$$\partial_2(\theta, \varphi_1, \varphi_2) = \min\left(1, \frac{\pi(\varphi_2)q_1(\varphi_2, \varphi_1)q_2(\varphi_2, \varphi_1, \theta)[1 - \partial_1(\varphi_2, \varphi_1)]}{\pi(\theta)q_1(\theta, \varphi_1)q_2(\theta, \varphi_1, \varphi_2)[1 - \partial_1(x, \varphi_1)]}\right)$$
$$= \min\left(1, \frac{N_2}{D_2}\right) \tag{7.37}$$

If q_i denotes the proposal at the i-th stage, the acceptance probability at that stage is as follows (Mira 2001):

$$\partial_i(\theta, \varphi_1, ..., \varphi_i)$$

$$= \min \left\{ \begin{array}{c} \left(1, \dfrac{\pi(\varphi_i)q_1(\varphi_i, \varphi_{i-1})q_2(\varphi_i, \varphi_{i-1}, \varphi_{i-2})\cdots q_i(\varphi_i, \varphi_{i-1}, ..., \theta)}{\pi(\theta)q_1(x, \varphi_1)q_2(\theta, \varphi_1, \varphi_2)}\right) \\[3mm] \left(\dfrac{[1 - \partial_1(\varphi_i, \varphi_{i-1})][1 - \partial_2(\varphi_i, \varphi_{i-1}, \varphi_{i-2})]\cdots[1 - \partial_{i-1}(\varphi_i, ..., \varphi_1)]}{[1 - \partial_1(\theta, \varphi_1)][1 - \partial_2(\theta, \varphi_1, \varphi_2)]\cdots[1 - \partial_{i-1}(\theta, \varphi_1, ..., \varphi_{i-1})]}\right) \end{array} \right\}$$

(7.38)

$$= \min\left(1, \frac{N_i}{D_i}\right)$$

The DR strategy demonstrated in Eqs. (7.36)–(7.38) has been shown to perform more efficiently than the MH chain with respect to Peskun–Tierney ordering (Mira 2001).

In an Adaptive Metropolis (AM), the covariance matrix of the Gaussian distribution is adapted spontaneously using the past chain. This adaptation destroys the Markovian property of the chain. However, it has been shown that the ergodicity of the generated sample remains. Starting from the initial covariance, C_0, the target covariance is updated at given intervals from the chain thus far generated. The initial non-adaptive period is defined as n_0 and the covariance of the Markov chain, X_n, is

$$C_n = \left\{ \begin{array}{cc} C_0 n \leq n_0 \\ S_d \text{Cov}(X_0, ..., X_{n-1}) + s_d \varepsilon I_d & n > n_0 \end{array} \right\}$$

(7.39)

where s_d is a parameter that depends only on the dimension (d) of the state space on which π is defined and $\varepsilon > 0$ is a small constant to prevent the sample covariance matrix from becoming singular, and $I d$ denotes the d-dimensional identity matrix.

If the DR method is adopted in the AM process, the DRAM method is obtained (Gilks et al. 1995). Here, the direct combination of DR and AM proposed by Haario et al. (2006) was utilized. With an m-stage DR algorithm, the proposal at the first stage of DR is adapted just as in AM, while the covariance of the proposal for the i-th stage ($i = 2,...m$) is always computed with a scaled-down version of the proposal of the first stage. More detail about MCMC-MLA method has been discussed in the paper of Ma et al. (2021b).

7.3.4.3 Case Study

Run 33 and Run 17 in the Prairie Grass experiment were selected to test the MCMC-MLA method. Figure 7.24 shows the Markov chain with the MLA model in the test cases. Compared with the results shown in Figure 7.21. The histograms of the Markov chains with the MLA model for different source parameters in runs 33 and 17 are shown in Figure 7.25. The results indicated that the mean estimated values of the location parameters with the MCMC–MLA model were closer to the real locations than those predicted using the MCMC–Gaussian method. As shown in Figure 7.26, the distance between the mean estimated and real positions using the MCMC–MLA model was 9.92 m in the test of Run 33 and 7.75 m for that in Run 17.

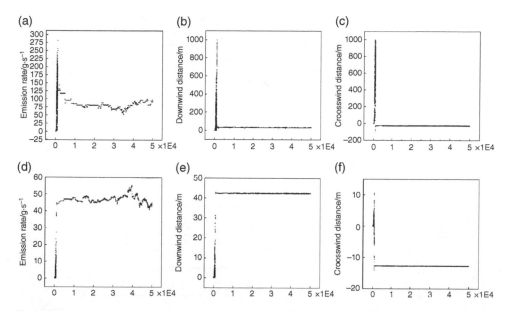

Figure 7.24 Markov chains using the proposed MCMC–MLA model for different parameters in: (a)–(c) Run 33 and (d)–(f) Run 17.

The skill scores of the MCMC method with Gaussian and Gaussian–MLA forward prediction models for Runs 33 and 17 are summarized in Table 7.4. According to these results, the location estimation using the MCMC–MLA approach was improved relative to the MCMC–Gaussian method, although the skill scores of the source strength in both runs using the MCMC–MLA method were larger than those using the MCMC–Gaussian method. The total average estimation skill score of MCMC–MLA was lower than that of the MCMC–Gaussian method.

The 95% CIs of location and source strength estimation for Run 33 and 17 were illustrated in Figure 7.27. It is shown that the CIs of location with MCMC–MLA are closer to the real location than that with MCMC–Gaussian method. Therefore, the location estimation of MCMC–MLA is more accurate than that of MCMC–Gaussian in test cases. But the CIs of source strength estimation with MCMC–MLA method seems not better than that of MCMC–Gaussian.

7.4 Conclusions and Future Work

7.4.1 Conclusions

In this chapter, different intelligent methods for recognizing the emission sources including abnormal signals, component, and source term were discussed.

For the gas identification methods, different emission scenarios were discussed. One is the case that the gases exist in the atmosphere, and another is that the emission gas is unknown.

For the gases that exist in the atmosphere, e.g. CO_2 from geosequestration project, the monitoring and identification may be a challenging issue due to complex variation of

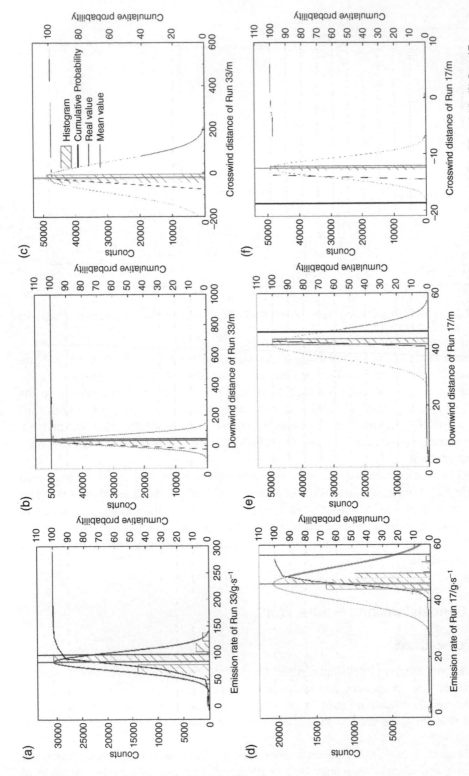

Figure 7.25 Histograms of the Markov chains for different source parameters using the MCMC–MLA model for: (a)–(c) Run 33 and (d)–(f) Run 17.

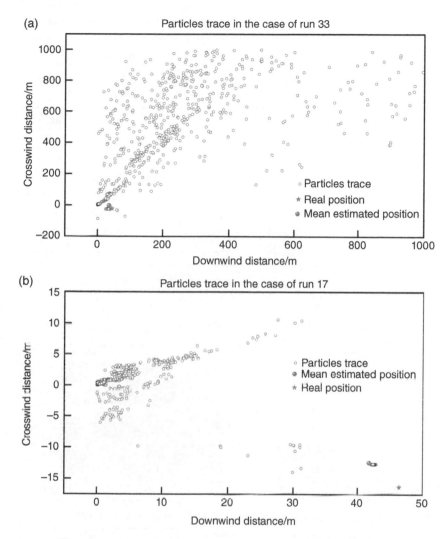

Figure 7.26 Distribution of real and estimated positions in the Markov chain using the MLA model for different test cases in: (a) Run 33 and (b) Run 17.

the atmosphere background. In this case, intelligent algorithm is a good tool to recognize abnormal signal. Time series network can provide an approximation for the daily natural variation of the natural gases. Therefore, the approximation error between the monitoring and the prediction can be utilized to distinguish the abnormal leakage signal due to CO_2 release in atmosphere. This method can also be applied to other fields that need to recognize abnormal signals in complex time series variations.

For the unknown emission gases or the VOCs that are not available in the atmosphere, AOS combined with sensor array and intelligent pattern recognition algorithms may be a good tool to improve the accuracy of the recognition. Compared with single sensor, sensor array can provide much more information about substances, thus the classification accuracy of machine learning model, e.g. SVM model with sensor array is much higher than that with single sensor. The dynamic response map of sensor array indicates the specific

Table 7.4 Skill scores of the MCMC method combined with different forward models for two runs.

	Run 33 case			Run 17 case		
	Real value	MCMC–Gaussian	MCMC–MLA	Real value	MCMC–Gaussian	MCMC–MLA
Emission rate (mg·m^{-3})	94.7	97.87	84.18	56.5	55.13	46.19
Downwind distance (m)	44.55	100.32	35.62	46.35	59.33	41.80
Crosswind distance (m)	−22.7	−26.81	−18.27	−18.73	−12.44	−12.45
S_q		0.0335	0.1111		0.0243	0.1825
S_x		1.2518	0.2003		0.2801	0.0982
S_y		0.1811	0.1951		0.3360	0.3351
S_l		1.2649	0.2798		0.4374	0.3492
S_a		0.4888	0.1689		0.2135	0.2057

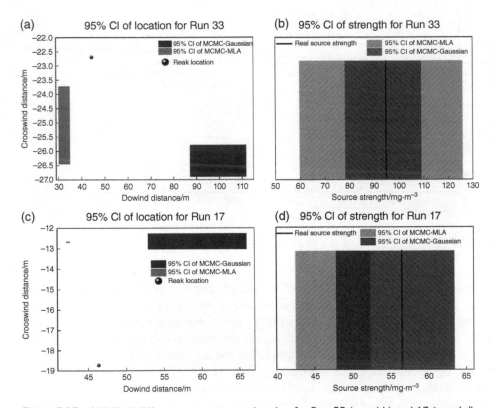

Figure 7.27 95%CI of different parameters estimation for Run 33 (a and b) and 17 (c and d).

features for certain substances even with different concentrations after normalizing, thus the dynamic response map of sensor array can be used to recognize the gases or VOCs. Therefore, data-driven models including machine learning and deep learning method could be utilized to identify different VOCs. Among different data-driven models, DLM performs better than simple machine learning model like SVM based on conventional extracted features. However, due to the limitation of sample size, pretrained transfer learning model could overcome the defect of overfitted simple deep learning. Hence, it is feasible to recognize different gases or VOCs with just one sensor array device combined with DLM with high accuracy.

Therefore, an appropriate method should be selected to monitor and identify the components of the emission gases. Intelligent algorithm like data-driven model has advantages in prediction with historical data.

For source term estimation problem, different intelligent algorithms including optimization, stochastic, and regularization method were discussed to identify the source parameters of the emission source in the atmosphere.

Intelligent optimization method such as GA and SA performed better than classic optimization methods such as NM and PS method in global searching. But the robustness and efficiency of the intelligent optimization methods were worse than that of classic optimization methods due to the group intelligent optimization finding the optimal parameters near the global area. Therefore, hybrid optimization method combining the classic and intelligent optimization methods seemed a good choice to estimate source parameters of the emission source with a best total performance among different tested optimization methods. However, optimization methods cannot provide a reasonable uncertainty analysis for estimation. MRE PSO method is a method that can provide probability estimation and has higher accuracy than conventional MCMC method. In spite of this, MRE method depends on some prior information such as measurement errors, parameter bounds, and expected inputs, which are hard to determine in real application. Therefore, Tikhonov-PSO regularization method was a method without prior error input and with reasonable uncertainty intervals output should be considered. One-step nonlinear Tikhonov-PSO regularization method performed better than two-step linear method. However, conventional forward dispersion model used in MRE and regularization method like Gaussian model is not fitted in some complex environment. Thus, the estimation accuracy is affected by the forward model. Machine learning model provides more accuracy than Gaussian model to predict gas dispersion as well as a similar computation time. Therefore, it is a potentially good tool to improve the performance of source term estimation. MCMC–MLA method was tested and compared with MCMC–Gaussian method to estimate the source parameters. The results showed that the location errors were reduced obviously with MCMC–MLA method.

Accordingly, different source estimation methods performed variably for different source parameters. Accuracy and efficiency are two key factors we should focus on. A suitable method should be selected for different application scenarios.

7.4.2 Limitations and Future Work

Although intelligent methods like machine learning methods could provide a good tool to predict the source components, recognize the leakage signal, and identify the source term, there are still some limitations.

First, for leakage signal recognition of the gases that exist in the atmosphere, e.g. CO_2, local historical data should be captured previously. The accuracy of the data source would affect the prediction results. On the other hand, the efficiency of the machine learning model to predict the atmospheric variation limits the application in online monitoring.

Moreover, for the emission event with just single component, AOS is a perfect method to monitor and identify the component of the source. However, the objectives of the AOS are limited in data source, which were used to train the machine learning model. It is difficult to recognize the component without the data source. Further, it is still impossible for identifying multiple components for mixture gases with AOS, but multiple components often exist together in chemical park.

Moreover, although intelligent methods such as optimization method, stochastic method, and machine learning have been utilized in source term estimation, most cases discussed in this chapter were the emission events with just one source. It more difficult to deal with the cases with multiple sources than that with single source. Forward dispersion model and inverse algorithm have great impact on the results of source estimation. Machine learning models to predict the gas dispersion still have limitations in the generalization due to limited data source. More experiments data in different scenarios should be combined together to build the model. The efficiency of enumeration or random search algorithms for the cases with uncertainty source number is still too high, which limits the application of source term estimation.

Therefore, there are still some challenging problems for emission source monitoring and identification. The algorithms to estimate leaking CO_2 concentrations should be further designed to monitor the leakage with the disturbance in the atmosphere. On the other hand, the intelligent algorithm of AOS to detect multiple VOCs components in mixture gases will be a challenging issue in future as well as the accurate gas detection method to identify one certain kind of gas under the environment with other gases. Further, the identification of emission sources under uncertain conditions including source components and source term will be another challenging issue. A method with both efficacy and accuracy should be investigated. Some new information methods such as transfer learning, kennel density function, complex system, and others like this should be induced in this field. It can be expected that a real-time emission map can be reconstructed for a large area scale like chemical park with more and more improved intelligent algorithms in future.

References

Addepalli, B., Sikorski, C., and Pardyjak, E.R. (2009). *Source Characterization of Atmospheric Releases Using Quasi-Random Sampling and Regularized Gradient Optimization*. Salt Lake City, USA: University of Utah.

Barad, M.L. (1958). *Project Prairie Grass, a Field Program in Diffusion. Volume 1*. Air Force Cambridge Research Labs Hanscom Afb MA.

Briggs, G. (1973). *Diffusion Estimation for Small Emissions ATDL, Contribution File No. 97*. NOAA, Oak Ridge, Tennessee: Air Resources Atmospheric Turbulence and Diffusion Laboratory.

Brooks, S., Gelman, A., Jones, G. et al. (2011). *Handbook of Markov Chain Monte Carlo*. CRC press.

Chow, F.K., Kosović, B., and Chan, S. (2008). Source inversion for contaminant plume dispersion in urban environments using building-resolving simulations. *Journal of applied meteorology and climatology* 47 (6): 1553–1572. https://doi.org/10.1175/2007JAMC1733.1.

Enting, I.G. (2002). *Inverse Problems in Atmospheric Constituent Transport*. Cambridge University Press.

Gilks, W.R., Best, N.G., and Tan, K.K. (1995). Adaptive rejection Metropolis sampling within Gibbs sampling. *Journal of the Royal Statistical Society: Series C (Applied Statistics)* 44 (4): 455–472. https://doi.org/10.2307/2986138.

Haario, H., Laine, M., Mira, A. et al. (2006). DRAM: efficient adaptive MCMC. *Statistics and computing* 16 (4): 339–354. https://doi.org/10.1007/s11222-006-9438-0.

Hanna, S.R., Briggs, G.A., and Hosker, R.P. Jr. (1982). *Handbook on atmospheric diffusion* (No. DOE/TIC-11223. In:. Oak Ridge, TN (USA): National Oceanic and Atmospheric Administration, Atmospheric Turbulence and Diffusion Lab.

Hastings, W.K. (1970). Monte Carlo sampling methods using Markov chains and their applications. *Biometrika* 57 (1): 97–109. https://doi.org/10.1093/biomet/57.1.97.

Haupt, S.E. (2005). A demonstration of coupled receptor/dispersion modeling with a genetic algorithm. *Atmospheric Environment* 39 (37): 7181–7189. https://doi.org/10.1016/j.atmosenv.2005.08.027.

Haupt, S.E., Pasini, A., and Marzban, C. (ed.) (2008). *Artificial Intelligence Methods in the Environmental Sciences*. Springer Science & Business Media.

Heemink, A.W. and Segers, A.J. (2002). Modeling and prediction of environmental data in space and time using Kalman filtering. *Stochastic Environmental Research and Risk Assessment* 16 (3): 225–240. https://doi.org/10.1007/s00477-002-0097-1.

Hsu, C.W. and Lin, C.J. (2002). A comparison of methods for multiclass support vector machines. *IEEE Transactions on Neural Networks* 13 (2): 415–425. https://doi.org/10.1109/72.991427.

Huchuk, B., Sanner, S., and O'Brien, W. (2019). Comparison of machine learning models for occupancy prediction in residential buildings using connected thermostat data. *Building and Environment* 160: 106177. https://doi.org/10.1016/j.buildenv.2019.106177.

Isakov, V. and Kindermann, S. (2000). Identification of the diffusion coefficient in a one-dimensional parabolic equation. *Inverse Problems* 16 (3): 665.

Kathirgamanathan, P. (2003). Source parameter estimation of atmospheric pollution from accidental releases of gas: a thesis presented in partial fulfilment of the requirements for the degree of Doctor of Philosophy in mathematics at Massey University, Palmerston North, New Zealand (Doctoral dissertation, Massey University).

Keats, A., Lien, F.S. and Yee, E. (2006). Source determination in built-up environments through Bayesian inference with validation using the MUST array and joint urban 2003 tracer experiments. In *Proc 14th Annual Conference of the Computational Fluid Dynamics Society of Canada*. 16–18.

Keats, A., Yee, E., and Lien, F.S. (2007). Bayesian inference for source determination with applications to a complex urban environment. *Atmospheric Environment* 41 (3): 465–479. https://doi.org/10.1016/j.atmosenv.2006.08.044.

Khlaifi, A., Ionescu, A., and Candau, Y. (2009). Pollution source identification using a coupled diffusion model with a genetic algorithm. *Mathematics and Computers in Simulation* 79 (12): 3500–3510. https://doi.org/10.1016/j.matcom.2009.04.020.

Lauret, P., Heymes, F., Aprin, L. et al. (2016). Atmospheric dispersion modeling using artificial neural network based cellular automata. *Environmental Modelling and Software* 85: 56–69. https://doi.org/10.1016/j.envsoft.2016.08.001.

Long, K.J., Haupt, S.E., and Young, G.S. (2010). Assessing sensitivity of source term estimation. *Atmospheric Environment* 44 (12): 1558–1567. https://doi.org/10.1016/j.atmosenv.2010.01.003.

Ma, D. and Zhang, Z. (2016). Contaminant dispersion prediction and source estimation with integrated Gaussian-machine learning network model for point source emission in atmosphere. *Journal of Hazardous Materials* 311: 237–245. https://doi.org/10.1016/j.jhazmat.2016.03.022.

Ma, D., Deng, J., and Zhang, Z. (2013). Correlation analysis for online CO_2 leakage monitoring in geological sequestration. *Energy Procedia* 37: 4374–4382. https://doi.org/10.1016/j.egypro.2013.06.340.

Ma, D., Wang, S., Zhang, Z. et al. (2014). Hybrid algorithm of minimum relative entropy-particle swarm optimization with adjustment parameters for gas source term identification in atmosphere. *Atmospheric Environment* 94: 637–646. https://doi.org/10.1016/j.atmosenv.2014.05.034.

Ma, D., Tan, W., Zhang, Z. et al. (2017). Parameter identification for continuous point emission source based on Tikhonov regularization method coupled with particle swarm optimization algorithm. *Journal of Hazardous Materials* 325: 239–250. https://doi.org/10.1016/j.jhazmat.2016.11.071.

Ma, D., Tan, W., Zhang, Z. et al. (2018). Gas emission source term estimation with 1-step nonlinear partial swarm optimization–Tikhonov regularization hybrid method. *Chinese Journal of Chemical Engineering* 26 (2): 356–363. https://doi.org/10.1016/j.cjche.2017.07.022.

Ma, D., Gao, J., and Zhang, Z. (2019). An improved firefly algorithm for gas emission source parameter estimation in atmosphere. *IEEE Access* 7: 111923–111930. https://doi.org/10.1109/ACCESS.2019.2935308.

Ma, D., Gao, J., Gao, Z. et al. (2020). Gas leakage recognition for CO_2 geological sequestration based on the time series neural network. *Chinese Journal of Chemical Engineering* 28 (9): 2343–2357. https://doi.org/10.1016/j.cjche.2020.06.014.

Ma, D., Gao, J., and Zhang, Z. (2021a). Gas recognition method based on the deep learning model of sensor array response map. *Sensors and Actuators B: Chemical* 330: 129349. https://doi.org/10.1016/j.snb.2020.129349.

Ma, D., Gao, J., Zhang, Z. et al. (2021b). Identifying atmospheric pollutant sources using a machine learning dispersion model and Markov chain Monte Carlo methods. *Stochastic Environmental Research and Risk Assessment* 35 (2): 271–286. https://doi.org/10.1007/s00477-021-01973-7.

Martin, J.R. (ed.) (1979). *Recommended Guide for the Prediction of the Dispersion of Airborne Effluents*, 3e. Avenue NY: American Society of Mechanical Engineers.

Metropolis, N., Rosenbluth, A.W., Rosenbluth, M.N. et al. (1953). Equation of state calculations by fast computing machines. *The journal of chemical physics* 21 (6): 1087–1092. https://doi.org/10.1063/1.1699114.

Mira, A. (2001). On Metropolis-Hastings algorithms with delayed rejection. *Metron* 59 (3–4): 231–241.

Neal, R.M. (2011). MCMC using Hamiltonian dynamics. *Handbook of markov chain monte carlo* 2 (11): 2.

Neupauer, R.M. and Borchers, B. (2001). A MATLAB implementation of the minimum relative entropy method for linear inverse problems. *Computers and Geosciences* 27 (7): 757–762. https://doi.org/10.1016/S0098-3004(01)00009-7.

Neupauer, R.M., Borchers, B., and Wilson, J.L. (2000). Comparison of inverse methods for reconstructing the release history of a groundwater contamination source. *Water Resources Research* 36 (9): 2469–2475. https://doi.org/10.1029/2000WR900176.

Rao, K.S. (2007). Source estimation methods for atmospheric dispersion. *Atmospheric Environment* 41 (33): 6964–6973. https://doi.org/10.1016/j.atmosenv.2007.04.064.

Thomson, L.C., Hirst, B., Gibson, G. et al. (2007). An improved algorithm for locating a gas source using inverse methods. *Atmospheric Environment* 41 (6): 1128–1134. https://doi.org/10.1016/j.atmosenv.2006.10.003.

Tierney, L. (1994). Markov chains for exploring posterior distributions. *The Annals of Statistics* 24: 1701–1728.

Vapnik, V.N. (1998). *Statistical Learning Theory*. New York: Wiley.

Woodbury, A.D. and Ulrych, T.J. (1993). Minimum relative entropy: forward probabilistic modeling. *Water Resources Research* 29 (8): 2847–2860. https://doi.org/10.1029/93WR00923.

Woodbury, A.D. and Ulrych, T.J. (1996). Minimum relative entropy inversion: theory and application to recovering the release history of a groundwater contaminant. *Water Resources Research* 32 (9): 2671–2681. https://doi.org/10.1029/95WR03818.

Woodbury, A.D. and Ulrych, T.J. (1998). Minimum relative entropy and probabilistic inversion in groundwater hydrology. *Stochastic Hydrology and Hydraulics* 12 (5): 317–358. https://doi.org/10.1007/s004770050024.

Yang, R., Zhang, J., Shen, S. et al. (2007). Numerical investigation of the impact of different configurations and aspect ratios on dense gas dispersion in urban street canyons. *Tsinghua science and technology* 12 (3): 345–351. https://doi.org/10.1016/S1007-0214(07)70051-2.

Yee, E. (2008). Theory for reconstruction of an unknown number of contaminant sources using probabilistic inference. *Boundary-Layer Meteorology* 127 (3): 359–394.

Yin, H.M. (1988). On a Class of Nonclassical Parabolic Problems (Doctoral dissertation, Washington State University).

8

Machine Learning and Deep Learning Applications in Medical Image Analysis

Pingfan Hu[1], Changjie Cai[2], Yu Feng[3], and Qingsheng Wang[1]

[1] Artie McFerrin Department of Chemical Engineering, Texas A&M University, College Station, TX, USA
[2] Department of Occupational and Environmental Health, Hudson College of Public Health, The University of Oklahoma, Oklahoma City, OK, USA
[3] School of Chemical Engineering, Oklahoma State University, Stillwater, OK, USA

8.1 Introduction

8.1.1 Machine Learning in Medical Imaging

Machine learning algorithms are very effective at using medical imaging to study specific diseases. Numerous machine learning methods have been used to analyze medical images, such as linear discriminant analysis, support vector machines (SVMs), decision trees, and random forests. Pixel/voxel-based machine learning (PML) model emerged in medical image analysis, which uses pixel/voxel values in images directly instead of features calculated from segmented objects as input information (Suzuki 2012). Machine learning algorithms can also combine with the computational aerosol dynamics method for lung disease diagnosis. The exhaled aerosol patterns that were simulated by computational fluid dynamics (CFD) with different asthma conditions were categorized using fractal analysis and SVMs classification as well as random forest (Xi et al. 2015; Xi and Zhao 2019).

8.1.2 Deep Learning in Medical Imaging

Both the 2-dimensional and 3-dimensional structures of an organ being studied are crucial in order to identify what is normal versus abnormal. By maintaining these local spatial relationships, convolutional neural networks (CNNs) are well suited to perform image recognition tasks (Ker et al. 2018). CNNs can also be used for classification, localization, detection, and segmentation (Jiao et al. 2020).

Machine Learning in Chemical Safety and Health: Fundamentals with Applications, First Edition.
Edited by Qingsheng Wang and Changjie Cai.
© 2023 John Wiley & Sons Ltd. Published 2023 by John Wiley & Sons Ltd.

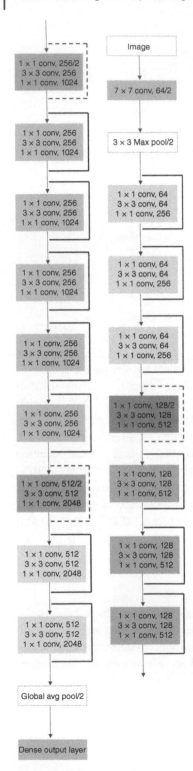

Figure 8.1 Modified ResNet50 architecture.

8.2 CNN-Based Models for Classification

8.2.1 ResNet50

As shown in Figure 8.1, ResNet50 is a 50-layer residual network. The main goal is to build a deeper neural network based on a modified ResNet50 without encountering the vanishing gradient problem (Hochreiter et al. 2001). The error gradients are computed at the end of the network. Backpropagation (Goodfellow et al. 2016) is used to propagate the error gradients backward through the network. Using the chain rule (Goodfellow et al. 2016), terms are multiplied with the error gradients and have to be kept as the networks go backward. However, in the long chain of multiplication, the gradient becomes very small as networks approach the earlier layers in a deep architecture. This small gradient is an issue because network parameters cannot be updated by a large enough amount and the training is very slow. To avoid the vanishing gradient problem, ResNet50 stacks these residual blocks together where an identity function is used to preserve the gradient. It is also called skip connection since the origin input is added to the output of the convolution block directly. The structure of the skip connection is shown in Figure 8.2 (He et al. 2016).

As shown in Figure 8.1, the input image goes through the first layer with 64 filters, with a filter size of 7×7. Next, it goes through the max-pooling layer, which helps reduce the spatial size of the convolved features and helps reduce the over-fitting problem. Then, it goes through 48 convolutional layers with skip connection and finally reaches the fully connected layer that helps learn nonlinear combinations of the high-level features outputted by previous layers. In the modified ResNet50 model employed in this study, parameters of the pre-trained convolutional layers on the ImageNet dataset (Deng et al. 2009) were used. The final pooling and fully connected layer in the original ResNet50 model were replaced by global average pooling and a dense output layer, in order to connect the dimensions of the previous layers with the new layers for classification of our own dataset. Regularization methods (e.g. batch normalization and dropout) and optimizers were used to avoid over-fitting and reduce computational time (He et al. 2016).

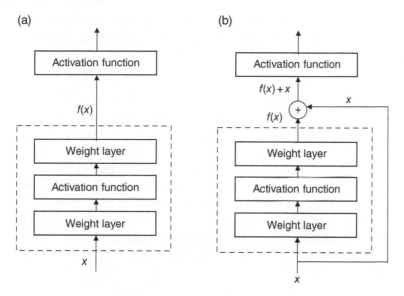

Figure 8.2 The residual learning building blocks: (a) regular block and (b) residual block.

8.2.2 YOLOv4 (Darknet53)

One popular state of the art CNN-based model for detecting objects in an image is "You Only Look Once" or YOLO (Redmon et al. 2016). YOLO version 3 (YOLOv3) expands on its previous version, YOLOv2, by utilizing a Darknet53 (53 convolutional layers) as its backbone in contrast to YOLOv2, which utilized Darknet19 (19 convolutional layers) (Redmon and Farhadi 2017). Although the precision has been greatly improved in YOLOv3 compared with YOLOv2 due to the increased number of convolutional layers, the resultantly increased computational complexity also makes YOLOv3 more computationally expensive. To optimize the balance between precision and computational efficiency, YOLOv4 has been developed to improve both the precision and speed of YOLOv3. YOLOv4 is considered as one of the most accurate real-time neural network detectors to date (Bochkovskiy et al. 2004). YOLOv4 has been successfully applied in various industries, including autonomous driving, agriculture, electronics, public health, etc. (Li et al. 2020; Cai et al. 2021; Kajabad et al. 2021; Sozzi et al. 2021). In this study, YOLOv4 was employed and tested for classifying the lung obstruction locations, including left lung, right lung, and both lungs.

YOLOv4 consists of three main blocks, including the "backbone," "neck," and "head" (Bochkovskiy et al. 2004). The "backbone" implements feature extraction. The model implements the Cross Stage Partial Network (CSPNet) backbone method to extract features (Wang et al. 2020), containing 53 convolutional layers for accurate image classification, also known as CSPDarknet53. The "neck" is a layer between the "backbone" and "head" that implements feature aggregation. Specifically, YOLOv4 uses the Path Aggregation Network (PANet) for feature aggregation (Liu et al. 2018) and Spatial Pyramid Pooling (SPP) method to set apart the important features obtained from the "backbone" (Liu et al. 2018). The "head" used in YOLOv4 is the same as the one in YOLOv3, which uses dense prediction for anchor-based detection that helps divide the image into multiple cells and inspect each cell to find the probability of having an object using the post-processing techniques (Redmon et al. 2016).

8.2.3 Grad-CAM

In practice, deep learning models are treated as "black box" methods. To enhance the fundamental understanding of where the CNN-based models are "looking" in the input image, a simple modification of the global average pooling layer combined with Grad-CAM (Rajasegarar et al. 2007) allows the classification-trained CNN to both classify the image and localize class-specific image regions. The gradient of the chosen convolutional layer is converted to weight. Then the 1D vector that stored the number of filters is reshaped to the image shape. After the layer output and weight are computed and normalized, the heat map showing the highly correlated regions of input for predictions is created. By generating such visual explanations, Grad-CAM makes the CNN-based model more transparent and insightful.

8.3   Case Study

8.3.1   Background

According to the National Vital Statistics Report (Hamilton et al. 2013), chronic obstructive pulmonary disease (COPD) is the third leading cause of death in America. COPD causes severe breathing difficulty due to airway stiffening, loss of airway deformation capability, and airway blockage induced by inflammation especially in small airways, which are regarded as the silent zone in the respiratory system (Yi et al. 2021; Rajendran and Banerjee 2020; Pramanik et al. 2021). Inhalation of therapeutic nano-/microparticles is the standard COPD treatment, but delivering a sufficiently high dose of therapeutic nano-/microparticles to obstruction sites in small airways has remained the long-standing barrier preventing the desired therapeutic outcomes.

To overcome such a barrier, it is important to detect the obstruction locations in small airways of COPD patients at an early stage and optimize the inhalation therapy to achieve targeted drug delivery to designated obstruction sites, instead of healthy airway tissues, for better therapeutic outcomes and reduced side effects. However, there is strong evidence to suggest that most patients are not aware of their small airway obstruction conditions at the early stage, due in part to the invasive nature of conventional diagnostic methods (Jindal 2012; Burgel et al. 2013; Deepak et al. 2017). Specifically, traditional methods to diagnose pulmonary diseases involve costly and invasive procedures such as X-ray screening and bronchoscope. Thus, it is imperative and beneficial to detect the obstruction locations in the peripheral lung precisely with noninvasive diagnostic methods.

This case study proposes and tests the feasibility of a new diagnostic methodology using both CFD and CNN. The new methodology is able to identify the obstruction location in the left lung, right lung, or both lungs using hyperpolarized magnetic resonance imaging (MRI)

(Salerno et al. 2001; Walkup and Woods 2014; Roos et al. 2015; Walkup et al. 2016). The method was driven by a central hypothesis enlightened by existing preliminary studies (Sul et al. 2018). The small airway obstruction will lead to detectable velocity distribution pattern shift of the expiratory airflow in the trachea. Specifically, based on the training and test data generated using the CFD simulation results of expiratory airflows in a subject-specific human tracheobronchial tree (trachea to G6), two CNN-based classification models were developed using open-source codes, e.g. ResNet50 and YOLOv4 (Darknet53) (Bochkovskiy et al. 2004). The modified ResNet50 model is a 50-layer residual network, and it was the winner of the ImageNet Large Scale Visual Recognition Challenge (ILSVRC) in 2015. The main goal of the residual network is to build a deeper neural network without the problem of vanishing gradients. To further analyze which regions suggest the obstruction locations, Gradient-weighted Class Activation Mapping (Grad-CAM) was applied to produce a coarse localization map highlighting the important regions (Daibo 2017). The results have also been validated by Darknet53, which acts as a backbone for the YOLOv4 object detection approach. (Li et al. 2020; Cai et al. 2021; Kajabad et al. 2021; Sozzi et al. 2021).

Two CNN-based models, e.g. modified ResNet50 and YOLOv4, are trained by CFD expiratory velocity contours in a subject-specific 3D tracheobronchial (TB) tree with 990 obstruction conditions at small airway terminals to automatically classify COPD airway obstruction locations. Grad-CAM and hue-value-saturation (HSV) thresholding techniques were employed to classify COPD obstruction locations and velocity contour pattern shifts in the lung and highlight the highly correlated regions in the contours for locating the obstruction sites.

8.3.2 Study Design

Based on the central hypothesis, the workflow of the training and test of the two CNN classification models are shown in Figure 8.3a and b. A subject-specific TB tree from the trachea to G6 was employed for the expiratory flow simulations using CFD. An experimentally validated CFD model (Feng et al. 2018) was employed to predict expiratory intrathoracic flow velocity distributions through the TB tree with 1 normal airway and 990 airway obstruction conditions. Using the airflow velocity distribution data labeled by the obstruction locations, two CNN-based classification models were trained and tested.

8.3.3 Training and Testing Database Preparation

As shown in Figure 8.3a and b, the preparation of the training and test database used the CFD simulation results for the expiratory flow field predictions with different obstruction conditions in the subject-specific TB tree. Specifically, only 1 or 2 of the 44 small airway openings were blocked for each simulation case in order to mimic the minimum changes in obstruction conditions in the human lung compared with the obstructions of multiple openings in left, right, or both lungs. The structure of the labeled training and test images is shown in Figure 8.4.

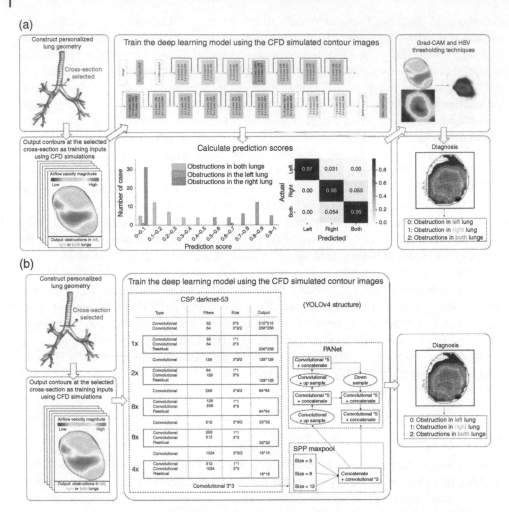

Figure 8.3 Workflows of CNN development to diagnose obstructed locations in the human lung based on expiratory flow patterns: (a) ResNet50 and (b) YOLOv4 (Darknet53).

The velocity contours used for training and testing the two CNN-based models were acquired at a selected cross-section ($x = 0.1$ m) for all CFD simulation cases. The cross-section was selected based on two rationales: (i) the available locations in the chest where the airflow velocity distributions can be measured by hyperpolarized MRI and (ii) the location that is closer to the obstruction sites at 44 small airway terminals. Specifically, the closer the selected cross-section and the obstruction sites are, the more negligible the dissipation effect will be, and the more identifiable shifts of the airflow velocity distributions can be maintained due to the variation in deeper-lung expiratory flow conditions induced by the obstruction. An example of the expiratory velocity contour at $x = 0.1$ m can be found in Figure 8.3a and b. The images were partitioned into training and testing sets for each obstruction class, with an approximately 80 to 20% split and fivefold cross-validation.

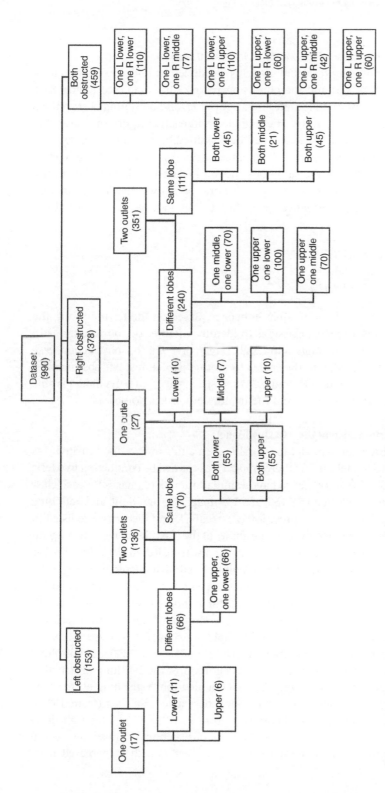

Figure 8.4 Data structure of the training and test images prepared using CFD

8.3.4 Results

8.3.4.1 Classification Performance of the Modified ResNet50 Model

The performance of the modified ResNet50 model is visualized by both the histogram of prediction scores (see Figure 8.5) and the confusion matrix heat map (see Figure 8.6). The prediction scores of all test cases for left lung, right lung, and both lung obstructions are shown in Figure 8.5a–c, respectively. The high score in each category indicates that the model has a high certainty, and the case will be classified based on the highest prediction score's class. As shown in Figure 8.5a–c, most of the test cases have the highest prediction scores in the class with the same obstruction locations, which indicates the reliability of the prediction score system. Furthermore, as shown in Figure 8.6 and Table 8.1, the testing dataset performance quantified by the average precision (AP) at a threshold of 0.5 is:

- Left lung obstructed: AP = 89.27%
- Right lung obstructed: AP = 93.89%
- Both lungs obstructed: AP = 98.43%

The total accuracy converges to 95.1% after 20 epochs. Based on the testing result, the modified ResNet50 model is a reliable classifier to identify whether the obstruction is in the left lung, right lung, or both lungs with high sensitivity. The no obstruction case (e.g. normal airways) was studied, and the scores for both obstructions, left obstructions, and right obstructions are 0.3785, 0.3337, and 0.2878, respectively. Thus, the model is able to distinguish between obstruction cases and the healthy no-obstruction case.

8.3.4.2 Classification Performance of the YOLOv4 Model

To validate the modified ResNet50 model for classification and compare the sensitivity of AP to different CNN-based models, the YOLOv4 model was trained by conducting two tests (see Table 8.1). In the first test (e.g. Test 1), to have a similar number of images for each class during training, we randomly selected 153, 153, and 153 images for left, right, and both lung obstructions, respectively. The values of precision (P), recall (R), and $F1$ *score* are listed in Table 8.1. Specifically, precision (P) represents the ability of the classifier to identify relevant data points that were classified as true and that were actually true. Recall (R) is described as the ability of the classifier to find all relevant data points. Maximizing P often comes at the expense of R and vice versa. The $F1$ *score* is considered as a parameter that can reflect both P and R more objectively. Determining the $F1$ *score* is useful in this assessment to ensure optimal precision (P) and recall scores (R) can be achieved. As shown in Table 8.1, the precision, recall, and $F1$ score are 0.93, 0.94, and 0.93 at a threshold of 0.5 for Test 1, respectively. By checking the AP for each class, the recognition of right lung ($AP = 96.78\%$) and both lung ($AP = 93.85\%$) obstructions were not as good as recognition of the left lung obstruction ($AP = 100\%$). Therefore, the second test (e.g. Test 2) doubled the "right lung obstructed" images (from 153 to 378) and tripled the "both lung obstructed" images (from 153 to 459). APs slightly increased for the two classes (e.g. from 96.78 to 97.74% for right lung obstructions and from 93.85 to 96.29% for both lung obstructions). Thus, the overall YOLOv4 model trained in Test 2 is slightly better than Test 1, with P increased from

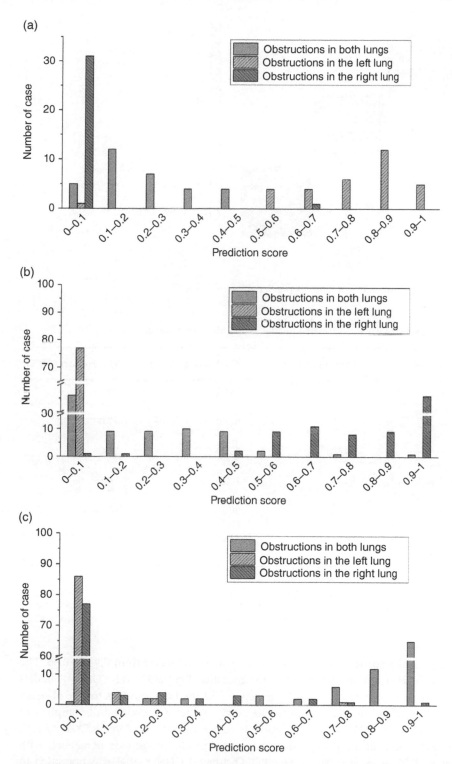

Figure 8.5 Prediction scores for (a) left lung obstructions, (b) right lung obstructions, and (c) both lungs obstructions.

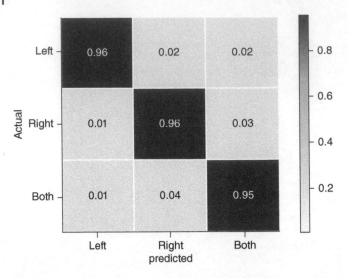

Figure 8.6 Confusion matrix heat map for prediction.

Table 8.1 Evaluation metrics of modified Resnet50 and YOLOv4 models.

Evaluation metrics	ResNet50 test[a](%)	YOLOv4 test 01[b](%)	YOLOv4 test 02[c](%)
Macro precision $P = TP/(TP + FP)$	93.86	93.00	96.00
Macro recall $R = TP/(TP + FN)$	95.27	94.00	97.00
F1 score $F1 = 2PR/(P + R)$	94.56	93.00	96.00
Average precision (AP) (Left lung obstructed)	89.27	100%	100%
Average precision (AP) (Right lung obstructed)	93.89	96.78	97.74
Average precision (AP) (Both lungs obstructed)	98.42	93.85	96.29

TP, true positive; FP, false positive; FN, false negative.
[a] ResNet50 Test : 153 images for left, 378 images for right, and 459 images for both lung obstructions.
[b] YOLO Test 01 : 153 images for left, 153 images for right, and 153 images for both lung obstructions.
[c] YOLO Test 02 : 153 images for left, 378 images for right, and 459 images for both lung obstructions.

0.93 to 0.96, *R* increased from 0.94 to 0.97, and *F1* score increased from 0.93 to 0.96. The comparison of evaluation results between the modified ResNet50 and YOLOv4 models summarized in Table 8.1 shows that both models can be used as classifiers for the obstruction location identifications. The modified ResNet50 model was coded and compiled in Keras 2.4.3. It was run on Windows Operating System with GPU (GeForce RTX 2080 with 16 GB-VRAM). The training computation is approximately 12 seconds per epoch with 19 minutes and 10 seconds in total. The YOLOv4-based CNN model was compiled in Microsoft Visual Studio 2019 (Microsoft Corporation, Albuquerque, NM, USA) and run on Windows Operating System with GPU (GeForce GTX 1660 Ti with 16 GB-VRAM), CUDNN_HALF, and OpenCV for accelerating the training. For test02, the training computation is approximately 10 seconds per iteration with batch-size of 64, subdivisions (or mini batch-size) of 64, and 6000 iterations in total; therefore, the training computation performance is approximately 130 seconds per epoch with 18 hours and 17 minutes in total. Considering the computational costs, the modified ResNet50 is more efficient than YOLOv4.

8.3.4.3 Post-Processing Via Grad-CAM Model and HSV

As shown in Figure 8.7, the Grad-CAM model is combined with the modified ResNet50 model to output the heat map visualization of the important region. The HSV thresholding technique is applied to the heat map plot of the Grad-CAM model. The HSV color space is a cylindrical coordinate representation of points in an RGB color model. It represents the human perception using Hue (the dominant color as perceived by an observer), Saturation (the amount of white light mixed with a Hue), and Value (the chromatic notion of intensity). The highlighted regions were detected by filtering out the color that suggests a low correlation with the obstructions. As shown in Figure 8.8, the highlighted regions for cases within the same obstruction class (e.g. left lung, right lung, or both lungs) were blended with the same weight to assist in finding the most important regions to identify different obstructions. The brightness of the color shows the importance of the region for classification. It can be observed that the area of the bright red region for both lung obstructions is larger than those for left or right lung obstruction cases. It indicates that the determination of both lung obstructions requires more information

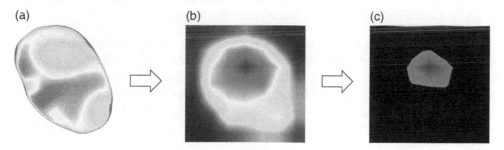

Figure 8.7 HSV thresholding technique procedure: (a) CFD model output, (b) Grad-CAM model output, (c) high correlation region for obstruction.

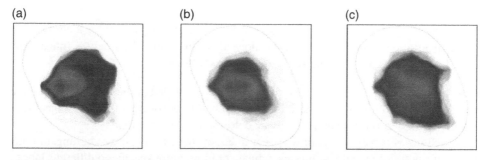

Figure 8.8 Blended highlight regions of construction for (a) left lung, (b) right lung, and (c) both lungs.

8.3.5 Conclusion

The prototype of a novel, effective, and noninvasive tool for early diagnosis of deeper airway obstructions has been developed using CFD and two CNN-based models. The important regions have been determined by Grad-CAM and HSV thresholding techniques and confirmed by the Pearson correlation coefficient calculations between CFD velocity contours. The reason for the falsely classified cases was analyzed based on the comparisons of fundamental airflow dynamics in the intrathoracic region. Key conclusions are summarized as follows:

1) The two CNN-based models (e.g. modified ResNet50 and YOLOv4) can detect small airway obstruction locations based on measurable expiratory airflow patterns in the intrathoracic region well with all evaluation scores higher than 93%, including precision, recall, *F1* score, and average precisions. YOLOv4 is slightly better in classification performance than the modified ResNet50 but requires a higher computational cost. Furthermore, the results of the two CNN-based models validate each other very well. The bagging or boosting ensemble method can be applied to achieve better overall prediction accuracy.
2) The comparisons of expiratory velocity contours show minor flow field pattern shifts with the variations of obstruction sites and demonstrate the necessity of employing CNN algorithms for the effective diagnosis of obstructions. In addition, the Pearson correlation coefficients show much lower similarities of the velocity contours in the highly correlated region identified by CNN-based models than the velocity contours of the whole cross-section, which explains why the Grad-CAM and HSV techniques rely more on the high-correlated regions to identify the obstruction locations.

This prototype of the diagnostic algorithms paves the way for the development of noninvasive and effective diagnostic tools with classification algorithms to effectively diagnose COPD at an early stage and provide high-resolution information for precise treatment (e.g. targeted drug delivery to the identified obstruction sites) with better therapeutic outcomes. No CFD knowledge is needed for users (e.g. physicians) to use the classification algorithm, which increases the transformative impact of the CNN-based models for clinical practice.

8.4 Limitations and Future Work

The present study developed a prototype of the diagnostic algorithm to identify lung obstruction locations via the expiratory airflow distributions in the trachea, integrating CFD and CNN. Limitations of the study are listed as follows:

1) Only one subject-specific TB tree configuration was employed in the CFD simulation for the preparation of training and test images, which did not consider the intersubject variability effect and the influence of airway deformation kinematics.
2) Obstructions were assumed to only appear in either the left lung, the right lung, or both lungs in the training and test images, which could be more specific to different lobes.

3) The airway was assumed to be rigid, which neglected the effect of airway expansion and contraction in the real-world breathing process.
4) The input images were produced by the CFD instead of real-world MRI images.

To address the limitations of the present study and further develop a diagnosis algorithm that can be ready for clinical practice, future work includes:

1) More subject-specific airway configurations with airway deformation kinematics will be obtained and employed in the CFD simulations to prepare the training and test images with the effect of intersubject variabilities, which will enhance the generalized predictability of the CNN algorithm.
2) Lobe-specific obstruction diagnosis will be achieved by improving the training process of the two CNN-based models.
3) Noises and missing parts could be added to mimic the real-world MRI images first and then replaced by hyperpolarized MRI images to further improve the accuracy and realism of the CNN algorithms.

The long-term goal is to provide physicians with a computationally efficient diagnosis algorithm, which can identify the obstruction locations in human lungs based on the pulmonary airflow velocity distributions that are measurable using hyperpolarized MRI (Salerno et al. 2001; Walkup and Woods 2014; Roos et al. 2015; Walkup et al. 2016).

References

Bochkovskiy, A., Wang, C.-Y., and Liao, H.-Y.M. (2004). Yolov4: optimal speed and accuracy of object detection. *arXiv preprint arXiv:2004.10934, 2020.*

Burgel, P.-R., Bergeron, A., De Blic, J. et al. (2013). Small airways diseases, excluding asthma and COPD: an overview. *European Respiratory Review* 22 (128): 131–147.

Cai, C., Nishimura, T., Hwang, J. et al. (2021). Asbestos detection with fluorescence microscopy images and deep learning. *Sensors* 21 (13): 4582.

Daibo, M. (2017). Toroidal vector-potential transformer. *2017 Eleventh International Conference on Sensing Technology (ICST),* Sydney, Australia (4–6 December 2017). IEEE.

Deepak, D., Prasad, A., Atwal, S.S., and Agarwal, K. (2017). Recognition of small airways obstruction in asthma and COPD – the road less travelled. *Journal of Clinical and Diagnostic Research: JCDR.* 11 (3): TE01–TE05.

Deng, J., Dong, W., Socher, R., et al. (2009). Imagenet: a large-scale image database. *2009 IEEE Conference on Computer Vision and Pattern Recognition,* Miami, FL (20–25 June 2009). IEEE.

Feng, Y., Zhao, J., Kleinstreuer, C. et al. (2018). An in silico inter-subject variability study of extra-thoracic morphology effects on inhaled particle transport and deposition. *Journal of Aerosol Science.* 123: 185–207.

Goodfellow, I., Bengio, Y., and Courville, A. (2016). *Deep Learning.* MIT Press.

Hamilton, B.E., Hoyert, D.L., Martin, J.A. et al. (2013). Annual summary of vital statistics: 2010–2011. *Pediatrics* 131 (3): 548–558.

He, K., Zhang, X., Ren, S., and Sun, J. (2016). Deep residual learning for image recognition. Proceedings of the IEEE Conference on Computer Vision and Pattern Recognition, Las Vegas, NV, USA (27–30 June 2016).

Hochreiter, S., Bengio, Y., Frasconi, P., and Schmidhuber, J. (2001). Gradient flow in recurrent nets: the difficulty of learning long-term dependencies. In: *A field guide to dynamical recurrent neural networks*. IEEE Press.

Jiao, Z., Hu, P., Xu, H., and Wang, Q. (2020). Machine learning and deep learning in chemical health and safety: a systematic review of techniques and applications. *ACS Chemical Health & Safety* 27 (6): 316–334.

Jindal, S.K. (2012). COPD: the unrecognized epidemic in India. *The Journal of the Association of Physicians of India* 60 (Suppl): 14–16.

Kajabad, E.N., Begen, P., Nizomutdinov, B., and Ivanov, S. (2021). YOLOv4 for urban object detection: case of electronic inventory in St. Petersburg. *2021 28th Conference of Open Innovations Association (FRUCT),* Moscow, Russia (27–29 January 2021). IEEE.

Ker, J., Wang, L., Rao, J., and Lim, T. (2018). Deep learning applications in medical image analysis. *IEEE Access.* 6: 9375–9389.

Li, Y., Wang, H., Dang, L. et al. (2020). A deep learning-based hybrid framework for object detection and recognition in autonomous driving. *IEEE Access.* 8: 194228–194239.

Liu, S., Qi, L., Qin, H., Shi, J., and Jia, J. (2018). Path aggregation network for instance segmentation. Proceedings of the IEEE Conference on Computer Vision and Pattern Recognition, Salt Lake City, UT, USA (18–23 June 2018).

Pramanik, S., Mohanto, S., Manne, R. et al. (2021). Nanoparticle-based drug delivery system: the magic bullet for the treatment of chronic pulmonary diseases. *Molecular Pharmaceutics* 18 (10): 3671–3718.

Rajasegarar, S., Leckie, C., Palaniswami, M., and Bezdek, J. C. (2007). Quarter sphere based distributed anomaly detection in wireless sensor networks. *2007 IEEE International Conference on Communications,* Glasgow, Scotland (24–28 June 2007). IEEE.

Rajendran, R.R. and Banerjee, A. (2020). Effect of non-Newtonian dynamics on the clearance of mucus from bifurcating lung airway models. *Journal of Biomechanical Engineering* 143 (2): 021011.

Redmon, J. and A. Farhadi. (2017). YOLO9000: better, faster, stronger. Proceedings of the IEEE Conference on Computer Vision and Pattern Recognition, Honolulu, HI, USA (21–26 July 2017).

Redmon, J., Divvala, S., Girshick, R., and Farhadi, A. (2016). You only look once: unified, real-time object detection. Proceedings of the IEEE Conference on Computer Vision and Pattern Recognition, Las Vegas, NV, USA (27–30 June 2017).

Roos, J.E., McAdams, H.P., Kaushik, S.S., and Driehuys, B. (2015). Hyperpolarized gas MR imaging: technique and applications. *Magnetic Resonance Imaging Clinics of North America* 23 (2): 217–229.

Salerno, M., Altes, T.A., Brookeman, J.R. et al. (2001). Dynamic spiral MRI of pulmonary gas flow using hyperpolarized 3He: preliminary studies in healthy and diseased lungs. *Magnetic Resonance in Medicine: An Official Journal of the International Society for Magnetic Resonance in Medicine.* 46 (4): 667–677.

Sozzi, M., Cantalamessa, S., Cogato, A. et al. (2021). Grape yield spatial variability assessment using YOLOv4 object detection algorithm. In: *Precision Agriculture'21*, 193–198. Wageningen Academic Publishers.

Sul, B., Oppito, Z., Jayasekera, S. et al. (2018). Assessing airflow sensitivity to healthy and diseased lung conditions in a computational fluid dynamics model validated in vitro. *Journal of Biomechanical Engineering* 140 (5).

Suzuki, K. (2012). Pixel-based machine learning in medical imaging. *International Journal of Biomedical Imaging* 2012: 1.

Walkup, L.L. and Woods, J.C. (2014). Translational applications of hyperpolarized 3He and 129Xe. *NMR in Biomedicine* 27 (12): 1429–1438.

Walkup, L.L., Thomen, R.P., Akinyi, T.G. et al. (2016). Feasibility, tolerability and safety of pediatric hyperpolarized 129 Xe magnetic resonance imaging in healthy volunteers and children with cystic fibrosis. *Pediatric Radiology* 46 (12): 1651–1662.

Wang, C.-Y., Liao, H. Y. M., Wu, Y. H., et al. (2020). CSPNet: a new backbone that can enhance learning capability of CNN. Proceedings of the IEEE/CVF Conference on Computer Vision and Pattern Recognition Workshops, Virtual (13–19 June 2020), 671.

Xi, J. and Zhao, W. (2019). Correlating exhaled aerosol images to small airway obstructive diseases: a study with dynamic mode decomposition and machine learning. *PLoS One* 14 (1): e0211413.

Xi, J. et al. (2015). Detecting lung diseases from exhaled aerosols: non-invasive lung diagnosis using fractal analysis and SVM classification. *PLoS One* 10 (9): e0139511.

Yi, H., Wang, Q., and Feng, Y. (2021). Computational analysis of obstructive disease and cough intensity effects on the mucus transport and clearance in an idealized upper airway model using the volume of fluid method. *Physics of Fluids* 33 (2): 021903.

9

Predictive Nanotoxicology: Nanoinformatics Approach to Toxicity Analysis of Nanomaterials

Bilal M. Khan[1,3] and Yoram Cohen[2,3]

[1] *Department of Computer Science and Engineering, California State University San Bernardino, San Bernardino, CA, USA*
[2] *Department of Chemical and Biomolecular Engineering, University of California, Los Angeles, CA, USA*
[3] *Institute of the Environment and Sustainability, University of California, Los Angeles, CA, USA*

9.1 Predictive Nanotoxicology

9.1.1 Introduction

Nanomaterials possess at least one and usually two dimensions that are less than 100 nm. Their functionality is linked to their salient characteristics including their size range, configurations, and surface modifications. Engineered nanomaterials (ENMs) have found use in a variety of industrial applications such as cosmetics, therapeutics, electronics, manufacturing, and healthcare (Bao et al. 2013; Thiruvengadam et al. 2018; Siddiquee et al. 2019; Sahoo et al. 2021). New applications of ENMs include, for example, new therapeutic delivery vectors, contrast agents, and biological research probes (Wolfbeis 2015; Bao et al. 2013; El-Hack et al. 2017; Siddiquee et al. 2019). The compositions of nanomaterials vary according to the products and brands, thus the drive to categorize them into a suitable new class of materials. Given the rapid developments of new nanomaterials and their applications (Roco 2011; CPI 2016; Sahoo et al. 2021), nanotechnology is among the most rapidly growing technologies of the twenty-first century and has already reached an annual global market of over $30 billion (Global Industry Analysts 2021).

It is stressed that along with the growth of the nanotechnology industry, there have been mounting concerns regarding the impact of ENMs on the environment and human health, thus necessitating detailed characterization of ENMs' physicochemical properties, their multimedia environmental distribution, and analysis of their potential toxicity. The study of the toxicity of ENMs in various settings (Haase et al. 2012) has led to the rapid expansion of the field of nanotoxicology. Toxic effects associated with the ENMs are diverse and this is expected given their size and diverse physicochemical properties. The concentrations of ENMs in the multimedia environment along with their toxicity profiles govern their potential environmental and health impacts. Despite significant concerns raised about the potential toxicity of ENMs, mere knowledge of the effects of specific ENMs on selected biological

Machine Learning in Chemical Safety and Health: Fundamentals with Applications, First Edition.
Edited by Qingsheng Wang and Changjie Cai.
© 2023 John Wiley & Sons Ltd. Published 2023 by John Wiley & Sons Ltd.

systems is insufficient (Stahura and Bajorath 2004; Mballo and Makarenkov 2010; Puzyn et al. 2010; Bosetti and Vereeck 2011; Sansone et al. 2012; Kleandrova et al. 2014; Bhattacharjee and Brayden 2015; CPI 2016; Joshi 2016; Najafi-Hajivar et al. 2016; Garner et al. 2017; Hua et al. 2018; Ponce et al. 2018; van der Merwe and Pickrell 2018; Berggren et al. 2020; Muratov et al. 2020; Bencsik and Lestaevel 2021; Herrera-Ibatá 2021; Wheeler and Lower 2021). There is an endless array of combinations of ENMs' chemical compositions, architectures, and shapes; thus, there is a wide range of resulting physicochemical properties, all of which must also be considered in conjunction with assessments of various exposure scenarios. Therefore, predictive nanotoxicology is of immense importance for providing a robust evidence- and mechanistic-based understanding of the environmental distribution and toxicological impacts of ENMs.

The ultimate goal is the tuned design of ENMs with respect to their composition, colloidal stability, surface functionality, self-assembly capabilities, and surface passivation or coating for a variety of applications. In this regard, the development and availability of fundamental and data-driven computational models can accelerate the development of engineered safe-by-design nanomaterials for their intended applications. For example, spatiotemporal models of the relationship of ENM surface properties with their bioactivities can be designed in order to dynamically represent various ENMs' physicochemical characteristics based on data collected at different time scales. Examples of a modular series of computational and analytical methods and simulation tools for the design and modeling of nanotherapeutics and disease applications appear in Table 9.1.

The availability of a host of models/tools for nanotoxicology is desirable for the nanotechnology community to assess the unique nanoscale characteristics and structure–activity relationships (SARs) that support the important field of nano environmental health and safety (EHS), including the development of sustainable technology. These models and tools are dependent on the characterization of intrinsic and extrinsic ENM properties as illustrated in Figure 9.1 (Nel et al. 2013b; Wang et al. 2017). A major way to arrive at such predictive models is (quantitative) SARs ((Q)SARs) (Puzyn et al. 2010) that relate biological activity (e.g. toxicity) of ENMs to their physicochemical/structural properties (Aillon et al. 2009; Waters et al. 2009; Cattaneo et al. 2010; Okuda-Shimazaki et al. 2010). The scientific foundation for (Q)SARs is the underlying premise that samples (e.g. chemicals or ENMs) of similar physicochemical/structural properties are likely to exhibit similar bioactivity. (Q)SARs for ENMs (i.e. nano-(Q)SARs) should not only be useful for predicting toxicity for untested ENMs but also provide an insight into their toxicity mechanisms. Nano-(Q)SAR-predicted toxicity can be used to inform ENM's environmental risk assessment in support of formulating rational environmental and health regulatory policies and guiding the selection and/or design of safe ENMs (Nel et al. 2013a).

9.1.2 Nano Quantitative Structure–Activity Relationship (QSAR)

Quantitative SARs for nanomaterials (nano-(Q)SARs) are a class of models that relate toxicity/bioactivity (quantified by suitable metrics) induced by ENMs to their physicochemical properties, chemical composition, and architecture, in addition to exposure scenarios (Puzyn et al. 2010; Toropov et al. 2013; Muratov et al. 2020). The development of (nano-(Q)SARs) relies on experimental bioactivity metrics derived from cellular, organism, or

Table 9.1   Examples of biomedical areas focusing on computational models/analysis tools.

A. Atomistic, biomolecular, and particle-based modeling and simulation

Modeling focus	Modeling/simulation approach
Protein corona	Spatiotemporal modeling (e.g. as a function of NP surface properties), modeling of enzymatic function
Nanoparticle (NP) colloidal stability	Modeling of Derjaguin-Landau-Verwey-Overbeek (DLVO) theory, molecular dynamics (MD), Monte Carlo (MC) simulations, coarse-grained (CG) simulations
NP drug-loading efficiency and stability	Interior, surface area/packing space calculations, supramolecular chemistry, MD, and MC simulations
NP drug release	Imaging, spectroscopy, and chromatography-based analytics that also allow controlled release modeling and simulation (e.g. pH, temperature, and molecular machines)

B. Cellular and subcellular modeling and simulation

Cell membrane wrapping (promotive/resistive forces) and inner genetics	Energy-based CG methods, MD, and simulations of NP interactions with the lipid membrane and glycocalyx
Endocytic and other update pathways	Imaging-based and fluorescence resonance energy transfer (FRET) analysis and modeling
NP intracellular fate and transport, as well as interactions with the cytoskeleton	Image- and function-based modeling and simulation, with and without targeting considerations
Cellular dosimetry models	"Particles in a box" simulation model; in vitro sedimentation, diffusion, and dosimetry (ISDD) models
NP interactions with organelles and the nucleus	Modeling of NP interactions with DNA and the nucleus: modeling of lysosomal and mitochondrial responses

C. Systemic and circulatory nano-bio interface modeling and simulation

Circulatory half-life, blood concentrations, area under the curve (AUC) of the NP carrier and its cargo	Pharmacokinetic (PK) modeling of drug substance and drug product to formulate calculations of drug dose
Design of synergistic drug combinations (e.g. cancer)	Development of software for synergy calculations and ratiometric design of dual delivery NP drug carriers
Biodistribution to deceased tissue (e.g. tumor site), allowing delivered dose calculations	Image-based NP design and tuning of physicochemical features for electron paramagnetic resonance (EPR) modeling, and targeting ligands and PK-based calculations of intratumoral drug biodistribution and dose
Controlled biodistribution to the reticuloendothelial system to enhance biodistribution and PK	Modeling and simulation of NP design to control opsonization and PEGylation

(Continued)

Table 9.1 (Continued)

D. Tissue organ and disease site modeling and simulation	
Biophysical properties of the target tissue	Modeling of vascular density, perfusion, permeability, interstitial pressure, stiffness, and intratumor diffusivity
Dosimetry at the disease site	Hierarchical compilation models spanning multiple temporal–spatial scales (in combination with PK and imaging models)
Vascular access, including vascular barriers (e.g. blood–brain and stromal vascular barriers)	Continuum mechanics approaches, dissipative particle dynamics, Lattice-Boltzmann methods, simulated NP design for stromal vascular engineering

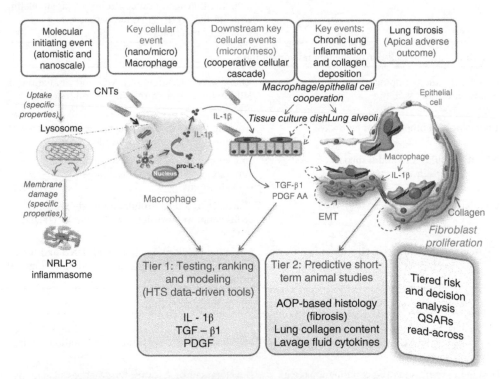

Figure 9.1 Adverse outcome pathway (AOP) modeling for multiwall and single-wall CNTs. The schematic depicts the initiation of a molecular initiating event that is tied to two sets of physicochemical properties of CNTs, one of which determines bioavailability and uptake by macrophages in the lung, and the second set determines the lysosomal damage. Lysosomal damage leads to a pro-inflammatory cascade that subsequently involves other cell types, the march of events leading to collagen deposition and fibrosis in the lung. High throughput screening (HTS), imaging, and QSAR tools can be used to categorize CNTs in terms of specific properties leading to hazard generation at the cellular level. Animal testing and predictive toxicological tools can then be used for a tiered risk assessment approach suitable for read-across, regulatory decision-making and, ultimately, safer design of CNTs and preventative measures to prevent lung damage. *Source:* Adapted from Wang et al. (2017); Nel et al. (2013b).

animal toxicity studies. A QSAR model for ENMs can include structure representation, descriptive analysis, and modeling of the correlation of their properties with biological activities from simple linear to complex nonlinear models. Typically, a molecular descriptor (Section 9.3.2) is referred to as a chemical structural feature, which is closely related to the target property of the ENM.

Among the experimental data generated regarding the toxicity of ENMs, considerable effort has focused primarily on various cell lines (e.g. human cells, macrophages, and pancreatic cells) (Tamayo et al. 1999; Braydich-Stolle et al. 2009; Ehrenberg et al. 2009; Waters et al. 2009; Fourches et al. 2010c; Zhang et al. 2011; Liu et al. 2011b; Ghorbanzadeh et al. 2012; Toropov et al. 2013; Nikota et al. 2015) and bacteria, with limited studies with simple organisms (e.g. zebrafish (King-Heiden et al. 2009; Asharani et al. 2011; Truong et al. 2011; Duan et al. 2013; Liu et al. 2013e; Karcher et al. 2016)) and even fewer animal studies (Krewski et al. 2010; Truong et al. 2011; Zhang et al. 2012; Chang et al. 2013). Reported data on toxicity outcomes have also included qualitative information (e.g. experimentation across multiple assays, cell lines/types with ENMs of different surface modifiers and core chemical compositions), and at different levels of confidence and consistency across different toxicological studies. Based on these datasets, efforts to arrive at the generalization of ENM toxicity behavior via data-driven models, which relate ENM characteristics to their induced bioactivity, have generally been based on single studies rather than the collective body of published evidence (Aillon et al. 2009; Ehrenberg et al. 2009; King-Heiden et al. 2009; Okuda-Shimazaki et al. 2010; Zhang et al. 2011; Wang et al. 2011; Ghorbanzadeh et al. 2012; Duan et al. 2013). ENMs at the nano-bio interface can trigger events at various levels of sublethal functions (Shaw et al. 2008b; Xiao-feng et al. 2009; Cheng et al. 2013; Liu et al. 2013e; Wang et al. 2016). Clearly, understanding the relationships between the structural and physicochemical properties of ENMs and the biological responses, they induce at sublethal and lethal levels, as well as associations among such responses are critical to arriving at causal relationships and nano-(Q)SARs of higher confidence level. Accordingly, in order to develop effective nano-(Q)SARs as models for predicting the toxicity of ENMs, knowledge discovery can serve to first identify critical biological pathways that can potentially lead to adverse outcomes.

Nano-(Q)SARs are also indispensable elements of nanoinformatics (Liu and Cohen 2015) emerging as "the science and practice of determining which information is relevant to the nanoscale science and engineering community, and then developing and implementing effective mechanisms for collecting, validating, storing, sharing, analyzing, modeling, and applying that information." Simple nano-(Q)SARs (e.g. linear regression, partial least squares (PLS), logistic regression (LR), and linear discriminant analysis (LDA)) provide valuable information regarding the correlation of specific ENM descriptors with their toxicity response. However, direct mechanistic justification or causal relationships present in the experimental data is/are typically not achievable via such models. The advantages of models that provide mechanistic insight into even a rudimentary level can allow model interpretation and identification of the significance of its descriptors so as to increase the level of confidence in the model. In this regard, it is stressed that, in general, nano-(Q)SARs that are based on simple regression models are only one-directional in that they relate ENM properties to toxicity outcome and without providing a clear path of causality. On the other hand, models based on advanced ML algorithms (e.g. decision trees and Bayesian Networks

(BNs)) provide a model structure that allows one to trace complex relationships between model input and outcomes.

Although there is no universal model that can depict all possible SARs, simple and transparent models are preferred. A simple model is one of a simple analytical form with a few descriptors to model the SAR. Transparency is usually dependent on the modeling approach itself; thus, linear regression analysis can be considered to be highly transparent, i.e. the algorithm is available, and predictions can be easily made. Moreover, multivariate and nonlinear modeling techniques (e.g. a neural network) are generally of lower transparency. In reality, (Q)SARs span the range from simple models to highly complex multivariate models. The requirements for model simplicity are highly dependent on the context and application of the model. There is no reason to exclude a model where the mechanism is not known or if there are multiple mechanisms. However, the advantages of a strong mechanistic-based model are the model interpretability (i.e. understanding the relationships of ENM properties and exposure conditions with their toxicity profile through the model's structure) and explainability (i.e. identifying causal relationships while predicting ENM bioactivity with higher accuracy given the evidence of ENM descriptors).

9.1.3 Importance of Data for Nanotoxicology

The utilization of existing datasets and data curated from the published literature on ENM toxicity is essential to developing robust and generalizable nanotoxicology models based on correlated toxicity endpoints as part of the health and environmental impact assessment of ENMs. Existing and emerging data sources, although increasingly posing data management challenges (Klaessig 2018), are the core to developing and promoting a sustainable nanoinformatics platform, which should advance the current state of the art in silico nanotoxicology. To this end, rapid developments in nanotechnology have led to the generation of various datasets of ENM characterization with respect to their physicochemical properties, fate and transport behavior, bioactivity at the cellular and organism level, and simulations data of environmental exposure concentrations and releases of ENMs throughout their lifecycle (Klaessig 2018). These generated datasets, which are particularly suited for nano-EHS, consist of both literature curated and raw data on ENM toxicity from controlled laboratory experiments and HTS[1] investigations. For example, OCHEM (Klaessig 2018) contains experimental data on ENMs with provisions for generating descriptors for model building, NanoMiner (Kong et al. 2013) contains data on 634 instances (including omics data) for ENMs, nanomaterial-biological interactions knowledgebase contains over 200 nanomaterial toxicological evaluations for embryonic zebrafish exposed to metal and metal-oxide ENMs, and NanoDatabank (http://Nanoinfo.org 2018) contains data for more than 1000 nanomaterial types with 900 investigations regarding ENM toxicity (including metal oxides, quantum dots (QDs), CNTs, and more) and 150 investigations regarding characterizations and fate and transport studies.

---

1 HTS is a widely adopted scientific method used in drug discovery and relevant fields of biology for rapid testing of model organism or cellular pathways via automated equipment (i.e. plates with wells), whereby multiple cells or simple organisms are exposed to different ENMs at different concentrations and exposure periods.

Early nanoinformatics efforts focused on data organization into *structured* datasets (i.e. reside in fixed fields) (Sansone et al. 2012; Thomas et al. 2013; Doganis et al. 2015). However, a significant portion of the publicly available data is *unstructured* datasets (i.e. not easily organized in a predefined manner and may reside in various file formats) such as text documents, PDFs, images, and MS excel sheets with embedded graphs). The available data are often scattered across multiple literature and online sources. Given the often-limited datasets from single (or few) assays and the typically limited number of organisms in reported studies, toxicity datasets from different sources need to be combined. Such datasets invariably include toxicity outcomes derived from multiple and varying assays, different organisms and conducted at varying ranges of ENM concentrations and exposure times (Nikota et al. 2015; Oh et al. 2016; Ha et al. 2018; Shin et al. 2018; Bilal et al. 2019).

It should be recognized that raw data (free from preprocessing choices by data curators) that can be processed and explored are most useful for nano-(Q)SAR development. This is due to the inherent requirement for developing a nano-(Q)SAR of a predefined data structure. Thus, efforts have also been devoted in recent years to promote the development of databases (along with appropriate metadata) and data management systems for nanomaterials that allow for: (i) the incorporation of data from multiple sources, (ii) data security, (iii) effective data sharing, (iv) intelligent data queries, and (v) integration utilities (Hendren et al. 2015; Robinson et al. 2016; Karcher et al. 2018). It is noted that the joint effort by the European Union (EU) and the United States (US) through the EU-US Roadmap Nanoinformatics 2030 (Klaessig 2018) has called for guidelines for the development of structured nanoinformatics datasets. It is stressed that such data should be organized according to the controlled ontology for the properties and bioactivity induced by ENMs, along with suitable data transfer formats for data exchange with other databases and for model development.

In general, irrespective of the level of complexity of the model being developed, the application of data mining and knowledge extraction approaches are critical for machine learning (ML)-based nano-(Q)SARs. Such approaches should consider the value of available information to: (i) evaluate the relevance and significance of various ENM characteristics (structural and chemical descriptors and physicochemical parameters) and experimental conditions, which can be considered as model input parameters affecting toxicity metrics; (ii) develop data-driven models for correlating toxicity metrics with identified quantitative and qualitative model input attributes; and (iii) identify causal pathways as well as related data-driven hypotheses. Indeed, various studies have recently been undertaken that consider model parameters that are relevant for the identification of causal relationships between ENM bioactivity and descriptors of importance in correlating those outcomes (Nikota et al. 2015; Oh et al. 2016; Ha et al. 2018; Shin et al. 2018; Bilal et al. 2019).

9.2 Machine Learning Modeling for Predictive Nanotoxicology

9.2.1 Overview

The application of ML algorithms in nanotoxicology has gained momentum over the past two decades due to two major reasons. First, given the complexity of the structural properties of ENMs, exposure scenarios, and a myriad of bioactivities, ML can utilize statistical approaches to cluster/group similar biological behaviors to enhance the applicability domain (AD) of nano-(Q)SARs. Second, the development of simple in-silico models is often infeasible due to the inherent complexity and nonlinearity of the dependence of bioactivity endpoints on complex ENM characteristics and exposure scenarios. Basic linear nano-(Q) SARs can be developed based on the existing voluminous body of evidence that ENM size impacts on ENM toxicity. However, complex nonlinear models that incorporate the impact of structural and chemical properties of ENMs on nanotoxicity require ML algorithms that can handle nonlinearities and complex multiparameter associations.

ML is a major branch of artificial intelligence (AI) consisting of diverse sets of statistical algorithms that allow machines (computers) to utilize empirical training data to build models based on such data. Model performance can be improved as additional data become available and through added domain knowledge necessary for the given task. An ML-based model can be regarded as providing both a hypothesis about the environment/task and uncovering patterns in the data so as to solve targeted problems. The factors included in developing ML-based models for nano-(Q)SARs require: (i) the predefined task to be accomplished (e.g. predict cytotoxicity of an ENM given its physicochemical properties and relevant exposure conditions), (ii) domain knowledge for ML model development, and (iii) experimental data that include toxicity outcomes. Given the above information, ML-based nano-(Q)SARs can be built to infer the correlation of ENM-induced toxicity with the relevant ENM and target bioreceptor properties and exposure conditions, thereby allowing the discovery of patterns, and identification of factors that impact ENM toxicity.

Knowledge learned from an ML model can, in some cases, be transferred from one domain (one ENM type) to another to accelerate the learning process, even with a lower level of data availability. This technique is referred to as *transfer learning*. Model building from an existing set of observations (samples) in the form of general fitted parameters (rules) is known as *induction*. The outcome of induction (model predictions) may not be consistent due to the uncertainty caused by variance or missing values. ML approaches that employ reasoning via probabilistic models (e.g. BNs) allow both induction and *deduction* where diagnostic reasoning can also be applied to find the likelihood of the causes given an effect.

ML-based model development often requires data preprocessing, cleaning, and transformation (e.g. normalization and standardization) including factored representation for data visualization. Along with factored data representation (i.e. tabular data where each row/ sample is a vector of descriptor values), a mapping from input to output is required to properly describe data from controlled experiments. In such cases, data attributes (input and output) can either have a finite discrete set of values (e.g. shell coating and core composition) or continuous numerical values (e.g. ENM size, $IC_{50}/EC_{50}/LC_{50}$, exposure temperature, and duration).

ML models for nano-(Q)SARs have extensively been developed and categorized as *regression* (Mballo and Makarenkov 2010; Okuda-Shimazaki et al. 2010; Zhang et al. 2011; Chau and Yap 2012; Ghorbanzadeh et al. 2012; Orimoto et al. 2012; Liu et al. 2013b; Toropov et al. 2013; Oh et al. 2016) or *classification* models (Almuallim and Dietterich 1994; Czermiński et al. 2001; Liu and Yu 2005; Park et al. 2009; Liu et al. 2011b) (Figure 9.2, Table 9.2). For continuous toxicity endpoints, regression models can be developed using a variety of statistical methods (e.g. linear regression and PLS regression), while classification methods

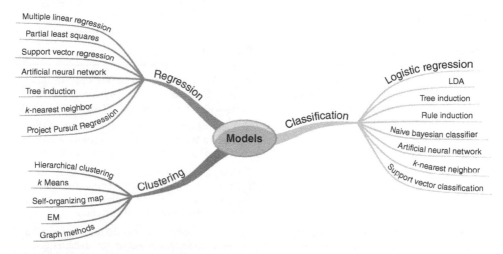

Figure 9.2 ML models categorized into classification, regression models, and clustering approaches applicable to nano-SAR development.

Table 9.2 Summary of model types and their descriptions.

Model type	Model name	Description
Classification	LR Chau and Yap (2012), Zhang et al. (2011)	Regression method that utilizes a logistic function to fit a binary target variable
	LDA Mballo and Makarenkov (2010)	Statistical approach to find linear combination of variables that characterize or separate two or more classes in the dataset
	Tree induction Svetnik et al. (2003)	Top-down recursive algorithm based on "divide and conquer" rule to find if-else relationships between model input parameters and the target outcome
	Rule induction Judson (1994)	Data mining process to extract "if-then-else" rules from dataset
	Naïve Bayes classifier Chau and Yap (2012)	Classification technique based on Bayes theorem that assumes that model input parameters are unrelated to each other
Applicable to both classification and regression	Artificial Neural Network (ANN) Webb (2002)	Inspired by biological neural networks, which are a collection of connected nodes that simulate a biological neuron
	k-Nearest Neighbor Fourches et al. (2010a)	Nonparametric classification approach that stores all data samples and classifies new samples based on k number of existing samples that are the closest (similar) based on a distance measure
	SVM Cristianini and Shawe-Taylor (2000), Webb (2002)	Aims at devising a computationally efficient way of segregating different data classes with a hyperplane that best differentiates the classes

(Continued)

Table 9.2 (Continued)

Model type	Model name	Description
Regression	Multiple linear regression Halder and Dias Soeiro Cordeiro (2021)	Regression algorithm that models two or more input parameters with the target outcome by fitting a linear equation to samples in the dataset
	PLS Mballo and Makarenkov (2010), Jonathan et al. (2000)	Transforms the original dimensionality of the data into a smaller set and performs least squares regression on the transformed dataset
	Project pursuit regression Judson (1994)	Projects given attributes in the dataset in an optimal transformed direction and then applies a smoothing function (e.g. radial, ridge, harmonic mean) on the transformed variables
Clustering	Hierarchical clustering Chon (2011), Shaw et al. (2008a)	Clustering analysis method that groups similar samples in clusters based on the distance between samples
	k-means Chau and Yap (2012)	A vector quantization method that partitions samples into k clusters based on the distance between samples
	SelfOMs Rallo et al. (2005), Liu et al. (2014)	An unsupervised learning approach that transforms high dimensional data into a lower dimension for feature selection
	Expectation Maximization Pham and Ruz (2009)	Finds maximum likelihood estimate of the model parameters with missing data

(Czermiński et al. 2001; Webb 2002; Stahura and Bajorath 2004; Liu and Yu 2005; Medlock 2008; Park et al. 2009; Liu et al. 2013d; Ha et al. 2018) (e.g. LR, support vector machines (SVMs), LDA, and naïve Bayesian classifier) can be used for modeling categorical toxicity endpoints. Indeed, ML methods have been applied to developing both regression (e.g. linear/nonlinear regression (Fourches et al. 2010c; Oh et al. 2016), SVMs (Czermiński et al. 2001; Liu et al. 2013a; Fourches et al. 2010c; Ghorbanzadeh et al. 2012), neural networks (Libotean et al. 2009; Orimoto et al. 2012)) and classification-based (e.g. self-organizing maps (SOMs) (Zollanvari et al. 2009; Rallo et al. 2011; Cohen et al. 2013; Liu et al. 2013d), hierarchical clustering, SVM (Czermiński et al. 2001; Fourches et al. 2010c; Ghorbanzadeh et al. 2012; Liu et al. 2013a)) nano-(Q)SARs. Other approaches that integrate both continuous and categorical information have also been proposed such as decision trees (Ha et al. 2018), random forests (RFs) (Oh et al. 2016; Ha et al. 2018), *k*-nearest neighbor, and BNs (Fourches et al. 2010c; Marvin et al. 2017; Bilal et al. 2019). Although (Q)SARs have a long history of success in drug and chemical-property discovery, modeling the bioactivities of ENMs presents challenges due to the novel nano-scale-governed physicochemical/structural properties of ENMs. Over the last two decades, over two dozen nano-(Q)SARs have been developed but most have been based on small datasets (< 500 samples) and multiple models have utilized the same datasets. An example of published nano-(Q)SARs since 2008 is provided in Table 9.3. It is noted that these nano-(Q)SAR models are mostly based on the data generated from controlled laboratory studies. The toxicity of ENMs is typically reported

Table 9.3 Summary of recently published nano-(Q)SARs.

ENMs	Bioactivity data	Property	Model type	Performance	References
TiO$_2$ (30–125 nm)	Lactic dehydrogenase (LDH) of co-cultures of immortalized rat L2 lung epithelial cells and rat lung alveolar macrophages.	Size (in different media), concentration, zeta potential	Linear regression	$r^2 = 0.77$ – resubstitution (i.e. training data is fed as test data for prediction accuracy)	Sayes and Ivanov (2010)
ZnO (50–1500 nm)	LDH of co-cultures of immortalized rat L2 lung epithelial cells and rat lung alveolar macrophages	Size (in different media), concentration, and zeta potential	LDA classification	100% – resubstitution	Sayes and Ivanov (2010)
44 iron oxide core (20–74 nm)	4 assays of 4 cell lines	Size, zeta potential, and spin–lattice and spin–spin relaxivities	SVM classification	73% – fivefold cross-validation (CV)	Fourches et al. (2010b)
44 iron oxide core NPs	Normalized signal-to-noise ratio (SNR) with respect to unexposed (control) cell responses	Primary size, spin–lattice, and spin–spin relaxivities, and zeta potentials	Naïve bayes classifier (NBC) classification	78% – repeated fivefold CV	Liu et al. (2013c)
109 NPs of the same iron oxide core (an overall size of 38 nm in aqueous solution) but different coatings	Cellular uptake of NPs	~150 molecular operating environment (MOE)	k-nearest neighbor (kNN) regression	$r^2 = 0.72$ – fivefold CV	Fourches et al. (2010b)
	Cellular uptake of NPs	10 selected from ~150 MOE	RVM regression	$r^2 = 0.77$ – fivefold CV	Liu et al. (2013b)
105 NPs with a single metal core by pancreatic cancer cells	Cellular uptake of NPs	367 PaDEL-Descriptor	Consensus model based on local Gaussian regression (LGR), NBC, kNN, and SVM	sensitivity and specificity values of 86.7 and 67.3%	Chau and Yap (2012)

(Continued)

Table 9.3 (Continued)

ENMs	Bioactivity data	Property	Model type	Performance	References
109 magnetofluorescent NPs in PaCa2	Cellular uptake of NPs	6 selected from 256 Dragon Descriptors	Multi-layer perceptron – neural network (MLP-NN)	$r^2 = 0.655$ – fivefold CV	Ghorbanzadeh et al. (2012)
109 NPs of the same iron oxide core (an overall size of 38 nm in aqueous solution) but different coatings	Cellular uptake of NPs	SMILES-based optimal descriptors (calculated with the CORAL software)	QSAR developed as a random event	$r^2 = 0.8043–0.9341$	Toropov et al. (2013)
17 metal oxide (15–90 nm)	Log($1/EC_{50}$) of *Escherichia coli* cells	ΔH_{Me+} (enthalpy of the formation of a gaseous cation having the same oxidation state as that in the metal oxide structure	Linear regression	$r^2 = 0.77$ – Leave one out (LOO) CV	Puzyn et al. (2011)
9 metal oxides (8–19 nm)	PI uptake of BEAS-2B cells	Primary size, particle volume fraction, period of particle metal, and atomization energy of metal oxide	Local Gaussian Regression (LGR)	100% – Leave one out cross-validation	Liu et al. (2011b)
23 metal oxide (10 to 70 nm and two particles of sizes 140 and 190 nm)	7 assays for 2 cell lines	Conduction band energy and ionic index of metal cation	SVM classification	94% – E632 estimator	Liu et al. (2013d)
29 surface-modified multiwalled carbon nanotubes (MWCNT) (diameter ~40 nm and length ~200–1000 nm.)	6 toxicity endpoints	Combinations of MOE, VolSurf (a method for the modeling and prediction of PK properties based on computed molecular interaction fields and multivariate statistics), and 4D-fingerprint descriptors	Linear regression	Q2 (prediction accuracy of regression model when applied on test dataset): 0.735–0.909	Shao et al. (2013)

in terms of nominal values such as toxicity categories based on reported lethal concentrations or continuous numerical values of concentrations at (or above) which a significant effect is observed.

9.2.2 Unsupervised Learning

A data instance with a known toxicity response is referred to as a *label*. Based on the labeled data, ML models are then trained to learn these mappings in a *supervised* learning environment and then serve to predict the toxicity of new samples unseen by these models. In contrast, *unsupervised* learning is typically based on clustering/grouping of data samples to discover patterns in the data without any explicit feedback on input–output mapping. Unsupervised learning in nano-(Q)SAR development has mainly focused on identifying groups or clusters of nanomaterials that exhibit similar behaviors with their given physicochemical properties under a range of exposure conditions (Russel and Norvig 2012). This approach is taken to explore relationships among descriptors for dimensionality reduction and feature extraction/selection. Unsupervised learning is also useful for visual data exploration to identify hidden data groups for a better understanding of the underlying problem (e.g. correlation of ENM attributes with their bioactivity) without predictions or testing underlying hypotheses. Techniques for such exploratory analyses include interrogation via histograms, scatter plots, principal component analysis (PCA), and statistical distributions. Other advanced clustering algorithms such as SOMs and hierarchical clustering have also been utilized for data visualization, as well as dimensionality reduction (i.e. feature selection) in ENM toxicity data analysis (Russel and Norvig 2012).

9.2.2.1 Data Exploration Via Self-Organizing Maps (SOMs)

Cluster analysis techniques such as SOMs have proven useful for discovering relationships in various complex multidimensional toxicity datasets. (Tamayo et al. 1999; Törönen et al. 1999; Chon 2011). The utility of SOMs for data visualization, as well as feature selection, has also been demonstrated for exploratory data analyses (Tamayo et al. 1999; Törönen et al. 1999; Bullinaria 2004; Chon 2011). SOM analysis has also been shown useful for the development of nano-(Q)SAR (Rallo et al. 2005) and structure–property relationships (QSPR) (Giralt et al. 2004). An example of the application of SOM is provided in Figure 9.3 exploring a dataset on the cellular toxicity of 44 iron-based NPs (Liu et al. 2013c) used for molecular imaging and nano-sensing. The bioactivity of these NPs, obtained from HTS toxicity profiles, was analyzed and consensus clustering was performed via SOMs. This enabled the identification of NP bioactivity classes based on cluster membership. NPs in this dataset were evaluated in four cell response assays, four different cell lines, and over NPs' concentration range of 0.01–0.3 (mg mL^{-1} of Fe) via HTS (Damoiseaux et al. 2011). Three major NP clusters were identified via co-clustering indices (a measure of cluster validity). NPs grouped in Cluster I showed significant response in mitochondrial membrane potential (Mito) and reducing equivalent resazurin (Red) assays, as well as a measurable response in adenosine triphosphate (ATP) assay. Cluster III included NPs that showed high bioactivity as measured by ATP and a few NPs (i.e. NP36, NP38, and NP44) also showed significant bioactivity

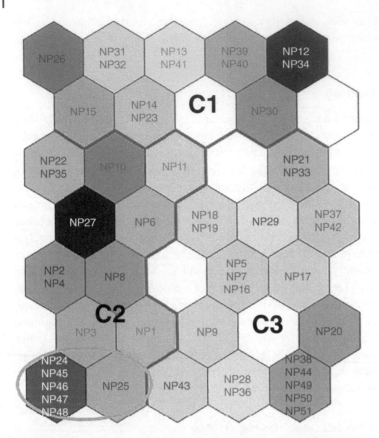

Figure 9.3 SOM-based consensus cluster analysis of the NP HTS bioactivity profiles for RAW 264.7 luciferase reporter cell lines, demonstrating three major clusters (NPs of similar bioactivity profiles). Note: NP24 and NP25 in Cluster II are approved for human use and this cluster was found to consist of NPs with very few hits suggesting the possibility of establishing a class definition for NPs of no or marginal bioactivity based on membership in this cluster. *Source:* Liu et al. (2013c).

in the apoptosis (APO) assay. NPs in Cluster II contained two "safe NPs" (NP24 and NP25) approved for medical use (Liu et al. 2013b). This suggests the hypothesis that NPs of similar characteristics within the same cluster may also likely be declared as "safe."

Another example of SOM clustering analysis is shown in Figure 9.4a for cell signaling pathways RAW 267.4 luciferase reporter cell lines exposed to metal and metal-oxide NPs. SOM clustering of various NPs was accomplished according to their signaling pathway response profile. Evaluation of the impact of different NPs within different clusters can be assessed by following the different cell signaling pathway responses as shown, for example, in Figure 9.4b, for SOM cluster II (Figure 9.4a). In this example, the strictly standardized mean difference (SSMD) of the normalized luminescence for the different cell signaling pathways (relative to a population of untreated cells) shows that signaling pathways E2 Transcription Factor (E2F) and p53 had the highest response within the same cluster; this observation is consistent with the suggestion that significant DNA damage induced by ZnO NPs triggers cell cycle arrest (Liu et al. 2014).

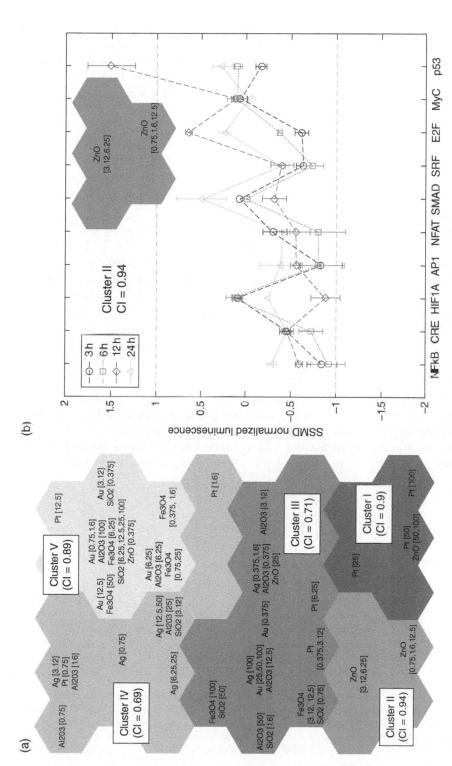

Figure 9.4 (a) The left figure portrays SOM clusters obtained from HTS data of toxicity-related cell signaling pathways of RAW 264.7 luciferase reporter cell lines for metal and metal-oxide NPs. (b) The right figure shows the SSMD-normalized luminescence at different exposure times. *Source:* Liu et al. (2014).

9.2.2.2 Evaluating Associations among Sublethal Toxicity Responses

Mapping the relationships (i.e. associations) among biological responses induced by exposure to ENMs can be helpful in revealing toxicity-response patterns, streamlining the development of robust nano-(Q)SARs. For example, association rule mining has been applied in nanoinformatics (Guzzi et al. 2014; Oellrich et al. 2014; Naulaerts et al. 2015) and bioinformatics (Guzzi et al. 2014; Oellrich et al. 2014; Naulaerts et al. 2015) to extract complex protein-to-protein (Nafar and Golshani 2006; Park et al. 2009), cellular responses (Liu et al. 2014), and gene expression data (Martinez et al. 2008; Alves et al. 2010; Mallik et al. 2013) relationships. Association rule mining of HTS of NPs' toxicity can be used to identify multiple-to-multiple relationships and thus can be regarded as data-driven hypotheses generators. NP triggering of one or more pathways can imply triggering of other associated pathways, consistent with biological crosstalk between cellular signal transduction and transcriptional regulation pathways.

An example of association rules, mined from data on cell signaling pathways for RAW264.7 cell lines exposed to six metal and metal-oxide ENMs, is provided in Figure 9.5 (Liu et al. 2014). Here, E2F signaling suggests definite detection of p53 and Myc (networks of genes used to mediate cellular stress responses), while Myc signaling suggests that p53 may also be detected (92% probability). Confirmatory SOM analysis was carried out showing that p53, E2F, and Myc were within the same cluster. The association rules revealed, as validated experimentally (Liu et al. 2014), that if Mito is encountered with Myc, then it is definite that p53 will also be encountered, which confirmed the above data-driven hypothesis.

In another example, the conditional dependence of cellular toxicity (IC_{50} and cell viability as toxicity metrics) of cadmium containing semiconductor QDs was assessed by clustering analysis using RFs-based regression models (Section 9.2.2.1) to identify QD samples with similar attribute–toxicity correlations (Oh et al. 2016). It is noted that QDs are of interest in applications in medicine, biosensing, and as probes/contrast agents (Michalet et al. 2005; Rosenthal et al. 2011; Petryayeva et al. 2013). In the above study on the cellular toxicity

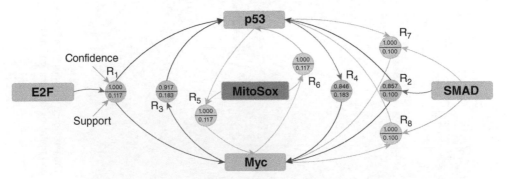

Figure 9.5 Nonredundant association rules derived from data on cell signaling pathways for RAW264.7 cell lines exposed to six different metal and metal-oxide NPs (Liu et al. 2014). Each cellular response is denoted by a box (4 outer boxes for the signaling pathway and inner box (MitoSox) for cytotoxicity). Each association rule is represented as a node labeled by rule as "R#" and Confidence/Support given inside the node. *Source:* Adapted from Liu et al. (2014).

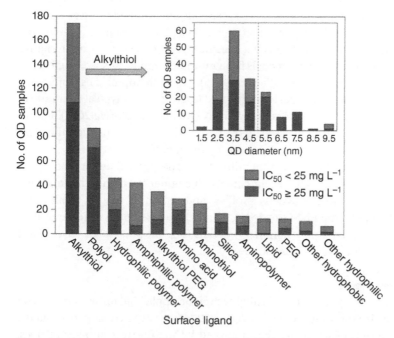

Figure 9.6 The conditional dependence, quantified via the distribution of the number of QD samples with respect to the surface ligand and with surface ligand = alkylthiol and QD diameter. *Source:* Oh et al. (2016).

of QDs, the identified clusters served to establish conditional dependencies of a reasonable level of generality and accuracy. From 307 publications, 1741 cell viability-related data samples were compiled, each with 24 qualitative and quantitative attributes describing the material properties and experimental conditions. Based on IC_{50} data, it was shown that >80% of the QDs with either lipid, amphiphilic polymer, or aminothiol *surface ligand* were associated with $IC_{50} < 25$ mg L^{-1} (highly toxic), while >80% of QDs with a polyol *surface ligand* demonstrating $IC_{50} \geq 25$ mg L^{-1} (less toxic) (Figure 9.6) (Oh et al. 2016). On the other hand, alkylthiol, as the prevalent *surface ligand* category (accounting for ~34% of the IC_{50} data), did reveal a noticeable correlation with IC_{50}. However, if one also considers *QD diameter* > 5 nm, then it is revealed that ~87% of QDs with alkylthiol *surface ligand* were associated with $IC_{50} \geq 25$ mg L^{-1}. It should be noted, however, that while the above example was restricted to a two-to-one toxicity-attribute relationship, a complex coupling of attributes in terms of their impact on observed toxicity often exists in multidimensional datasets and thus significant data are required to uncover such complex relationships.

9.2.3 Supervised Learning

Supervised learning is the discovery of a model function h from a hypothesis space (i.e. a set of candidate models) to approximate a true unknown function f that generates the input–output pairs $(x_1, y_1), (x_2, y_2), ..., (x_n, y_n)$ of size N. With structured formats of the data for supervised learning, each input data sample x_i is represented by an n-dimensional vector and y_i is the labeled output. The hypothesis space is generally determined using exploratory data analysis (simple bar charts, histograms, scatter, and box plots) to analyze patterns and correlations among data variables that satisfy a specific hypothesis space. An alternative approach consists of experimenting with models from multiple hypothesis spaces and determining model(s) of the best performance (Section 9.3.5). For both approaches, supervised learning divides a given dataset into model training and test sets with a typical split ratio of 80% and 20% for training and test datasets, respectively (Russel and Norvig 2012). Performance evaluation of the models is then based on the best-fit function that predicts the highest possible accuracy for test data that are not seen by the model.

Various supervised ML algorithms have been utilized for nano-(Q)SAR development with varying levels of interpretability and explainability (i.e. understanding the mapping between ENM descriptors and their toxicity profiles, and reasoning for the significance of ENM descriptors for predicting ENM's toxicity). RF models, based on decision tree algorithms, BNs, SVMs, and other simple nonlinear models (e.g. LR and Naïve Bayes) have notably been most representative in these studies.

9.2.3.1 Random Forest Models

The RF approach, which is an ensemble learning technique applicable for both regression and classification, consists of a multitude of decision trees constructed as part of model training phase. The output of RF is the class returned by the greatest number of trees, and the average/mean value predicted by each tree in the forest, for classification and regression tasks, respectively (Breiman 2001; Liaw and Wiener 2002; Svetnik et al. 2003; Bishop 2006; Han et al. 2012).

9.2.3.2 Support Vector Machines

SVMs have been adopted for the development of both regression and classification-based nano-(Q)SAR. SVM aims at devising a computationally efficient way of segregating different data classes with a hyperplane that best differentiates the classes (Jain et al. 2000; Bishop 2006; Russel and Norvig 2012). SVM selects this hyperplane (i.e. the maximum margin separator) from the hypothesis space that maximizes the distance to data points of either class (Figure 9.7). The support vectors are the data points that are closest to the separator constructed by the SVM model (three circled points in Figure 9.7).

9.2.3.3 Bayesian Networks

BNs are probabilistic graphical models that represent a set of variables and their conditional dependence via directed acyclic graphs (DAGs). BNs are used to represent the uncertainty in domain knowledge in the form of probability distributions (Russel and Norvig 2012). BNs, which are based on probability and graph theories, are a suitable tool to handle uncertainty and complexity via probability distributions and can function as an expert system to predict outcomes given a set of variables. BNs are capable of deriving conditional independence properties within the structure, where the conditional independence of a node of its predecessors, given only its parents, is represented by the non-descendants' property. The non-descendants' property defines each variable to be conditionally independent of its non-descendants, given its parents (Figure 9.8a). Given the non-descendants' property, defining joint distributions within the BN becomes computationally feasible (i.e. only a set of

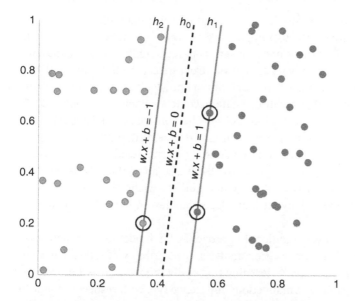

Figure 9.7 SVM classification of two-class problems (data points on the left and right of h_0 represent two classes, class 1 and class 2, respectively) with three candidate linear separators (h_0, h_1, and h_2). The maximum margin separator (h_0) is the middle point of the margin (area between h_1 and h_2). In the dual representation equation (right), α is the Lagrange dual variable associated with each data point, y is the class variable which can be +1 and −1 for samples with positive (e.g. toxic) and negative (e.g. nontoxic) examples, respectively. x are the data samples that represent a sample class, w is the weight vector, b is the intercept, and circled data points are the support vectors that touch the boundary of the margin.

(a)   (b)

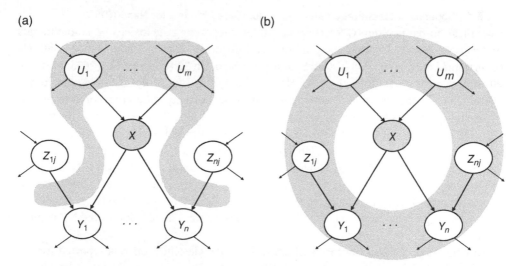

Figure 9.8 (a) BN representing non-descendants' property where a variable X is shown conditionally independent of its non-descendants (Z_{ij}s) given its parents (U_js) in the shaded area. (b) BN representing the Markov Blanket property where a node X is conditionally independent of all other nodes in the network given its parents (U_js), children (Y_js), and children's parents (Z_{ij}s) in the shaded area. *Source:* Adapted from: Russel and Norvig (2012).

conditional independence properties are defined instead of a full network joint distribution). The non-descendants' property further implies a Markov blanket in which "a variable is conditionally independent of all other nodes, given its parents, children, and children's parents" (Figure 9.8b) (Russel and Norvig 2012). BNs' structure can be initially developed based on deterministic models (if available) and/or domain knowledge, along with available data. The BNs can then be updated as additional data become available (Neapolitan 2003; Chan and Darwiche 2004; Jensen and Nielsen 2007). The constructed BN model, once trained with the presented data, can function as an expert system that allows intelligent queries while providing useful insight even when only partial evidence is provided (Money et al. 2012; Murphy et al. 2016; Bilal et al. 2017). It is also noted that BN-based models can be used for attribute sensitivity analysis (Chan and Darwiche 2004; Money et al. 2012; Bilal et al. 2017) conditional toxicity-attribute dependencies, and quantitative toxicity predictive models.

BNs offer several advantages over traditional approaches for model development that include: (i) suitability for multi-criteria decision analysis, (ii) capability of robust handling of uncertainties, (iii) iterative and adaptive learning (i.e. easily refined as new knowledge/evidence is acquired), (iv) capability of bidirectional inference (i.e. with respect to input to output or output to input relationships), (v) ability to incorporate both quantitative and qualitative information, (vi) graphical cause-effect representation of a system (high model interpretability), (vii) modular structure that can be easily expanded to incorporate new information (e.g. expanding a BN node into submodules), and (viii) extendibility to create decision networks by adding decision and utility nodes to arrive at optimal decisions, control, or plans corresponding to a maximum or a minimum total utility (Neapolitan 2003; Chan and Darwiche 2004; Jensen and Nielsen 2007).

9.2.3.4 Supervised Classification and Regression-Based Models for Nano-(Q)SARs

Classification-based nano-(Q)SARs are useful for categorizing the toxicity of nanomaterials in terms of nominal values such as toxicity categories. Various classification methods (Czermiński et al. 2001; Webb 2002; Stahura and Bajorath 2004; Liu and Yu 2005; Medlock 2008; Park et al. 2009; Liu et al. 2013d; Ha et al. 2018) (e.g. LR, SVM, LDA, and naïve Bayes classifier) have been proposed for modeling categorical toxicity endpoints. For example, an LR-based nano-(Q)SAR classifier was developed (Liu et al. 2011b) based on HTS data for BEAS2B cells exposed to nine metal oxide NPs. In a controlled setting, data for this model were first processed to identify and label toxic versus nontoxic outcomes via statistical analysis of the standardized difference in the biological responses of cells exposed to NPs relative to an unexposed control cell population. The best-performing model (nearly 100% classification accuracy) was based on three NP descriptors (i.e. the period of the NP metal, atomization energy of the metal oxide, and the NP primary size) in addition to the NP volume fraction.

Another meta-analysis approach which utilized RF classification was reported for the analysis of cellular toxicity of metal oxide NPs (Ha et al. 2018). The study, which was based on data extracted from 216 published studies, utilized 14 attributes (including NPs' physicochemical properties and experimental conditions), and demonstrated that cytotoxicity of NPs was highly correlated with the administered NP dose, assay type, exposure time, and NP surface area. RF classification models for the preprocessed dataset revealed average prediction accuracy of up to 89% in terms of F_1 score measure (i.e. a measure of model predictive accuracy calculated by the number of correctly predicted true positive samples divided by all true positive samples in the dataset including misclassified samples) (Ha et al. 2018).

A classification nano-(Q)SAR was also reported for cellular toxicity of metal-oxide NPs, based on toxicity class definition derived from both dose–response analysis and consensus SOM clustering (Liu et al. 2013d). An initial pool of thirty NP descriptors was utilized for a generalizable candidate model and several nano-(Q)SAR models were evaluated. The best performing nano-(Q)SAR (built with SVM) was based on two descriptors, namely conduction band energy (E_C) and ionic index (Z^2/r) with a classification accuracy of ~94%. Consistent with mechanistic understanding, NPs were found to be toxic at a higher probability with decreasing ionic index and within the range of the biological band gap.

Advanced ML algorithms (RF and BN) have also been utilized to evaluate the feasibility of meta-analysis for comprehensive ENM impact assessment. For example, the study presented in the data exploration section (Figure 9.6) (Oh et al. 2016) demonstrated RF models for analyzing and extracting pertinent knowledge from published studies on the cellular toxicity of QDs. Using RF regression models, QD toxicity was shown to closely correlate with QD surface properties (including *shell*, *ligand*, and *surface modifications*), *diameter*, *assay type*, and *exposure time*. A subset of attributes for developing an RF model was selected via an exhaustive search based on increasing RF model predictive accuracy (correlation of coefficient – R^2). Using the RF model based on the subset of the seven most significant attributes, it was then shown that a robust ENM nano-(Q)SAR workflow can be developed for interrogating the wide range of toxicity data from multiple sources.

Classification and regression-based nano-SARs (Fourches et al. 2010a), based on a large dataset of ENMs, were proposed for in vitro biological effects of two distinct NP datasets: (i) a set of 109 NPs of similar core (monocrystalline magnetic NPs, with a 3-nm core of $(Fe_2O_3)n(Fe_3O_4)m$) and diverse surface modifiers (dataset A) (Weissleder et al. 2005), and (ii) a second set (dataset B) containing 44 iron oxide core NPs of different polymeric coatings used in molecular imaging and nano-sensing (Shaw et al. 2008a). A kNN-based regression model (Zheng and Tropsha 2000) for NP uptake by pancreatic cancer cells (PaCa-2) was developed for dataset A. The model was formulated using a set of molecular descriptors for the organic coatings without explicit considerations of the intrinsic NP characteristics (e.g. primary or aggregate size or core ENM properties). Model performance was quantified by a square correlation coefficient of 0.72 (assessed via fivefold cross-validation Section 9.3.3) (Kohavi 1995; Jonathan et al. 2000; Zhang 2007). A classification model for dataset B was developed using bioactivity profiles for each NP (comprised of bioactivity measures for four assays for four cell types at four NP concentrations). The classification endpoint for the nano-SAR was defined based on the average (for each NP) of multiple different bioactivity measures. The NPs were then categorized into two subsets (each of the same number of NPs) based on the median of the average bioactivity endpoint for the NP dataset. The nano-SAR was constructed based on an SVM model (Cristianini and Shawe-Taylor 2000) using four NP descriptors (*primary size* and *zeta potential*) and two magnetic properties (*spin–lattice relaxivity* and *spin–spin relaxivity*) (Thorek et al. 2006; Shaw et al. 2008a), demonstrating a classification accuracy of 73% in a fivefold cross-validation. Although this model demonstrated the feasibility of correlating fundamental NP properties with their bioactivity, the use of an arbitrary threshold of average bioactivity seems to have deviated from OECD principles (OECD 2007) of SAR development that require the use of a meaningful and definitive biological endpoint.

A log-linear regression model (Puzyn et al. 2011) was proposed based on the toxicity of 17 metal oxide NPs based on EC_{50} for *Escherichia coli*. A simple nano-(Q)SAR regression model was constructed with a single descriptor (formation enthalpy of a gaseous cation having the same oxidation state as that in the metal oxide structure). The nano-SAR demonstrated reasonable performance indicated by a correlation coefficient of 0.77 and 0.83 in cross-validation (Kohavi 1995; Jonathan et al. 2000; Zhang 2007) and external validation, respectively. Although size-specific behavior of ENMs is often expected or the NP size range is too narrow, such behavior may be obscured when there are other factors that dominate ENM toxicity. Inclusion of the effect of ENM primary size, exposure concentration (as volume fraction) and other ENM descriptors (e.g. metal-oxide atomization energy and period of the NP metal) was proposed in a later study on classification of nano-SAR cytotoxicity (i.e. loss of membrane integrity) of BEAS-2B cell line exposed to nine different metal-oxide NPs (Liu et al. 2011b). The reported nano-SAR demonstrated 100% classification accuracy in both internal and external validations.

9.2.4 Predictive Nano-(Q)SARs for the Assessment of Causal Relationships

An important undertaking in the development of nano-(Q)SAR is the need to address model interpretability and explainability (i.e. causal relationships of ENM toxicity with the reported ENM attributes). The above traits are not the intrinsic capabilities of most traditional techniques such as SVM, LR, Naïve Bayes, or ANNs. Additionally, for datasets with a large number of ENM descriptors (both qualitative and quantitative), increased model complexity (e.g. higher number of model input parameters, ranges of parameter values with high variance, models with high computational cost) poses challenges for model-based reasoning of descriptor-toxicity relationships. For example, although decision trees are highly interpretable, trees in RF models can grow to a scale where the decision path from the root node to leaf (decision) nodes can become exceedingly difficult to interpret for datasets of high dimensionality. Additionally, such models typically rely on single value outputs adding further challenge in terms of building confidence from a decision-making perspective.

Handling of uncertainties poses a significant challenge in nano-(Q)SAR development due to the ENM attributes and experimental conditions that may differ across multiple studies. For example, based on RF models, developed based on meta-analysis of QD cellular toxicity (Oh et al. 2016), a similarity network was constructed via hierarchical clustering of QD cellular toxicity with respect to six significant attributes (determined by assessment of RF attribute significance). The approach demonstrated the heterogeneity of the 1741 data samples collected from 307 publications (Figure 9.9) based on a proximity matrix (Svetnik et al. 2003) that quantified attribute similarity according to the frequency with which QD samples appear in the same leaf node of an RF model tree. To visualize heterogeneity with respect to major structures in the QD cell viability data, the network was further partitioned into 30 clusters of cell viability via hierarchical clustering. Nodes in the network represent QDs which are connected (links in the network) if the similarity with other QDs (based

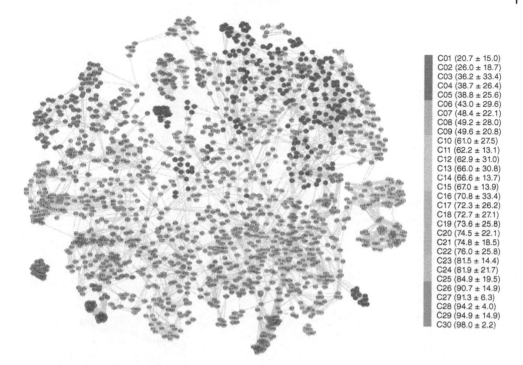

C01 (20.7 ± 15.0)
C02 (26.0 ± 18.7)
C03 (36.2 ± 33.4)
C04 (38.7 ± 26.4)
C05 (38.8 ± 25.6)
C06 (43.0 ± 29.6)
C07 (48.4 ± 22.1)
C08 (49.2 ± 28.0)
C09 (49.6 ± 20.8)
C10 (61.0 ± 27.5)
C11 (62.2 ± 13.1)
C12 (62.9 ± 31.0)
C13 (66.0 ± 30.8)
C14 (66.6 ± 13.7)
C15 (67.0 ± 13.9)
C16 (70.8 ± 33.4)
C17 (72.3 ± 26.2)
C18 (72.7 ± 27.1)
C19 (73.6 ± 25.8)
C20 (74.5 ± 22.1)
C21 (74.8 ± 18.5)
C22 (76.0 ± 25.8)
C23 (81.5 ± 14.4)
C24 (81.9 ± 21.7)
C25 (84.9 ± 19.5)
C26 (90.7 ± 14.9)
C27 (91.3 ± 6.3)
C28 (94.2 ± 4.0)
C29 (94.9 ± 14.9)
C30 (98.0 ± 2.2)

Figure 9.9 QD similarity network based on the RF model for cell viability using the six most significant attributes with decreasing order of R^2 through exhaustive search for the attributes. QDs are represented as nodes identified by colors corresponding to average viability ranges. A total of 30 clusters were identified via hierarchical clustering, and these are colored in the network according to the averaged cell viability of QDs in each cluster. Connected QDs are of proximity larger than the average within cluster proximity. Sparse connections among clusters in the similarity network indicate heterogeneity of the QD toxicity studies, while clusters having little, or no connectivity suggest that these studies have little commonality in terms of QD properties and/or experimental conditions. *Source:* Adapted from: Oh et al. (2016).

on their reported descriptors) was greater than the average within cluster similarity. The sparse connection in the similarity network demonstrated the heterogeneity of QD toxicity studies, with some clusters having little or no connectivity. Moreover, isolated QD clusters that are not connected to any other clusters (e.g. C14, C28, and C30) represent distinct attribute–cell viability correlations. Visualization of heterogeneity via such approach can be helpful to assess potential skewness in the data that could introduce model bias due to one or more specific data sources. For example, the network (Figure 9.9) shows that isolated cluster C30, consisting of data from single study is of the highest average cell viability among all clusters.

In order to link causes (i.e. descriptors) and effects (i.e. toxicity metrics), predictive models have to account, in addition to ENM properties, for reported bioactivities resulting from differences in experimental conditions (e.g. assay types, exposure concentrations, exposure period, organism and more). In developing predictive models, it is critical to: (i) quantify the relevance and significance of both continuous and categorical (e.g. assay type) attributes in

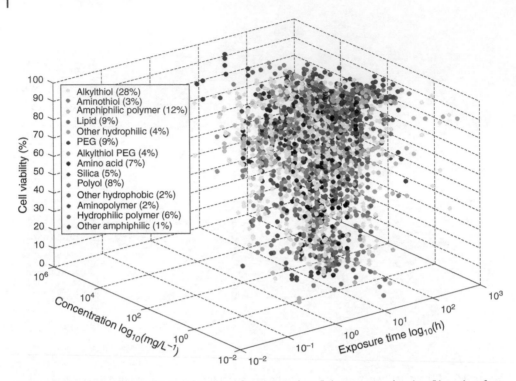

Figure 9.10 Cell viability (%) as a function of exposure time (hr), concentration (mg/L), and surface ligand showing high scatter and no clear pattern. *Source:* Adapted from Bilal et al. (2019).

affecting the determined toxicity metrics, (ii) account for data uncertainties, particularly if collected from multiple sources, and (iii) assess conditional dependencies (i.e. association rules) of toxicity metrics on various attribute combinations which can then serve to identify causal relationships (Oh et al. 2016; Bilal et al. 2019). For example, in an extended study of QD cellular toxicity (Bilal et al. 2019), exploratory data analysis via a scatter plot (Figure 9.10) demonstrated the challenge posed by data heterogeneity. Consistent with previous work (Oh et al. 2016), data heterogeneity was also found in the above study as visualized by a scatter plot of cell viability as a function of exposure time, QD concentration and *surface ligand*. The analysis did not reveal any discernible patterns to suggest significant correlation of the above attributes with higher or lower cell viability. However, BN-based model proved to be useful for intelligent query of the available body of evidence and to determine attribute significance via sensitivity analysis, and to represent data variance in the form of probability distributions (Uusitalo 2007; Aguilera et al. 2011; Money et al. 2014; Bilal et al. 2017; Marvin et al. 2017).

BN based models described in the above studies were constructed by establishing a network architecture of attributes and outcomes in a visual representation in the form of links that allow one to follow causal relationships (Neapolitan 2003; Jensen and Nielsen 2007). The linkages represent conditional dependences of target outcomes on attributes which serve to interpret "if/then" causal relationships where the parent (antecedent) and child (descendent) nodes are at the outgoing and incoming link in the BN, respectively. The set of model attributes and their conditional dependencies represent knowledge from the dataset(s) of ENM attributes and toxicity outcomes in the form of probability distributions. Variations at parent nodes are recorded along with the resulting conditional probability distributions at the children nodes (e.g. toxicity outcomes or any selected ENM descriptor) which are affected by the observed state of each parent node. Following such an approach, one may identify, for example, the conditional dependence (i.e. association rule) that would lead to a toxicity outcome within a specific range.

It is noted that most previous NP toxicity studies have demonstrated the potential benefit of BNs for developing hazard ranking based on qualitative "toxicity/hazard" classes (Murphy et al. 2016; Marvin et al. 2017). However, given the lack of sufficient experimental data, these studies have relied primarily on expert opinion in constructing the Conditional Probability Tables (CPTs) (discretization and probability assignment). Nonetheless, the developed BN based models demonstrated a capability for identifying the most significant parameters that are likely to impact the hazard attributed to specific NPs. For example, literature curated data can be visualized as illustrated in Figure 9.10 for exploring cellular toxicity of Cd-containing QD via with BNs (Bilal et al. 2019). BN models were developed based on the above dataset which was curated from 517 publications providing 3028 cell viability data samples and 837 IC_{50} values. The BN QD toxicity models were developed using both continuous (i.e. numerical) and categorical attributes. IC_{50} correlated with the following attributes: QD diameter, exposure time, surface ligand, shell, assay type, surface modification, and surface charge, with the addition of QD concentration for modeling cell viability. The BN models performed with predictive accuracy (R^2) of > 0.8 and could be conveniently updated as new knowledge is acquired. In addition, data exploration via BN models enabled extraction of association rules for QDs cellular toxicity (i.e. mappings of ENM descriptors with IC_{50}).

BN models were also developed as nano-(Q)SARs for both regression and classification-based models (Murphy et al. 2016; Marvin et al. 2017) with cause–effect relationships for toxicity outcomes (e.g. cytotoxicity, genotoxicity, immunological effects, cardio pulmonary effects and others) associated with exposure to different ENM types (TiO_2, SiO_2, Ag, CeO_2, ZnO) based on data compiled from the published literature. For example, a BN approach for predicting toxicity metrics for eight toxicity groups was proposed for different metal and metal oxide NPs based on data compiled from 32 published studies (Marvin et al. 2017). Despite significant data gaps for the various NPs (e.g. missing data in the range of 59–72% for surface charge, surface reactivity, and administration route, 33% for cytotoxicity and 74–99% for seven other toxicity outcomes), the derived BN model identified the NP properties most relevant for correlating toxicity outcomes (*NP surface area, surface coatings* and *surface charge*). Another approach for ranking hazard due to inhalation exposure to NPs was proposed based on a BN constructed specifically for Ag, TiO_2, and CNTs NPs relying on a data extracted from NIOSH and EU research reports (Murphy et al. 2016). A BN model was constructed to predict the NIOSH upper boundary with respect to the acceptable doses and thus the occupational exposure concentrations. An approach of control banding was proposed in the above study but with a significant uncertainty of up to two orders of magnitude and higher for prediction of exposure concentrations.

9.3 Development of Machine Learning Based Models for Nano-(Q)SARs

9.3.1 Overview

ML has been proven useful for identifying nanomaterial properties and exposure conditions that affect cellular and organism toxicity, thereby providing information needed for risk analysis and for safe-by-design approaches for the development of new nanomaterials. Continuous endpoints regression models can be developed using methods such as linear regression and PLS regression. Classification methods such as LR, LDA, and naïve Bayesian classifier can be used to develop models suitable to predict categorical endpoints. As described in Section 9.2.2, a number of other sophisticated ML methods have also been applied, such as SVM, ANN, decision tree, RF, BN, and k-nearest neighbor, for developing both regression and classification-based toxicity models.

Workflows, using tools that are compliant with standard guidelines for model development (Directorate et al. 2007) and assessment of their domain of applicability (Liu et al. 2013c) can be depicted by the general workflow shown in Figure 9.11. The development of these robust nano-(Q)SARs requires screening for the target outcomes in relation to various ENM characteristics and biological settings. This initial screening is integral to performance and metadata assessments and evaluation of subsequent models. Developing nano-(Q)SARs is based on the available experimental/literature data or proven theoretical foundations.

9.3.1.1 Data-Driven Models

Understanding of the complexity of ML tools for developing a generalized predictive data-driven model is critically dependent on data-scrubbing and preprocessing activities. Preprocessing of both unstructured and/or heterogeneous data of ENMs is critical (Gesellschaft für

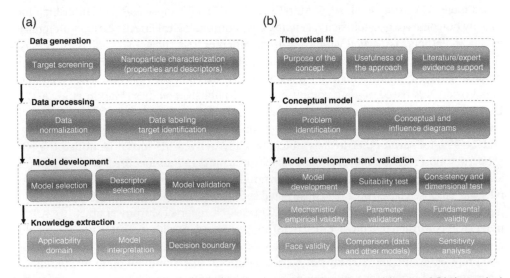

Figure 9.11 The nano-(Q)SAR model development workflow for: (a) data-driven, and (b) theoretical models.

Chemische Technik und Biotechnologie 2014; Morris et al. 2015; National Cancer Institute 2016; NIOSH 2016; http://Re3data.org 2016; http://NanoHUB.org 2016; ENanoMapper 2016; National Institute for Public Health and the Environment 2016; Karcher et al. 2018) given the diversity of toxicity studies and endpoints, data complexity, and problem definition. Features' or descriptors' selection, therefore, is an important undertaking when ML tools and/or algorithms are developed for complex datasets of a large number of features (i.e. ENM characteristics and environmental settings). In this regard, the strengths of applying a robust feature selection approach for a range of datasets are important as demonstrated in various studies (Liu et al. 2011a; Liu and Shi 2013; UCLA 2016) (Figure 9.11a). In this regard, it is noted that most ML development of data-driven nanotoxicology models relies on the stationarity assumption, whereby unseen test data samples are assumed to follow the same distribution as that of model training data.

9.3.1.2 Mechanistic/Theoretical Models

Due to the complexity of nano-bio interactions and data scarcity in various toxicity areas of interest, developing mechanistic/theoretical models are sought to assess cause-and-effect relationships. Development of theoretical models via well-established mathematical approaches (Figure 9.11b) requires an abstract description of a given system including conceptual/influence diagrams to highlight different aspects of the system (Consumer Product Safety Commission, NNI 2016). However, to develop theoretical/mechanistic models and understand underlying layers of complex theoretical systems, theoretical models require multiple levels of validation (i.e. face validity, parameter validation, sensitivity analysis, mechanistic/empirical validity, and comparison with other tools and datasets).

9.3.2 Data Generation, Collection, and Preprocessing

ML-based models require data for model training. Such data can be acquired from various sources (e.g. data generated from experiments, data compiled from the published literature, qualitative data from surveys/interviews, and theoretical simulations). When the body of evidence is insufficient, transfer learning (Section 9.2) may prove useful for model building. However, such an approach hinges on having models that have been pretrained using specific data and can then be retrained with newly acquired data (Russel and Norvig 2012). For meta-analysis approaches, crowdsourcing can also be carried out by collecting and processing data from multiple sources (e.g. published literature, surveys, and phone interviews) (Gernand and Casman 2014; Kaweeteerawat et al. 2015; Murphy et al. 2016; Ha et al. 2018; Wheeler and Lower 2021).

Data integrity and understanding of dataset dimensions are important undertakings for nano-(Q)SAR development. Data-driven ML-based nano-(Q)SAR models are typically supported by data provenance strategies (e.g. definitions and descriptions of each model descriptor, source of the data, statistical summaries of model descriptors, information of the contributor(s), and evolution of data sources over time for time series comparison) (Oh et al. 2016; Bilal et al. 2019). In planning nano-(Q)SAR development, one should consider the following questions:

1) Should a supervised or unsupervised ML model be developed?
2) Is the nano-(Q)SAR used for classification or regression?

3) Is a stationary model sufficient or does it require time-series toxicity data?
4) Is the dataset size sufficient for robust model development?
5) Is the number of attributes sufficient for developing the nano-(Q)SAR model?
6) Can the dimensionality of the dataset be reduced via descriptor selection (Section 3.2)?
7) Can the impact of outliers (data samples having values that are significantly different from the majority of other samples) be reduced/handled via appropriate transformation/normalization?
8) Can missing data fields be imputed without introducing bias or detrimental impacts on data fairness?
9) Has the data been cleaned to avoid human errors (e.g. correction of spelling errors in categorical attributes)?

In addition to this, a class balance in the dataset is an additional important factor to consider for an unbiased and robust nano-(Q)SAR model development. Bias in the dataset due to class imbalance (e.g. the class representing *nontoxic* outcomes has a significantly higher number of samples than the class representing *toxic* outcomes) can severely impact ML model performance. Data bias can be usually overcome via undersampling the majority class (i.e. removing some samples from the majority class) or oversampling the minority class (i.e. imputing synthetic samples based on a statistical approach or newly collected samples for the minority class) (Pears et al. 2014). ML methods for imbalanced classes such as gradient boosting, random forests, or ensemble learning approaches can successfully handle outliers and overcome the class imbalance issue via regularization/penalization approaches. Thus, such approaches are suitable for models built based on datasets with imbalanced classes (Torgo et al. 2013; Pears et al. 2014).

Data normalization techniques are often used as part of data preprocessing, including, for example: (i) data normalization with respect to the minimum and maximum values in the dataset (e.g. $\frac{(x_i - x_{\min})}{(x_{\max} - x_{\min})}$ where x_i, $x_{\min}$, and $x_{\max}$ are the ith data sample, the minimum and maximum values of attribute x, respectively), (ii) standardization of descriptor values to have an established standard deviation, (iii) variable encoding (i.e. the conversion of categorical attributes into separate Boolean attributes, particularly for ANN- and SVM-based models, (iv) discretization of continuous attribute values into fixed bins, and (v) dimensionality reduction and feature extraction by constructing alternative data dimensions from an existing set of features (e.g. PCA and multidimensional scaling (MDS)).

9.3.3 Descriptor Selection

Current nano-(Q)SAR developments typically rely on an initial reasonably large pool of descriptors (features) that are then pruned to retain the most significant descriptors. This approach is followed to avoid missing potentially relevant features that correlate with ENM-induced bioactivity. Thus, feature selection is for establishing the most suitable features that correlate with toxicity endpoints (Puzyn et al. 2010). Feature selection methods that have been utilized for nano-(Q)SAR development range from the simple sequential forward selection method to more sophisticated approaches such as sequential forward floating selection (SFFS) and genetic algorithms (GAs). Among nano-(Q)SARs of similar performance, it is generally preferred to select the one having the smallest number of features for the sake of model simplicity. As a rule of thumb, it is recommended that the number of QSAR descriptors should be below 1/10–2/10 of the total training compounds (or ENMs) (Puzyn et al. 2010).

The "curse of dimensionality" is a critical problem for any modeling method that utilizes data and applies statistical approaches to predict the ENM descriptor–bioactivity relationship with higher confidence (Ravi Kanth et al. 1998; Pagel et al. 2000; Korn et al. 2001). Particularly, when using data-mining methods for knowledge discovery, learning quality and efficiency collapse dramatically at an excessively high level of data dimensionality (Blum and Langley 1997). Such a collapse can be ascribed to features that do not provide useful information (e.g. irrelevant or redundant features) for modeling toxicity endpoints for high-dimensional datasets (Guyon and Elisseeff 2003). It is possible that the intrinsic dataset dimension (i.e. dimensionality of the object represented by the dataset) (De Sousa et al. 2007; Rozza et al. 2012) is much smaller than its embedding dimension (i.e. the number of features) due to linear or nonlinear feature correlations (Pagel et al. 2000; Roweis and Saul 2000; Tenenbaum et al. 2000). Moreover, the presence of redundant or irrelevant features could complicate data analyses and compromise modeling results (John et al. 1994; Liu and Yu 2005). Therefore, data dimensionality (feature) reduction methods (Webb 2002; Liu and Yu 2005; Amayri and Bouguila 2013) are required in order to prune the initial feature pool to the most relevant set for model development with improved generalization capability.

In principle, approaches for feature reduction can be categorized into two types: feature extraction/transformation (Jain et al. 2000; Webb 2002) and feature selection (Liu and Yu 2005). Feature extraction approaches (e.g. PCA) construct a new set of features by projecting (transforming) the original feature space to a lower dimensional one. Compared to feature selection methods, the former approach can provide a more compact representation (greater reduction in dimensionality) of the original feature set. However, the projection/transformation may lead to a loss of physical interpretation pertaining to the original features (Wei and Billings 2007; Liu et al. 2009). Unlike feature extraction, feature selection aims to identify an optimal subset of the original features according to an objective, such as minimizing redundancy and/or maximizing relevance (Liu and Yu 2005; Liu and Motoda 2008).

A feature selection method typically includes a search strategy that generates a subset from a full feature and a quality measurement that quantifies the "goodness" of the feature subset generated by the search strategy (Figure 9.12) (Liu and Yu 2005; Liu and Motoda 2008). The quantified feature quality is usually fed back to the search strategy to guide the generation of new candidate feature subsets. Feature selection methods are generally categorized into wrappers (Kohavi and John 1997) and filters (Liu and Setiono 1996) according to whether the performance of a predetermined learning algorithm (model) is used for evaluating the quality of candidate feature subsets. Wrappers assess the quality of a feature subset with the performance of a model developed using the subset, while filters evaluate a feature subset based on intrinsic data characteristics, such as information (Jebara and Jaakkola 2000; Yu and Liu 2004), distance/similarity (Pudil and Hovovicova 1998; Dash and Gopalkrishnan 2008; Robnik-Šikonja and Kononenko, n.d.; Zhao et al. 2013), consistency (Almuallim and Dietterich 1994; Dash and Liu 2003), and dependence (Modrzejewski 1993;

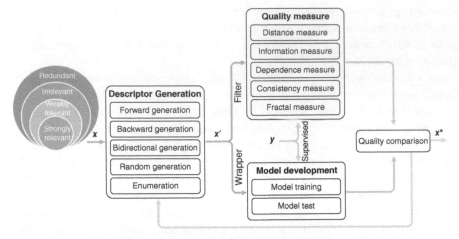

Figure 9.12 Wrapper approach with exhaustive (or partially exhaustive) search.

Hall 2000). There are also hybrid methods (Das 2001; Xing et al. 2001) that combine the advantages of wrappers and filters. Models developed with features selected by wrappers are tuned to the specific interaction between the chosen model and the given dataset. Therefore, such models are of higher accuracy relative to those developed based on features selected by filters. To date, the majority of reported nano-(Q)SAR studies have resorted to using wrappers for the identification of suitable descriptors. The availability of wrappers for nano-(Q)SAR development is, in part, due to the limited physicochemical characterization data for ENMs. However, there is a higher computational cost associated with wrappers due to the embedding of the entire process of development and performance evaluation of a nano-(Q)SAR. Also, wrappers are usually biased toward the predetermined models, which could lead to impaired generality (Loughrey and Cunningham 2005). As more descriptors and ENM samples become available, filters may become the preferred choice for identifying feature subsets of high quality within a reasonable computational cost. Indeed, filter methods have been increasingly utilized for feature selection, both due to their low computational cost and unbiased result toward a particular model. Examples of various feature selection approaches adopted for nano-(Q)SAR development are listed in Table 9.4.

From a mathematical viewpoint, feature selection is a combinatorial optimization problem of identifying the global optimum from the 2^{N-1} subset (i.e. feature space) of a full set of N features. An exhaustive evaluation of each feature subset (i.e. enumeration) becomes infeasible even when the number of features are not too large. Therefore, heuristic search strategies are often used to reduce the computational cost with the tradeoff of arriving at a potentially suboptimal feature subset due to partial exploration of the feature space. For example, the sequential forward/backward selection (SFS/SBS) strategy has been applied in the development of nano-(Q)SARs for computational ENM toxicology, whereby each step identifies the feature whose inclusion/exclusion leads to the greatest increase or least decrease in the quality of the selected feature subset (Puzyn et al. 2010; Ponzoni et al. 2017; Halder and Dias Soeiro Cordeiro 2021). Bidirectional search strategies have also been adopted for nano-(Q)SAR development, such as SFFS that conducts at each selection step.

Table 9.4 Feature selection approaches utilized for nano-(Q)SAR model development.

Technique	Definition	Reference
SDJ-FS	Spatial distance join (SDJ)-based feature selection	Liu and Shi (2013)
KLS-FS	Kernel least squares forward selection	Liu et al. (2011a)
DMIFS	Dynamic mutual information-based feature selection	Liu et al. (2009)
FFSEM	Fuzzy feature selection-based on min–max learning rule and extension matrix	Li and Wu (2008)
BDFS	Backward deletion feature selection	Huang and Chow (2005)
LS-SFM	Least squares support feature machine	Li et al. (2007)
MRMR	Minimum redundancy maximum relevance	Ding and Peng (2005), Schowe and Morik (2011)

Subsequently, a forward selection and then a backward elimination are used to evaluate whether previously selected features should be removed due to the addition of the newly selected one (Puzyn et al. 2010). It has been reported that with descriptors selected by SFFS, nano-(Q)SARs were developed with improved prediction accuracy over those selected by SFS. Random search strategies, as exemplified by GAs, have also been demonstrated to be useful for selecting descriptors of reasonable quality for nano-(Q)SAR development (Chakrabarty et al. 2016).

9.3.4 Model Selection and Training

Data processing and initial exploratory data analysis form the basis of the first steps for developing ML-based nano-(Q)SARs. Data exploration via simple exploratory analysis and statistical approaches are useful to gain an understanding of the data, detect missing/erroneous values or samples with unusual outcomes (outliers), and follow various data distributions (e.g. normally distributed or highly skewed descriptor values) to search for an appropriate model (Russel and Norvig 2012).

Subsequently, model development proceeds along several sequential steps that include choosing the model class (e.g. SVM, ANN, Naïve Bayes, LR, RF, and BN), model training with available training data, tuning of model parameters using unseen (validation) data, and evaluating model performance based on test data (Section 9.3.4). Model training and optimal fit metrics are obtained by splitting the dataset into training and test sets. ML model is then provided with training data to learn patterns and adjust its parameters to iteratively reduce prediction error. The test set is held out and locked until model training is completed along with parameter tuning and retraining and used for model performance evaluation for robustness and accuracy. For model selection, a third subset of the dataset, the "validation set," is typically used to evaluate multiple candidate models, which are followed by a final evaluation of the best-chosen model using the test dataset. In order to avoid model lack of robustness due to insufficient ENM toxicity experimental data, techniques such as k-fold-cross-validation (Zhang 1993; Kohavi 1995) and sampling with/without replacement (Oh et al. 2016) are often utilized to allow each data sample to be served with double duty (i.e. as training and validation data in iterations at different training/testing steps). In cross-validation, data are divided into k (typically set to 5 or 10) subsets of equal sizes, and k rounds of training are performed where at each round, $1/k$ of data is kept for model validation (validation set) while the remaining subsets are used for model training. The average model test score for all rounds has been shown to provide a better performance measure as opposed to only one test score (Russel and Norvig 2012).

Although there is no universal method for selecting an ML modeling approach (or a class of models), data types of model attributes available in hand and toxicity endpoints, data structure and domain knowledge can be helpful for selecting an appropriate model for attaining high predictive accuracy. In general, each ML modeling approach is embedded with its own assumption about hidden relationships between attributes and endpoints (Russel and Norvig 2012). Due to uncertainties when using data from multiple sources, variance in reported experimental settings (e.g. exposure concentrations and duration, cellular toxicity study, or experiments on organisms/species), and selection of models suitable for a given dataset pose a formidable challenge. To avoid missing possible relationships between ENM descriptors and their bioactivity, it is common in nano-(Q)SAR development to test and choose from different classes of models. The selection of an ML algorithm for nano-(Q) SAR development leads to model training based on sample data with the goal of an optimal fit (Section 9.3.4) by minimizing the prediction error. The prediction error is quantified via metrics that are most suitable for the type of nano-(Q)-SAR. For example, (i) correlation of determination (R^2) and root mean square error (RMSE) are typically utilized for regression-based models, and (ii) classification error rate (percent samples misclassified), true/false positive rate, and AUC are utilized for classification nano-(Q)SARs (Russel and Norvig 2012).

9.3.5 Model Validation

The goal of ML-based nano-(Q)SAR development is to select, among the available ML algorithms, the one that best fits test data samples. This is typically determined by minimization of the prediction error (e.g. for a classification-based model, the frequency of model misclassification of a data sample as belonging to an incorrect class). In addition to predicting biological response induced by ENMs with higher accuracy (Section 9.3.3), nano-(Q)SARs require verification and validation of model reliability, trust, AD, and robustness (Russel and Norvig 2012). Data dimensionality reduction (i.e. the descriptor selection) can be useful for the identification of the most significant descriptors that govern ENM-induced bioactivity/toxicity. It is also important to ensure that the nano-(Q)SAR is not a "chance-correlation" (i.e. the correlation is estimated as a random event by the model with weak statistical significance) due to high bias/variance. Therefore, it is recommended that the QSAR development process (including feature selection, parameter tuning/optimization, model development via a specific ML algorithm, and its evaluation using test data) should be repeated via y-randomization to assess the highest possible model performance for the given training dataset in order to identify potential chance correlations (Rücker et al. 2007). Irrespective of the increase in computational cost, chance correlation assessment can reduce the risk of selecting a "chance correlation" model.

9.3.5.1 Descriptor Importance

Feature ranking which constitutes the identification of attributes of higher significance is useful for robust model development. For example, with a simple linear model with normalized features, it may be possible to infer the relative significance of a model input parameter for its coefficient (weight). However, for nonlinear models or model features that are not normalized, it is difficult to directly infer relative parameter significance due to the inherent assumption that the input parameters of the linear model are uncorrelated. In this regard, the ranking of parameter significance via model performance assessment can be helpful. For example, one can employ feed forward feature selection (FFFS) by adding one parameter at a time and determining the step increase in R^2 or decrease in RMSE (Russel and Norvig 2012). Once a nano-(Q)SAR is developed (without feature selection), feature ranking can serve to assess the relative significance of the descriptors selected for the model. Feature ranking with relevant descriptors can be useful in providing an insight into the biological interpretation afforded by the developed models. More importantly, feature ranking can also serve as a basis for sensitivity analysis for selecting significant features for improved model explainability for a safer-by-design strategy (Zhang 1993; Kohavi 1995).

9.3.5.2 Applicability Domain

Nano-(Q)SARs developed based on experimental HTS toxicity data for various cell lines/species, or data from multiple published sources as part of meta-analysis are often associated with descriptors of mixed data types (qualitative and quantitative) of high dimensionality (e.g. including ENM physicochemical properties, surface modifications, biological conditions, experimental settings, and ENM exposure concentrations). In order to assess and enhance the generalizability of nano-(Q)SAR due to the diversity in data types of ENM descriptors, it is important to assess the model AD that can incorporate ENM physicochemical properties, environmental factors/exposure conditions, and attributes relevant to the biological activity of the studied ENMs, thereby providing robust predictions for new ENMs (or the samples from validation set).

Every model should have its own AD, which is derived not only based on the ENMs in the training set but also based on model parameters and (statistical) approach used to develop the model. Ideally, the AD should be clearly defined and documented by the model developer. The needed information should include, for example: (i) a statement of the unambiguous model algorithm, (ii) details of the training set (chemical identification, descriptors, and endpoint values), and (iii) details of the methods and workflow used to derive the model (Directorate et al. 2007). Nano-(Q)SARs with narrow AD are more likely to provide better model fit and predictive performance relative to those based on a wide AD. However, a model applicable to a narrow AD has the drawback of possible bias to training data and be applicable to only the studied ENM.

9.3.6 Model Diagnosis and Debugging

Nano-(Q)SARs should be designed to avoid the following: (i) high prediction error, (ii) lack of robustness, (iii) limited application domain due to a model that is overly complex or too simple, (iv) an inappropriate number of descriptors (too many or few), (v) irrelevant and/or redundant descriptors, and (vi) insufficient training samples. Therefore, model diagnosis

Table 9.5 Symptoms, potential causes, and possible solutions for common issues encountered with nano-QSARs.

Issue	Symptom	Potential cause	Possible colution
High bias	High training error	Model is too simple; irrelevant descriptors; poor endpoint quality (erroneous configuration of a predefined event/outcome).	Increase model complexity; refine descriptor selection; add additional (new) descriptors.
High variance	Low training errorbut high CV error	Model is too complex; insufficient training samples; too many descriptors;	Implement simpler model structure; add training samples; reduce the number of input attributes.
Lack of robustness	"Chance correlation"	Model is overly complex; excessive number of model attributes.	Implement simpler model structure; reduce the number of model input features.

and debugging are required at every model development step. Some of the common issues associated with nano-(Q)SARs development, as well as their cause, impact and potential solutions are summarized in Table 9.5 with additional details presented in the following subsections.

High model prediction error can arise from various model deficiencies. There are different remedies that can cure model deficiencies, and thus it is critical to identify the specific model deficiency responsible for high model prediction error. In this regard, it is helpful to analyze the composition of model prediction error. In general, the expected error (i.e. prediction error) of a model $\hat{y}(x)$ that is learned from a dataset to predict the true (i.e. observed) $y(x)$ can be expressed by Eq. 9.1 (Wasserman 2004; Bishop 2006):

$$E\left[(y(x)-\hat{y}(x))^2\right] = (y(x)-E[\hat{y}(x)])^2 + E\left[(\hat{y}(x)-E[\hat{y}(x)])^2\right] \tag{9.1}$$

The above expression is referred to as "bias-variance decomposition" (Wasserman 2004; Bishop 2006), (Figure 9.13), which shows that model prediction error is comprised of two components. The first component $(y(x)-E[\hat{y}(x)])^2$ measures (squared) model bias with its variance quantified by the second component $E\left[(\hat{y}(x)-E[\hat{y}(x)])^2\right]$. Intuitively, model variance is indicative of its complexity (or the smoothness in the improvement of the objective function). It is emphasized that one should not mistake model complexity to imply model analytical function complexity. In general, models that closely fit a large number of data points have low bias but may be associated with high variance (Figure 9.13).

The bias-variance relationship with model prediction error in Figure 9.13 illustrates that a high prediction error can stem from either high bias or variance, with different proposed approaches to handle each of them. Therefore, estimations of model bias and variance are required for diagnosing models of high prediction error in order to arrive at suitable solutions (i.e. "suite the remedy to the deficiency"). For a given model, its bias can be estimated by the re-substitution error $(\frac{1}{m}\sum_{i=1}^{m}(y_i-\hat{y}(x_i))^2)$, where m is the number of data

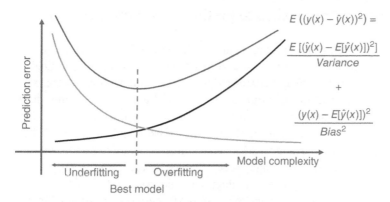

$$E\left((y(x) - \hat{y}(x))^2\right) =$$

$$\frac{E\left[(\hat{y}(x) - E[\hat{y}(x)])^2\right]}{Variance}$$

$$+$$

$$\frac{(y(x) - E[\hat{y}(x)])^2}{Bias^2}$$

Figure 9.13 Bias-variance-decomposition. Diagram of the relationship of model complexity with prediction error due to the problems imposed by model bias and variance.

samples, y_i is the observed outcome and $\hat{y}(x_i)$ is model predicted outcome for the given ith sample. The estimated model bias, together with the model prediction error, obtained via various model validation approaches (such as cross or bootstrapping validation), can provide a reasonable diagnostic of whether a model suffers high variance or bias.

It can be inferred that a model will suffer high variance if it demonstrates high cross-validation error but low training error with an increased training sample size. On the other hand, high cross-validation error together with high training error indicates high model bias. A high model variance may arise from an insufficient number of training samples, excessive number of descriptors, and high model complexity. Accordingly, remedies include increasing the number of training samples and reducing the number of model input descriptors to those of the highest relevance and low redundancy. This can be accomplished via feature selection using simpler models or regularization techniques (e.g. regularized linear regression, neural networks, and SVM). Highly biased models, on the other hand, may result from a lack of relevant input parameters, overly simple models, or poor endpoint quality (Table 9.5). Therefore, possible strategies to reducing high bias include adding additional and more relevant descriptors, increasing model complexity, and improving endpoint quality.

Y-randomization (Rücker et al. 2007) is a widely used diagnostics mechanism for assessing model robustness. A model that demonstrates a "chance correlation" in y-randomization can be diagnosed as lacking robustness. Similar to the occurrence of high model variance, the primary causes for lack of robustness are high model complexity and an excessive number of descriptors. An over-fitted (high variance) model may also lack robustness. Indeed, models of high variance are usually associated with a large number of descriptor coefficients and thus are less stable. Accordingly, the strategies for reducing high model variance, such as getting more training samples, selecting fewer descriptors, using simpler models or regularization technique, can also help improve model robustness.

9.4 Nanoinformatics Approaches to Predictive Nanotoxicology

Given the need for integrated information sources and in silico tools for the development of nanotoxicity models needed for environmental and health risk assessments, the field of nanoinformatics has emerged over the past decade as "the science and practice of determining which information is relevant to the nanoscale science and engineering community, and then developing and implementing effective mechanisms for collecting, validating, storing, sharing, analyzing, modeling, and applying that information" (Liu and Cohen 2015). A key challenge in nanoinformatics is in establishing: (i) interoperability of data repositories containing heterogeneous data sets (Panneerselvam and Choi 2014), (ii) common vocabulary (i.e. ontology) to unambiguously describe NPs (Thomas et al. 2011), (iii) standard formats for data exchange (e.g. ISA-TAB nano specification (Thomas et al. 2011); NCBO (Martínez-Romero et al. 2017); NBI (NBI Knowledgebase 2022)), (iv) criteria to arrive at the definition of the minimum set of nanomaterial characterization parameters (e.g. MINChar (Auffan et al. 2009)), and ENM databases and related information portals (e.g. EU Science HUB (EU Science HUB 2019); NCI caNanoLab (Morris et al. 2014), CEIN NDR (CEIN 2020)).

Assessing the environmental impact of ENMs (Figure 9.14) requires information and data regarding their physicochemical properties, bioactivity, and exposure of receptors of concern; the latter are governed by the fate and transport of ENMs post their releases to various environmental media (both outdoor and indoor). Once the above-required data are validated and organized, data mining and various ML approaches can be utilized to derive correlations (or dependence relations) that may exist between the various descriptors (or

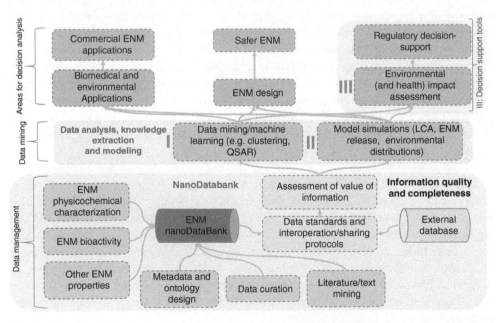

Figure 9.14 Nanoinformatics elements for the environmental and health impact assessment for nanomaterials.

features) that characterize ENMs (i.e. with respect to various structural, physicochemical, and fate and transport information/parameters) and the target toxicity outcomes associated with specific biological receptors under various exposure scenarios.

The computational tools, databases, and various utilities offered by ongoing efforts devoted to the development of nanoinformatics platforms are projected to enable various stakeholders (e.g. regulatory agencies, educational institutions, researchers, and various industrial entities) to formulate research strategies, prioritization of product development, testing objectives and needs for material improvements, as well as providing educational and training materials/tools. Nanoinformatics platforms can serve to facilitate multidisciplinary collaboration among researchers in both institutions of higher learning and industry, thereby accelerating the development of computational tools and construction of rich databases to serve the broad nanotechnology community. It is emphasized that experimental data are essential for supporting the development, validation, and refinement of models/simulators (mechanistic or data-driven). In this regard, sharing of data/information on the characterization and toxicity of nanomaterials will undoubtedly accelerate the development of accurate nano-(Q)SAR models that will benefit industry and regulators to arrive at rational decisions regarding risk management strategies and safe-by-design ENMs.

9.5 Summary

The main aspects of the development of nano-(Q)SARs and research needs reviewed in this chapter are summarized as follows:

- Quantitative SARs for ENMs (i.e. *Nano-(Q)SARs*) are *ML*-based models that relate biological activity (e.g. toxicity) at the single cell or multicellular organism levels induced by exposure to ENMs to the physicochemical/structural properties of the ENMs. ENM toxicity predicted by Nano-(Q)SARs can be used to inform ENMs' health and environmental risk assessment and guide the selection and/or design of safe ENMs;
- The development of robust nano-(Q)SAR requires *ML* model development workflows that include data management and preprocessing, model and *attribute selection*, training, testing, and various model optimization approaches to establish a meaningful model *AD*;
- Nano-(Q)SARs require data, which have to be acquired from various sources (e.g. published literature, experimental data, and qualitative surveys). Such data may be available at various levels of detail (along with appropriate metadata). In this regard, *nanoinformatics* efforts for nano-(Q)SAR development have promoted data organization and structuring via an appropriate data management system to allow data security, sharing, intelligent data queries, and integration utilities;
- The development of nano(Q)-SARs via *supervised learning* can be accomplished when the available data contains labels (i.e. known toxicity profiles and given ENM descriptors and experimental conditions such as cell lines, test organisms, assay types, exposure durations, and concentrations). ML algorithms are then applied to learn a best-fit function that accurately represents the relationships between ENM descriptors, experimental conditions, receptor characteristics, and the induced biological activity;

- The application of ML algorithms, as part of nano-(Q)SAR development, also includes *unsupervised ML learning* approaches. Such approaches can be utilized when: (i) the available data are unlabeled (i.e. the expected toxicity endpoints are unknown), or (ii) it is desired to discover relationships among the properties of ENMs and exposure attributes aiming at dimensionality reduction and *feature extraction/selection*. Unsupervised learning in nano-(Q)SAR development can also be useful for identifying groups or clusters of nanomaterials of similar attributes and also patterns of similar biological activities; and
- Nano-(Q)SARs can also be established via statistical learning approaches that extract from ENM toxicity data *conditional dependence relationships* (association rules) among ENM-induced bioactivities and ENM structural/physicochemical properties along with the experimental conditions.

Although there have been major strides forward in the development of nano-(Q)SARs, there remain challenges that need to be addressed in order to expand the range of applicability of nano-(Q)SARs as safe-by-design tools in nanotechnology. To achieve the above goal, efforts devoted to advancing the development of nano-(Q)SARs would benefit from addressing the following needs:

- Development of *shared databases* that provide detailed data on the physicochemical characteristics of ENMs and biological responses they induce at various exposure concentrations and exposure periods for a range of biological receptors from a single cell to whole organisms. Such databases should allow direct data utilization for model building and data exploration;
- *Data exploration* should always be taken as a first step to uncovering relationships among descriptors and biological response outcomes. In this manner, discovery and interpretability of relationships between model significant attributes and outcomes would serve to assess the utility of proposed nano-(Q)SARs and enhance their potential use as predictive tools;
- There is a need for nano-(Q)SARs that have the capability of revealing and/or supporting *improved mechanistic understanding of the response of biological receptors due to exposure to ENMs*. Nano-(Q)SARs could also be developed along strict adherence to response pathway analysis that incorporate fundamental domain knowledge;
- The reliability of developed nano(Q)SARs is likely to be enhanced through *considerations of data bias* by comparing data from different sources and assessing the impact of each data source (and associated experiments) on the resulting nano-(Q)SAR. The goal is to ensure that developed models have strong support from multiple data sources;
- As a standard practice, it is recommended that nano-(Q)SARs are developed and reported with a clearly stated *AD*. Such practice requires an understanding of the problem domain and the complete model development process (from managing data to model selection, tuning, and maintenance) to support model reproducibility and a clear *balance between model interpretability* (i.e. correlations of biological responses with ENMs' physicochemical properties over various exposure conditions) *and explainability* (i.e. strong causal analysis of the nano-(Q)SAR outcome with a given set of input attributes);
- A *repository of nano-(Q)SARs* would be most useful for rapid comparison of the performance of different nano-(Q)SAR-building approaches.

Acronyms

AI	Artificial Intelligence
AIC	Akaike Information Criterion
ANN	Artificial Neural Network
AOP	Adverse Outcome Pathway
APO	Apoptosis
ATP	Adenosine triphosphate
AUC	Area Under the Curve
BDFS	Backward Deletion Feature Selection
BEAS-2B	Human Bronchial Epithelial Cells
BIC	Bayesian Information Criteria
BN	Bayesian Network
CEIN	Center for Environmental Implications of Nanotechnology
CG	Coarse-Grained
CNT	Carbon Nanotube
CPT	Conditional Probability Table
CV	Cross-Validation
DAG	Directed Acyclic Graph
DLVO	Derjaguin-Landau-Verwey-Overbeek
DMIFS	Dynamic Mutual Information-based Feature Selection
DNA	Deoxyribonucleic Acid
E2F	E2 Transcription Factor
EC50	Effective Concentration
EHS	Environmental Health and Safety
EM	Expectation Maximization
ENM	Engineered Nanomaterial
EPR	Electron Paramagnetic Resonance
EU	European Union
FFFS	Feed Forward Feature Selection
FFSEM	Fuzzy Feature Selection Based on min–max Learning Rule and Extension Matrix
FRET	Fluorescence Resonance Energy Transfer
HTS	High Throughput Screening
IC	Inhibitory Concentration
IL-1	Interleukin-1
ISA-TAB	Investigation/Study/Assay (ISA) tab-delimited (TAB)
ISDD	In Vitro Sedimentation, Diffusion, and Dosimetry
KLS-FS	Kernel Least Squares Forward Selection
KNN	k-Nearest Neighbor
LC	Lethal Concentration
LCA	Lifecycle Assessment
LDA	Linear Discriminant Analysis
LDH	Lactic dehydrogenase

LS-SFM   Least Squares Support Feature Machine
MC   Monte Carlo
MD   Molecular Dynamics
MDS   Multidimensional Scaling
MINChar   Minimum Information on Nanoparticle Characterization
ML   Machine Learning
MRMR   Minimum Redundancy Maximum Relevance
NBI   Nanomaterial Biological Interactions
NCBO   National Center for Biomedical Ontology
NDR   NanoDatabank Repository
NIOSH   National Institute for Occupational Safety & Health
NLRP3   NOD-, LRR-, and pyrin domain-containing protein 3
NM   Nanomaterial
NP   Nanoparticle
NPO   Nanoparticle Ontology
OCHEM   Online Chemical Modeling Environment
OECD   Organization for Economic Co-operation and Development
OPS   Optimum Prediction Space
PCA   Principal Component Analysis
PDGF   Platelet Derived Growth Factor
PI   Propidium Iodide
PK   Pharmacokinetic
QD   Quantum Dot
QSAR   Quantitative Structure–Activity Relationship
QSPR   Quantitative Structure–Property Relationship
RF   Random Forest
RMSE   Root Mean Square Error
ROC   Receiver Operator Characteristics
SAR   Structure–Activity Relationship
SBS   Sequential Backward Selection
SDJ-FS   Spatial Distance Join (SDJ) based Feature Selection
SFFS   Sequential Forward Floating Selection
SFS   Sequential Forward Selection
SOM   Self-Organizing Maps
SSMD   Strictly Standardized Mean Difference
SVM   Support Vector Machine
TGF   Transforming Growth Factor
US   United States

References

Aguilera, P.A., Fernandez, A., Fernandez, R. et al. (2011). Bayesian networks in environmental modelling. *Environmental Modelling and Software* 36 (12): 1376–1388. https://doi.org/10.1016/j.envsoft.2011.06.004.

Aillon, K.L., Xie, Y., El-Gendy, N. et al. (2009). Effects of nanomaterial physicochemical properties on in vivo toxicity. *Advanced Drug Delivery Reviews* 61: 457–466. https://doi.org/10.1016/j.addr.2009.03.010.

Almuallim, H. and Dietterich, T.G. (1994). Learning Boolean concepts in the presence of many irrelevant features. *Artificial Intelligence* 69 (1): 279–305. http://dx.doi.org/10.1016/0004-3702(94)90084-1.

Alves, R., Rodriguez-Baena, D.S., and Aguilar-Ruiz, J.S. (2010). Gene association analysis: a survey of frequent pattern mining from gene expression data. *Briefings in Bioinformatics* 11 (2): 210–224. http://dx.doi.org/10.1093/bib/bbp042.

Amayri, O. and Bouguila, N. (2013). On online high-dimensional spherical data clustering and feature selection. *Engineering Applications of Artificial Intelligence* 26 (4): 1386–1398. http://dx.doi.org/10.1016/j.engappai.2012.10.009.

Asharani, P.V., Lianwu, Y., Gong, Z., and Valiyaveettil, S. (2011). Comparison of the toxicity of silver, gold and platinum nanoparticles in developing zebrafish embryos. *Nanotoxicology* 5: 43–54. https://doi.org/10.3109/17435390.2010.489207.

Auffan, M., Rose, J., Bottero, J.Y. et al. (2009). Towards a definition of inorganic nanoparticles from an environmental, health and safety perspective. *Nature Nanotechnology* https://doi.org/10.1038/nnano.2009.242.

Bao, G., Mitragotri, S., and Tong, S. (2013). Multifunctional nanoparticles for drug delivery and molecular imaging. *Annual Review of Biomedical Engineering* https://doi.org/10.1146/annurev-bioeng-071812-152409.

Bencsik, A. and Lestaevel, P. (2021). The challenges of 21st century neurotoxicology: the case of neurotoxicology applied to nanomaterials. *Frontiers in Toxicology.* https://doi.org/10.3389/ftox.2021.629256.

Berggren, K., Xia, Q., Likharev, K.K. et al. (2020). Roadmap on emerging hardware and technology for machine learning. *Nanotechnology* https://doi.org/10.1088/1361-6528/aba70f.

Bhattacharjee, S. and Brayden, D.J. (2015). Development of nanotoxicology: implications for drug delivery and medical devices. *Nanomedicine* https://doi.org/10.2217/nnm.15.69.

Bilal, M., Liu, H., Liu, R., and Cohen, Y. (2017). Bayesian network as a support tool for rapid query of the environmental multimedia distribution of nanomaterials. *Nanoscale* 9 (12): 4162–4174. https://doi.org/10.1039/C6NR08583K.

Bilal, M., Eunkeu, O., Liu, R. et al. (2019). Bayesian network resource for meta-analysis: cellular toxicity of quantum dots. *Small* 15 (34): 1900510. https://doi.org/10.1002/smll.201900510.

Bishop, C.M. (2006). *Pattern Recognition and Machine Learning.* New York: Springer.

Blum, A.L. and Langley, P. (1997). Selection of relevant features and examples in machine learning. *Artificial Intelligence* 97 (1–2): 245–271. https://doi.org/10.1016/S0004-3702(97)00063-5.

Bosetti, R. and Vereeck, L. (2011). Future of nanomedicine: obstacles and remedies. *Nanomedicine* https://doi.org/10.2217/nnm.11.55.

Braydich-Stolle, L.K., Schaeublin, N.M., Murdock, R.C. et al. (2009). Crystal structure mediates mode of cell death in TiO_2 nanotoxicity. *Journal of Nanoparticle Research* 11 (6): 1361–1374. https://doi.org/10.1007/s11051-008-9523-8.

Breiman, L. (2001). Random forests. *Machine Learning* 45 (1): 5–32. https://doi.org/10.1023/A:1010933404324.

Bullinaria, J.A. (2004). Self organizing maps: fundamentals. http://www.cs.bham.ac.uk/~jxb/NN/l16.pdf (accessed 24 July 2022).

Cattaneo, A.G., Gornati, R., Sabbioni, E. et al. (2010). Nanotechnology and human health: risks and benefits. *Journal of Applied Toxicology* 30 (8): 730–744. https://doi.org/10.1002/jat.1609.

CEIN 2020. CEIN nanomaterials data repository. https://nanoinfo.org/nanodatabank (accessed 17 April 2021).

Chakrabarty, A., Mannan, S., and Cagin, T. (2016). Chapter 3 – molecular-level modeling and simulation in process safety. In: *Multiscale Modeling for Process Safety Applications Cagin* (ed. A. Chakrabarty, S. Mannan and B.T. Tahir), 111–210. Boston: Butterworth-Heinemann https://doi.org/10.1016/B978-0-12-396975-0.00003-6.

Chan, H. and Darwiche, A. (2004). Sensitivity analysis in bayesian networks: from single to multiple parameters. *Proceedings of the 20th Conference on Uncertainty in Artificial Intelligence*, (7–11 July 2004), 67–75. UAI '04. Banff Canada: AUAI Press. http://dl.acm.org/citation.cfm?id=1036843.1036852.

Chang, X., Zhang, Y., Tang, M., and Wang, B. (2013). Health effects of exposure to nano-TiO_2: a meta-analysis of experimental studies. *Nanoscale Research Letters* 8 (1): 51. https://doi.org/10.1186/1556-276X-8-51.

Chau, Y.T. and Yap, C.W. (2012). Quantitative nanostructure-activity relationship modelling of nanoparticles. *RSC Advances* 2 (22): 8489–8496. https://doi.org/10.1039/C2RA21489J.

Cheng, L.-C., Jiang, X., Wang, J. et al. (2013). Nano-bio effects: interaction of nanomaterials with cells. *Nanoscale* 5 (9): 3547–3569. https://doi.org/10.1039/C3NR34276J.

Chon, T.S. (2011). Self-organizing maps applied to ecological sciences. *Ecological Informatics* 6 (1): 50–61. https://doi.org/10.1016/j.ecoinf.2010.11.002.

Cohen, Y., Rallo, R., Liu, R., and Liu, H.H. (2013). In silico analysis of nanomaterials hazard and risk. *Accounts of Chemical Research* 46 (3): 802–812. https://doi.org/10.1021/ar300049e.

Consumer Product Safety Commission, NNI (2016). Quantifying exposure to engineered nanomaterials (QEEN) from manufactured products. http://www.nano.gov/sites/default/files/pub_resource/qeen_workshop_report_2016.pdf. (accessed 13 June 2022).

CPI (2016). Consumer products inventory an inventory of nanotechnology-based consumer products introduced on the market. http://www.nanotechproject.org/inventories/consumer. (accessed 22 January 2020.

Cristianini, N. and Shawe-Taylor, J. (2000). *An Introduction to Support Vector Machines and Other Kernel-Based Learning Methods*. Cambridge University Press https://doi.org/10.1017/cbo9780511801389.

Czermiński, R., Yasri, A., and Hartsough, D. (2001). Use of support vector machine in pattern classification: application to QSAR studies. *Quantitative Structure-Activity Relationships* 20 (3): 227–240. https://doi.org/10.1002/1521-3838(200110)20:3<227::AID-QSAR227>3.0.CO;2-Y.

Damoiseaux, R., George, S., Li, M. et al. (2011). No time to lose-high throughput screening to assess nanomaterial safety. *Nanoscale* 3 (4): 1345–1360. https://doi.org/10.1039/C0NR00618A.

Das, S. (2001). Filters, wrappers and a boosting-based hybrid for feature selection. *ICML* 1: 74–81. Citeseer.

Dash, M. and Gopalkrishnan, V. (2008). Distance based feature selection for clustering microarray data. In: *Database Systems for Advanced Applications* (ed. J.R. Haritsa, R. Kotagiri and V. Pudi), 512–519. Heidelberg: Springer https://doi.org/10.1007/978-3-540-78568-2_41.

Dash, M. and Liu, H. (2003). Consistency-based search in feature selection. *Artificial Intelligence* 151 (1–2): 155–176. http://dx.doi.org/10.1016/S0004-3702(03)00079-1.

De Sousa, E.P.M., Caetano Traina, A.J.M., Traina, L.W., and Faloutsos, C. (2007). A fast and effective method to find correlations among attributes in databases. *Data Mining and Knowledge Discovery* 14 (3): 367–407. https://doi.org/10.1007/s10618-006-0056-4.

Ding, C. and Peng, H. (2005). Minimum redundancy feature selection from microarray gene expression data. *Journal of Bioinformatics and Computational Biology* 03 (02): 185–205. https://doi.org/10.1142/S0219720005001004.

Directorate, E., Joint Meeting, O.F., The, C.C. et al. (2007). Guidance document on the validation of (quantitative) structure-activity relationship [(Q)Sar] models. *Transport*.

Doganis, P., Fadeel, B., Grafström, R., et al. (2015). Deliverable report D3.1 technical specification and initial implementation of the protocol and data management web services. http://www.enanomapper.net/deliverables/d3/150131eNanoMapper-D3.1-IDEA-2015012701.pdf. (accessed 14 June 2022).

Duan, J., Yongbo, Y., Shi, H. et al. (2013). Toxic effects of silica nanoparticles on zebrafish embryos and larvae. Edited by Vishal Shah. *PLoS One* 8 (9): e74606. https://doi.org/10.1371/journal.pone.0074606.

Ehrenberg, M.S., Friedman, A.E., Finkelstein, J.N. et al. (2009). The influence of protein adsorption on nanoparticle association with cultured endothelial cells. *Biomaterials* 30: 603–610. https://doi.org/10.1016/j.biomaterials.2008.09.050.

El-Hack, M.E., Abd, M.A., Farag, M.R. et al. (2017). Nutritional and pharmaceutical applications of nanotechnology: trends and advances. *International Journal of Pharmacology* https://doi.org/10.3923/ijp.2017.340.350.

ENanoMapper (2016). ENanoMapper. http://www.enanomapper.net/about. (accessed 14 June 2022).

EU Science HUD (2019). NANOhub - nanomaterials data, IUCLID 5.6 files. https://data.jrc.ec.europa.eu/collection/id-00253. (accessed 14 June 2022).

Fourches, D., Dongqiuye, P., Tassa, C. et al. (2010a). Quantitative nanostructure – activity relationship modeling. *ACS Nano* https://doi.org/10.1021/nn1013484.

Fourches, D., Dongqiuye, P., Tassa, C. et al. (2010b). Quantitative anostructure – activity relationship modeling. *ACS Nano* 4 (10): 5703–5712. https://doi.org/10.1021/nn1013484.

Fourches, D., Dongqiuye, P., Tassa, C. et al. (2010c). Quantitative nanostructure – activity relationship (QNAR) modeling. *ACS Nano* 4 (10): 5703–5712. https://doi.org/10.1021/nn1013484.

Garner, K.L., Suh, S., and Keller, A.A. (2017). Assessing the risk of engineered nanomaterials in the environment: development and application of the nano fate model. *Environmental Science and Technology* https://doi.org/10.1021/acs.est.6b05279.

Gernand, J.M. and Casman, E.A. (2014). A meta-analysis of carbon nanotube pulmonary toxicity studies—how physical dimensions and impurities affect the toxicity of carbon nanotubes. *Risk Analysis* 34 (3): 583–597. https://doi.org/10.1111/risa.12109.

Gesellschaft für Chemische Technik und Biotechnologie (2014). Dechema. https://dechema.de/en. (accessed 14 June 2022).

Ghorbanzadeh, M., Fatemi, M.H., and Karimpour, M. (2012). Modeling the cellular uptake of magnetofluorescent nanoparticles in pancreatic cancer cells: a quantitative structure activity relationship study. *Industrial & Engineering Chemistry Research* 51 (32): 10712–10718. https://doi.org/10.1021/ie3006947.

Giralt, F., Espinosa, G., Arenas, A. et al. (2004). Estimation of infinite dilution activity coefficients of organic compounds in water with neural classifiers. *AIChE Journal* 50 (6): 1315–1343. https://doi.org/10.1002/aic.10116.

Global Industry Analysts (2021). Global nanotechnology market to reach $70.7 billion by 2026. https://www.reportlinker.com/p0326269/Global-Nanotechnology-Industry.html?utm_source=GNW (accessed 24 September 2021).

Guyon, I. and Elisseeff, A. (2003). An introduction to variable and feature selection. *Journal of Machine Learning Research* 3: 1157–1182. https://doi.org/10.1162/153244303322753616.

Guzzi, P.H., Milano, M., and Cannataro, M. (2014). Mining association rules from gene ontology and protein networks: promises and challenges. *Procedia Computer Science* 29: 1970–1980. http://dx.doi.org/10.1016/j.procs.2014.05.181.

Ha, M.K., Trinh, T.X., Choi, J.S. et al. (2018). Toxicity classification of oxide nanomaterials: effects of data gap filling and PChem score-based screening approaches. *Scientific Reports* 8 (1): 3141. https://doi.org/10.1038/s41598-018-21431-9.

Haase, A., Tentschert, J., and Luch, A. (2012). Nanomaterials: a Ccallenge for toxicological risk assessment? *Experientia. Supplementum* 101: 219–250. https://doi.org/10.1007/978-3-7643-8340-4_8.

Halder, A.K. and Cordeiro, M.N.D.S. (2021). QSAR-co-X: an open source toolkit for multitarget QSAR modelling. *Journal of Cheminformatics* 13 (1): 29. https://doi.org/10.1186/s13321-021-00508-0.

Hall, M.A. (2000). Correlation-Based Feature Selection for Discrete and Numeric Class Machine Learning. *Proceedings of the Seventeenth International Conference on Machine Learning*, (29 June 2000–2 July 2000) 359–66. ICML '00. San Francisco, CA, USA.: Morgan Kaufmann Publishers Inc.

Han, J., Kamber, M., and Pei, J. (2012). *Data Mining: Concepts and Techniques*. Morgan Kaufmann https://doi.org/10.1016/C2009-0-61819-5.

Hendren, C.O., Powers, C.M., Hoover, M.D., and Harper, S.L. (2015). The nanomaterial data curation initiative: a collaborative approach to assessing, evaluating, and advancing the state of the field. *Beilstein Journal of Nanotechnology* 6: 1752–1762. https://doi.org/doi:10.3762/bjnano.6.179.

Herrera-Ibatá, D.M. (2021). Machine learning and perturbation theory machine learning (PTML) in medicinal chemistry, biotechnology, and nanotechnology. *Current Topics in Medicinal Chemistry* https://doi.org/10.2174/1568026621666210121153413.

http://NanoHUB.org (2016). NanoBIO node. https://nanohub.org/groups/nanobio (accessed 14 June 2022).

http://Nanoinfo.org (2018) NanoDatabank. https://nanoinfo.org/nanodatabank. (accessed 13 October 2020).

http://Re3data.org (2016). Nanomaterial registry. https://www.re3data.org/repository/r3d100011129 (accessed 15 June 2022).

Hua, S., de Matos, M.B.C., Metselaar, J.M., and Storm, G. (2018). Current trends and challenges in the clinical translation of anoparticulate nanomedicines: pathways for translational development and commercialization. *Frontiers in Pharmacology* https://doi.org/10.3389/fphar.2018.00790.

Huang, D. and Chow, T.W.S. (2005). Efficiently searching the important input variables using bayesian discriminant. *IEEE Transactions on Circuits and Systems I: Regular Papers* https://doi.org/10.1109/TCSI.2005.844364.

Jain, A.K., Duin, R.P.W., and Mao, J. (2000). Statistical pattern recognition: a review. *IEEE Transactions on Pattern Analysis and Machine Intelligence* https://doi.org/10.1109/34.824819.

Jebara, T. and Jaakkola, T. (2000). Feature selection and dualities in maximum entropy discrimination. *Proceedings of the Sixteenth Conference on Uncertainty in Artificial Intelligence*, 291–300. San Francisco. (30 June 2000–3 July 2000). Morgan Kaufmann Publishers Inc.

Jensen, F.V. and Nielsen, T.D. (2007). Bayesian network and decision graph. *The Knowledge Engineering Review*. 19: https://doi.org/10.1007/978-0-387-68282-2.

John, G.H, Kohavi, R., Pfleger, K. (1994). Irrelevant features and the subset selection problem. *Machine Learning: Proceedings of the Eleventh International Conference*, (10–13 July 1994) 121–29. New Brunswick, NJ: Rutgers University.

Jonathan, P., Krzanowski, W.J., and McCarthy, W.V. (2000). On the use of cross-validation to assess performance in multivariate prediction. *Statistics and Computing* https://doi.org/10.1023/A:1008987426876.

Joshi, P.N. (2016). Green chemistry for nanotechnology: opportunities and future challenges. *Research & Reviews: Journal of Chemistry* 5 (1): 3–4.

Judson, P.N. (1994). Rule induction for systems predicting biological activity. *Journal of Chemical Information and Computer Sciences* 34 (1): 148–153. https://doi.org/10.1021/ci00017a018.

Karcher, S.C., Harper, B.J., Harper, S.L. et al. (2016). Visualization tool for correlating nanomaterial properties and biological responses in zebrafish. *Environmental Science: Nano* 3 (6): 1280–1292. https://doi.org/10.1039/C6EN00273K.

Karcher, S., Willighagen, E.L., Rumble, J. et al. (2018). Integration among databases and data sets to support productive nanotechnology: challenges and recommendations. *NanoImpact* 9: 85–101. https://doi.org/10.1016/j.impact.2017.11.002.

Kaweeteerawat, C., Ivask, A., Liu, R. et al. (2015). Toxicity of metal oxide nanoparticles in Escherichia Coli correlates with conduction band and hydration energies. *Environmental Science & Technology* 49 (2): 1105–1112. https://doi.org/10.1021/es504259s.

King-Heiden, T.C., Wiecinski, P.N., Mangham, A.N. et al. (2009). Quantum dot Nanotoxicity assessment using the zebrafish embryo. *Environmental Science & Technology* 43: 1605–1611. https://doi.org/10.1021/es801925c.

Klaessig, H. (2018). EU US roadmap Nanoinformatics 2030. http://doi.org/10.5281/zenodo.1486012 (accessed 25 July 2022).

Kleandrova, V.V., Luan, F., González-Díaz, H. et al. (2014). Computational ecotoxicology: simultaneous prediction of Ecotoxic effects of nanoparticles under different experimental conditions. *Environment International* https://doi.org/10.1016/j.envint.2014.08.009.

Kohavi, R. (1995). A study of cross-validation and bootstrap for accuracy estimation and model selection. *International Joint Conference on Artificial Intelligence* 14 (12): 1137–1143. https://doi.org/10.1067/mod.2000.109031.

Kohavi, R. and John, G.H. (1997). Wrappers for feature subset selection. *Artificial Intelligence* 97 (1–2): 273–324. http://dx.doi.org/10.1016/S0004-3702(97)00043-X.

Kong, L., Tuomela, S., Hahne, L. et al. (2013). NanoMiner — integrative human transcriptomics data resource for nanoparticle research. *PLoS One* 8 (7): e68414. https://doi.org/10.1371/journal.pone.0068414.

Korn, F., Pagel, B.U., and Faloutsos, C. (2001). On the "dimensionality curse" and the "self-similarity blessing". *IEEE Transactions on Knowledge and Data Engineering* https://doi.org/10.1109/69.908983.

Krewski, D., Acosta, D., Andersen, M. et al. (2010). Toxicity testing in the 21st century: a vision and a strategy. *Journal of Toxicology and Environmental Health. Part B, Critical Reviews* 13: 51–138. https://doi.org/10.1080/10937404.2010.483176.

Li, Y. and Wu, Z.-F. (2008). Fuzzy feature selection based on min–max learning rule and extension matrix. *Pattern Recognition* 41 (1): 217–226. http://dx.doi.org/10.1016/j.patcog.2007.06.007.

Li, J., Chen, Z., Wei, L. et al. (2007). Feature selection via least squares support feature machine. *International Journal of Information Technology and Decision Making* 06 (04): 671–686. https://doi.org/10.1142/S0219622007002733.

Liaw, A. and Wiener, M. (2002). Classification and regression by RandomForest. *R News* 2: 18–22.

Libotean, D., Giralt, J., Giralt, F. et al. (2009). Neural network approach for Modeling the performance of reverse osmosis membrane desalting. *Journal of Membrane Science* 326 (2): 408–419. http://dx.doi.org/10.1016/j.memsci.2008.10.028.

Liu, R. and Cohen, Y. (2015). Nanoinformatics for environmental health and biomedicine. *Beilstein Journal of Nanotechnology* 6 (December): 2449–2451. https://doi.org/10.3762/bjnano.6.253.

Liu, H. and Motoda, H. (2008). Chapter 10: Local feature selection for classification. In: *Computational Methods of Feature Selection* (ed. H. Liu and H. Motoda). New York: Chapman and Hall/CRC.

Liu, H. and Setiono, R. (1996). A probabilistic approach to feature selection-a filter solution. *ICML* 96: 319–327. Citeseer.

Liu, R. and Shi, Y. (2013). Spatial distance join based feature selection. *Engineering Applications of Artificial Intelligence* 26 (10): 2597–2607. http://dx.doi.org/10.1016/j.engappai.2013.08.016.

Liu, H. and Yu, L. (2005). Toward integrating feature selection algorithms for classification and clustering. *IEEE Transactions on Knowledge and Data Engineering* 17: 491–502. https://doi.org/10.1109/TKDE.2005.66.

Liu, H., Sun, J., Liu, L., and Zhang, H. (2009). Feature selection with dynamic mutual information. *Pattern Recognition* 42: 1330–1339. https://doi.org/10.1016/j.patcog.2008.10.028.

Liu, R., Rallo, R., and Cohen, Y. (2011a). Unsupervised feature selection using incremental least squares. *International Journal of Information Technology and Decision Making* 10 (06): 967–987. https://doi.org/10.1142/S0219622011004671.

Liu, R., Rallo, R., George, S. et al. (2011b). Classification NanoSAR development for cytotoxicity of metal oxide nanoparticles. *Small* 7: 1118–1126. https://doi.org/10.1002/smll.201002366.

Liu, R., Rallo, R., and Cohen, Y. (2013a). Quantitative structure-activity-relationships for cellular uptake of nanoparticles. *2013 13th IEEE International Conference on Nanotechnology (IEEE-NANO 2013)*. Beijing, China. (05–08 August 2013). IEEE. https://doi.org/10.1109/NANO.2013.6720861.

Liu, R., Rallo, R., Weissleder, R. et al. (2013c). Nano-SAR development for bioactivity of nanoparticles with considerations of decision boundaries. *Small* 9 (9–10): 1842–1852. https://doi.org/10.1002/smll.201201903.

Liu, R., Zhang, H.Y., Ji, Z.X. et al. (2013d). Development of structure-activity relationship for metal oxide nanoparticles. *Nanoscale* 5 (12): 5644–5653. https://doi.org/10.1039/C3NR01533E.

Liu, X., Tang, K., Harper, S. et al. (2013e). Predictive modeling of nanomaterial exposure effects in biological systems. *International Journal of Nanomedicine* 8 (Suppl 1): 31–43. https://doi.org/10.2147/IJN.S40742.

Liu, R., France, B., George, S. et al. (2014). Association rule Mining of Cellular Responses Induced by metal and metal oxide nanoparticles. *Analyst* 139 (5): 943–953. https://doi.org/10.1039/C3AN01409F.

Loughrey, J. and Cunningham, P. (2005). Overfitting in wrapper-based feature subset selection: The harder you try the worse it gets. In: *Research and Development in Intelligent Systems XXI. SGAI 2004* (ed. M. Bramer, F. Coenen and T. Allen), 33–43. London: Springer London.

Mallik, S., Mukhopadhyay, A., Maulik, U., and Bandyopadhyay, S. (2013). Integrated analysis of gene expression and genome-wide DNA methylation for tumor prediction: an association rule mining-based approach. *2013 IEEE Symposium on Computational Intelligence in Bioinformatics and Computational Biology (CIBCB)*. Singapore (16–19 April 2013). IEEE. https://doi.org/10.1109/CIBCB.2013.6595397.

Martinez, R., Pasquier, N., and Pasquier, C. (2008). GenMiner: mining non-redundant association rules from integrated gene expression data and annotations. *Bioinformatics* 24 (22): 2643–2644. http://dx.doi.org/10.1093/bioinformatics/btn490.

Martínez-Romero, M., Jonquet, C., O'Connor, M.J. et al. (2017). NCBO ontology recommender 2.0: an enhanced approach for biomedical ontology recommendation. *Journal of Biomedical Semantics*. https://doi.org/10.1186/s13326-017-0128-y.

Marvin, H.J.P., Bouzembrak, Y., Janssen, E.M. et al. (2017). Application of Bayesian networks for Hazard ranking of nanomaterials to support human health risk assessment. *Nanotoxicology* 11 (1): 123–133. https://doi.org/10.1080/17435390.2016.1278181.

Mballo, C. and Makarenkov, V. (2010). Using machine learning methods to predict experimental high-throughput screening data. *Combinatorial Chemistry & High Throughput Screening* 13: 430 441. https://doi.org/BSP/CCHTS/E-Pub/00064 [pii].

Medlock, B.W. (2008). Investigating classification for natural language processing tasks. https://www.cl.cam.ac.uk/techreports/UCAM-CL-TR-721.pdf (accessed 21 November 2021).

van der Merwe, D. and Pickrell, J.A. (2018). Toxicity of nanomaterials. In: *Veterinary Toxicology: Basic and Clinical Principles*, 3e (ed. R.C. Gupta). Hopkinsville, KY: United States: Elsevier Inc https://doi.org/10.1016/B978-0-12-811410-0.00018-0.

Michalet, X., Pinaud, F.F., Bentolila, L.A. et al. (2005). Quantum dots for live cells, in vivo imaging, and diagnostics. *Science* https://doi.org/10.1126/science.1104274.

Modrzejewski, M. (1993). Feature selection using rough sets theory. *Machine Learning: ECML-93*, 213–26. (5–7 April 1993). Veinna, Austria: Springer.

Money, E.S., Reckhow, K.H., and Wiesner, M.R. (2012). The use of Bayesian networks for nanoparticle risk forecasting: model formulation and baseline evaluation. *Science of the Total Environment* 426: 436–445. https://doi.org/10.1016/j.scitotenv.2012.03.064.

Money, E.S., Barton, L.E., Dawson, J. et al. (2014). Validation and sensitivity of the FINE Bayesian network for forecasting aquatic exposure to Nano-silver. *Science of the Total Environment* 473–474: 685–691. https://doi.org/10.1016/j.scitotenv.2013.12.100.

Morris, S.A., Gaheen, S., Lijowski, M. et al. (2014). CaNanoLab: a nanomaterial data repository for biomedical research. *Proceedings - 2014 IEEE International Conference on Bioinformatics and Biomedicine, IEEE BIBM 2014*. (02–05 November 2014). Belfast, United Kingdom. https://doi.org/10.1109/BIBM.2014.6999371.

Morris, S.A., Gaheen, S., Lijowski, M. et al. (2015). Experiences in supporting the structured collection of cancer nanotechnology data using CaNanoLab. *Beilstein Journal of Nanotechnology* 6 (July): 1580–1593. https://doi.org/10.3762/bjnano.6.161.

Muratov, E.N., Bajorath, J., Sheridan, R.P. et al. (2020). QSAR without borders. *Chemical Society Reviews* https://doi.org/10.1039/d0cs00098a.

Murphy, F., Sheehan, B., Mullins, M. et al. (2016). A tractable method for measuring nanomaterial risk using Bayesian networks. *Nanoscale Research Letters* 11 (1): 503. https://doi.org/10.1186/s11671-016-1724-y.

Nafar, Z. and Golshani, A. (2006). Data mining methods for protein-protein interactions. *2006 Canadian Conference on Electrical and Computer Engineering*. Ottawa, ON, Canada. (07–10 May 2006). IEEE. https://doi.org/10.1109/CCECE.2006.277746.

Najafi-Hajivar, S., Zakeri-Milani, P., Mohammadi, H. et al. (2016). Overview on experimental models of interactions between nanoparticles and the immune system. *Biomedicine and Pharmacotherapy* https://doi.org/10.1016/j.biopha.2016.08.060.

National Cancer Institute (2016). CaNanoLab. https://cananolab.nci.nih.gov/caNanoLab#/ (accessed 14 June 2022).

National Institute for Public Health and the Environment (2016). Centre for safety of substances and products. http://www.rivm.nl/en/About_RIVM/Organisation/Centres/Centre_for_Safety_of_Substances_and_Products (accessed 14 June 2022).

Naulaerts, S., Meysman, P., Bittremieux, W. et al. (2015). A primer to frequent Itemset mining for bioinformatics. *Briefings in Bioinformatics* 16 (2): 216–231. https://doi.org/10.1093/bib/bbt074.

NBI Knowledgebase (2022). NBI Knowledgebase. https://nbi.oregonstate.edu (accessed 15 June 2022).

Neapolitan, R.E. (2003). Learning Bayesian networks. *Molecular Biology* 6 (2): 674. http://www.amazon.com/Learning-Bayesian-Networks-Richard-Neapolitan/dp/0130125342.

Nel, A.E., Nasser, E., Godwin, H. et al. (2013a). A multi-stakeholder perspective on the use of alternative test strategies for nanomaterial safety assessment. *ACS Nano* 6422–6433. https://doi.org/10.1021/nn4037927.

Nel, A., Xia, T., Meng, H. et al. (2013b). Nanomaterial toxicity testing in the 21st century: use of a predictive toxicological approach and high-throughput screening. *Accounts of Chemical Research* 46 (3): 607–621. https://doi.org/10.1021/ar300022h.

Nikota, J., Williams, A., Yauk, C.L. et al. (2015). Meta-analysis of transcriptomic responses as a means to identify pulmonary disease outcomes for engineered nanomaterials. *Particle and Fibre Toxicology* 13 (May): 25. https://doi.org/10.1186/s12989-016-0137-5.

NIOSH (2016). Nanoparticle information library. http://nanoparticlelibrary.net (accessed 15 June 2022).

OECD (2007). *Guidance Document on the Validation of (Quantitative) Structure-Activity Relationship [(Q)Sar] Models*. OECD Transport.

Oellrich, A., Jacobsen, J., Papatheodorou, I., and Smedley, D. (2014). Using association rule mining to determine promising secondary phenotyping hypotheses. *Bioinformatics* 30 (12): i52–i59. http://dx.doi.org/10.1093/bioinformatics/btu260.

Oh, E., Liu, R., Nel, A. et al. (2016). Meta-analysis of cellular toxicity for cadmium-containing quantum dots. *Nature Nanotechnology* https://doi.org/10.1038/nnano.2015.338.

Okuda-Shimazaki, J., Takaku, S., Kanehira, K. et al. (2010). Effects of titanium dioxide nanoparticle aggregate size on gene expression. *International Journal of Molecular Sciences* 11: 2383–2392. https://doi.org/10.3390/ijms11062383.

Orimoto, Y., Watanabe, K., Yamashita, K. et al. (2012). Application of artificial neural networks to rapid data analysis in combinatorial nanoparticle syntheses. *The Journal of Physical Chemistry C* 116 (33): 17885–17896. https://doi.org/10.1021/jp3031122.

Pagel, B.U., Korn, F., and Faloutsos, C. (2000). Deflating the dimensionality curse using multiple fractal dimensions. *Data Engineering, 2000. Proceedings. 16th International Conference On*. San Diego, CA, USA. (29 February 2000–3 March 2000). IEEE. https://doi.org/10.1109/ICDE.2000.839457.

Panneerselvam, S. and Choi, S. (2014). Nanoinformatics: emerging databases and available tools. *International Journal of Molecular Sciences* https://doi.org/10.3390/ijms15057158.

Park, S.H., Reyes, J.A., Gilbert, D.R. et al. (2009). Prediction of protein-protein interaction types using association rule based classification. *BMC Bioinformatics* 10 (1): 36. https://doi.org/10.1186/1471-2105-10-36.

Pears, R., Finlay, J., and Connor, A.M. (2014). Synthetic minority over-sampling TEchnique (SMOTE) for predicting software build outcomes. *ArXiv Preprint ArXiv* 1407: 2330. http://arxiv.org/abs/1407.2330.

Petryayeva, E., Russ Algar, W., and Medintz, I.L. (2013). Quantum dots in bioanalysis: a review of applications across various platforms for fluorescence spectroscopy and imaging. *Applied Spectroscopy* https://doi.org/10.1366/12-06948.

Pham, D.T. and Ruz, G.A. (2009). Unsupervised training of Bayesian networks for data clustering. *Proceedings of the Royal Society A: Mathematical, Physical and Engineering Science* 465 (2109): 2927 LP–2948. http://rspa.royalsocietypublishing.org/content/465/2109/2927.abstract.

Ponce, A.G., Ayala-Zavala, J.F., Marcovich, N.E. et al. (2018). Nanotechnology trends in the food industry: recent developments, risks, and regulation. *In Impact of Nanoscience in the Food Industry*. https://doi.org/10.1016/B978-0-12-811441-4.00005-4.

Ponzoni, I., Sebastián-Pérez, V, Requena-Triguero, C. et al. (2017). Hybridizing feature selection and feature learning approaches in QSAR Modeling for drug discovery. *Scientific Reports* 7 (1): 2403. https://doi.org/10.1038/s41598-017-02114-3.

Pudil, P. and Hovovicova, J. (1998). Novel methods for subset selection with respect to problem knowledge. *IEEE Intelligent Systems and Their Applications*. https://doi.org/10.1109/5254.671094.

Puzyn, T., Leszczynski, J., and Cronin, M.T. (2010). *Recent Advances in QSAR Studies: Methods and Applications*. New York: Springer.

Puzyn, T., Rasulev, B., Gajewicz, A. et al. (2011). Using Nano-QSAR to predict the cytotoxicity of metal oxide nanoparticles. *Nature Nanotechnology* https://doi.org/10.1038/nnano.2011.10.

Rallo, R., Espinosa, G., and Giralt, F. (2005). Using an ensemble of neural based QSARs for the prediction of toxicological properties of chemical contaminants. *Process Safety and Environmental Protection* 83 (4): 387–392. https://doi.org/10.1205/psep.04389.

Rallo, R., France, B., Liu, R. et al. (2011). Self-organizing map analysis of toxicity-related cell Signaling pathways for metal and metal oxide nanoparticles. *Environmental Science & Technology* 45 (4): 1695–1702. https://doi.org/10.1021/es103606x.

Ravi Kanth, K.V., Agrawal, D., and Singh, A. (1998). Dimensionality reduction for similarity searching in dynamic databases. *ACM SIGMOD Record* 27: 166–176.

Robinson, M., Richard, L., Lynch, I. et al. (2016). How should the completeness and quality of curated nanomaterial data be evaluated? *Nanoscale* 8 (19): 9919–9943. https://doi.org/10.1039/c5nr08944a.

Robnik-Šikonja, M. and Kononenko, I. (n.d.). Theoretical and empirical analysis of ReliefF and RReliefF. *Machine Learning* 53 (1): 23–69. https://doi.org/10.1023/A:1025667309714.

Roco, M.C. (2011). The long view of nanotechnology development: the National Nanotechnology Initiative at 10 years. *Journal of Nanoparticle Research* https://doi.org/10.1007/s11051-010-0192-z.

Rosenthal, S.J., Chang, J.C., Kovtun, O. et al. (2011). Biocompatible quantum dots for biological applications. *Chemistry and Biology* https://doi.org/10.1016/j.chembiol.2010.11.013.

Roweis, S.T. and Saul, L.K. (2000). Nonlinear dimensionality reduction by locally linear embedding. *Science* 290 (5500): 2323–2326. https://doi.org/10.1126/science.290.5500.2323.

Rozza, A., Lombardi, G., Ceruti, C. et al. (2012). Novel high intrinsic dimensionality estimators. *Machine Learning* 89 (1): 37–65. https://doi.org/10.1007/s10994-012-5294-7.

Rücker, C., Rücker, G., and Meringer, M. (2007). Y-randomization and its variants in QSPR/QSAR. *Journal of Chemical Information and Modeling* 47 (6): 2345–2357. https://doi.org/10.1021/ci700157b.

Russel, S. and Norvig, P. (2012). *Artificial Intelligence—a Modern Approach*, 3e. Prentice Hall The Knowledge Engineering Review. https://doi.org/10.1017/S0269888900007724.

Sahoo, M., Vishwakarma, S., Panigrahi, C., and Kumar, J. (2021). Nanotechnology: current applications and future scope in food. *Food Frontiers*. https://doi.org/10.1002/fft2.58.

Sansone, S.-A., Rocca-Serra, P., Field, D. et al. (2012). Toward interoperable bioscience data. *Nature Genetics* 44 (January): 121.

Sayes, C. and Ivanov, I. (2010). Comparative study of predictive computational models for nanoparticle-induced cytotoxicity. *Risk Analysis* 30: 1723–1734. https://doi.org/10.1111/j.1539-6924.2010.01438.x.

Schowe, B. and Morik, K. (2011). Fast-ensembles of minimum redundancy feature selection. In: *Ensembles in Machine Learning Applications* (ed. O. Okun, G. Valentini and M. Re), 75–95. Berlin, Heidelberg: Springer Berlin Heidelberg https://doi.org/10.1007/978-3-642-22910-7_5.

Shao, C.-Y., Chen, S.-Z., Bo-Han, S. et al. (2013). Dependence of QSAR models on the selection of trial descriptor sets: a demonstration using Nanotoxicity endpoints of decorated nanotubes. *Journal of Chemical Information and Modeling* 53: 142–158. https://doi.org/10.1021/ci3005308.

Shaw, S.Y., Westly, E.C., Pittet, M.J. et al. (2008a). Perturbational profiling of nanomaterial biologic activity. *Proceedings of the National Academy of Sciences of the United States of America* 105 (21): 7387–7392. https://doi.org/10.1073/pnas.0802878105.

Shin, H.K., Seo, M., Shin, S.E. et al. (2018). Meta-analysis of daphnia magna Nanotoxicity experiments in accordance with test guidelines. *Environmental Science: Nano* 5 (3): 765–775. https://doi.org/10.1039/C7EN01127J.

Siddiquee, S., Melvin, G.J.H., and Rahman, M.M. (2019). *Nanotechnology: Applications in Energy, Drug and Food*. Nanotechnology: Applications in Energy, Drug and Food https://doi.org/10.1007/978-3-319-99602-8.

Stahura, F.L. and Bajorath, J. (2004). Virtual screening methods that complement HTS. *Combinatorial Chemistry & High Throughput Screening* 7: 259–269.

Svetnik, V., Liaw, A., Christopher Tong, J. et al. (2003). Random forest: a classification and regression tool for compound classification and QSAR modeling. *Journal of Chemical Information and Computer Sciences* https://doi.org/10.1021/ci034160g.

Tamayo, P., Slonim, D., Mesirov, J. et al. (1999). Interpreting patterns of gene expression with self-organizing maps: methods and application to hematopoietic differentiation. *Proceedings of the National Academy of Sciences of the United States of America* 96 (6): 2907–2912. https://doi.org/10.1073/pnas.96.6.2907.

Tenenbaum, J.B., de Silva, V., and Langford, J.C. (2000). A global geometric framework for nonlinear dimensionality reduction. *Science* 290 (5500): 2319–2323. https://doi.org/10.1126/science.290.5500.2319.

Thiruvengadam, M., Rajakumar, G., and Chung, I.M. (2018). Nanotechnology: current uses and future applications in the food industry. *3 Biotech* https://doi.org/10.1007/s13205-018-1104-7.

Thomas, D.G., Pappu, R.V., and Baker, N.A. (2011). NanoParticle ontology for Cancer nanotechnology research. *Journal of Biomedical Informatics* https://doi.org/10.1016/j.jbi.2010.03.001.

Thomas, D.G., Gaheen, S., Harper, S.L. et al. (2013). ISA-TAB-Nano: a specification for sharing nanomaterial research data in spreadsheet-based format. *BMC Biotechnology* 13 (1): 2. https://doi.org/10.1186/1472-6750-13-2.

Thorek, D.L.J., Chen, A.K., Czupryna, J., and Tsourkas, A. (2006). Superparamagnetic Iron oxide nanoparticle probes for molecular imaging. *Annals of Biomedical Engineering* https://doi.org/10.1007/s10439-005-9002-7.

Torgo, L., Ribeiro, R.P., Pfahringer, B., and Branco, P. (2013). SMOTE for regression. *Lecture Notes in Computer Science (Including Subseries Lecture Notes in Artificial Intelligence and Lecture Notes in Bioinformatics)* 8154 (LNAI): 378–389. https://doi.org/10.1007/978-3-642-40669-0_33.

Törönen, P., Kolehmainen, M., Wong, G., and Castrén, E. (1999). Analysis of gene expression data using self-organizing maps. *FEBS Letters* 451 (2): 142–146. https://doi.org/10.1016/s0014-5793(99)00524-4.

Toropov, A.A., Toropova, A.P., Puzyn, T. et al. (2013). QSAR as a random event: modeling of nanoparticles uptake in PaCa2 Cancer cells. *Chemosphere* 92 (1): 31–37. http://dx.doi.org/10.1016/j.chemosphere.2013.03.012.

Truong, L., Harper, S.L., and Tanguay, R.L. (2011). Evaluation of embryotoxicity using the Zebrafish model. *Methods in Molecular Biology* (Clifton, N.J.) 691: 271–279. https://doi.org/10.1007/978-1-60761-849-2_16.

UCLA (2016). Nanoinfo.Org: Nanoinformatics Platform for the Environmental and Health Impact Assessment of Nanomaterials. https://doi.org/https://nanoinfo.org (accessed 21 November 2020).

Uusitalo, L. (2007). Advantages and challenges of Bayesian networks in environmental modelling. *Ecological Modelling* 203 (3–4): 312–318. https://doi.org/10.1016/j.ecolmodel.2006.11.033.

Wang, X., Xia, T., Ntim, S.A. et al. (2011). Dispersal state of multiwalled carbon nanotubes elicits Profibrogenic cellular responses that correlate with Fibrogenesis biomarkers and fibrosis in the murine lung. *ACS Nano* 5: 9772–9787. https://doi.org/10.1021/nn2033055.

Wang, Q., Chen, B., Cao, M. et al. (2016). Response of MAPK pathway to Iron oxide nanoparticles in vitro treatment promotes osteogenic differentiation of HBMSCs. *Biomaterials* 86 (April): 11–20. http://dx.doi.org/10.1016/j.biomaterials.2016.02.004.

Wang, X., Sun, B., Liu, S., and Xia, T. (2017). Structure activity relationships of engineered nanomaterials in inducing NLRP3 Inflammasome activation and chronic lung fibrosis. *NanoImpact* 6 (April): 99–108. http://dx.doi.org/10.1016/j.impact.2016.08.002.

Wasserman, L. (2004). *All of Statistics: A Concise Course in Statistical Inference*. New York: Springer.

Waters, K.M., Masiello, L.M., Zangar, R.C. et al. (2009). Macrophage responses to silica nanoparticles are highly conserved across particle sizes. *Toxicological Sciences : An Official Journal of the Society of Toxicology* 107: 553–569. https://doi.org/10.1093/toxsci/kfn250.

Webb, A.R. (2002). *Statistical Pattern Recognition*, 2nde. Hoboken: Wiley https://doi.org/10.1198/tech.2003.s172.

Wei, H. and Billings, S.A. (2007). Feature subset selection and ranking for data dimensionality reduction. *IEEE Transactions on Pattern Analysis and Machine Intelligence* https://doi.org/10.1109/TPAMI.2007.250607.

Weissleder, R., Kelly, K., Sun, E.Y. et al. (2005). Cell-specific targeting of nanoparticles by multivalent attachment of small molecules. *Nature Biotechnology* https://doi.org/10.1038/nbt1159.

Wheeler, R.M. and Lower, S.K. (2021). A Meta-analysis framework to assess the role of units in describing nanoparticle toxicity. *NanoImpact*. https://doi.org/10.1016/j.impact.2020.100277.

Wolfbeis, O.S. (2015). An overview of nanoparticles commonly used in fluorescent bioimaging. *Chemical Society Reviews* https://doi.org/10.1039/c4cs00392f.

Xiao-feng, P., Bo, D., and Qiang, Z. (2009). Synthesis of magnetic Nanobiomaterials and its biological effects. *Journal of Nanoscience and Nanotechnology* 9 (2): 1369–1373.

Xing, E.P., Jordan, M.I., and Karp, R.M. (2001). Feature selection for high-dimensional genomic microarray data. *ICML* 1: 601–608. Citeseer.

Yu, L., and Liu, H. (2004). Redundancy based feature selection for microarray data. *Proceedings of the Tenth ACM SIGKDD International Conference on Knowledge Discovery and Data Mining*, 737–42. New York, NY, United States. (22–25 August 2004). Association for Computing Machinery. https://doi.org/10.1145/1014052.1014149.

Zhang, P. (1993). Model selection via multifold cross validation. *The Annals of Statistics* 21 (1): 299–313. https://doi.org/10.1214/aos/1176349027.

Zhang, P. (2007). Model selection via multifold cross validation. *Ann. Statist.* https://doi.org/10.1214/aos/1176349027.

Zhang, H., Xia, T., Meng, H. et al. (2011). Differential expression of Syndecan-1 mediates cationic nanoparticle toxicity in undifferentiated versus differentiated normal human bronchial epithelial cells. *ACS Nano* 5: 2756–2769. https://doi.org/10.1021/nn200328m.

Zhang, H., Burnum, K.E., Luna, M.L. et al. (2011). Quantitative proteomics analysis of adsorbed plasma proteins classifies nanoparticles with different surface properties and size. *Proteomics* 11 (23): 4569–4577. https://doi.org/10.1002/pmic.201100037.

Zhang, H., Ji, Z., Xia, T. et al. (2012). Use of metal oxide nanoparticle band gap to develop a predictive paradigm for oxidative stress and acute pulmonary inflammation. *ACS Nano* 6: 4349–4368. https://doi.org/10.1021/nn3010087.

Zhao, Z., Wang, L., Liu, H., and Ye, J. (2013). On similarity preserving feature selection. *Knowledge and Data Engineering, IEEE Transactions On* 25 (3): 619–632.

Zheng, W. and Tropsha, A. (2000). Novel variable selection quantitative structure-property relationship approach based on the k-nearest-neighbor principle. *Journal of Chemical Information and Computer Sciences* https://doi.org/10.1021/ci980033m.

Zollanvari, A., Cunningham, M.J., Braga-Neto, U., and Dougherty, E.R. (2009). Analysis and Modeling of time-course gene-expression profiles from nanomaterial-exposed primary human epidermal keratinocytes. *BMC Bioinformatics* 10 (11): S10. https://doi.org/10.1186/1471-2105-10-S11-S10.

10

Machine Learning in Environmental Exposure Assessment

Gregory L. Watson

Department of Biostatistics, University of California at Los Angeles, Los Angeles, CA, USA

10.1 Introduction

Environmental exposure assessment seeks to quantify exposure to potentially toxic environmental stressors, especially airborne pollutants. Health effects are typically estimated by regressing health outcome data on estimates of pollution exposure. While population health outcomes may often be obtained from administrative data, exposure data tends to be limited to observations recorded at sparsely distributed monitors or the output of numerical simulation models that computationally approximate atmospheric processes. The differing spatial resolution of these data sources is referred to as spatial misalignment (Gryparis et al. 2009).

To overcome this challenge, the analysis is often split into two steps. First, exposure is quantified by estimating the pollution surface across the entire spatial domain of interest using a statistical or machine learning model. Second, estimated exposure is linked to health outcome data to establish a dose–response relationship. This two-step approach is appealing because it splits a challenging problem into two easier tasks and allows for the production of exposure predictions as an intermediate output that may be used for multiple subsequent analyses. Uncertainty in the exposure modeling, however, is often ignored in the second step, and inaccuracies in the exposure predictions can bias subsequent epidemiological analyses (Szpiro and Paciorek 2013).

Environmental exposure assessment is concerned with the first step: measuring or estimating exposure for the population of interest. There is a long history of this in the air pollution literature. Traditionally pollution concentrations were recorded by monitors fixed at particular locations, requiring exposure to be estimated from that data across the region using spatial interpolation models. It is instructive to trace the development of these models to understand the contributions of the machine learning community to this field.

Machine Learning in Chemical Safety and Health: Fundamentals with Applications, First Edition.
Edited by Qingsheng Wang and Changjie Cai.
© 2023 John Wiley & Sons Ltd. Published 2023 by John Wiley & Sons Ltd.

10.2 Environmental Exposure Modeling

In their most basic form, exposure models simply interpolate measurements of pollution concentration across the geographic region of interest without the use of ancillary data.

At the heart of this approach is the assumption that spatially nearby quantities tend to be more similar than those farther apart. This assumption is something of a maxim and has been called the first rule of geography (Tobler 1970). From a statistical perspective, this tendency is an example of dependence: observations taken nearby exhibit spatial dependence, e.g. knowledge of the value of a quantity at one spatial location is informative of the value at spatially nearby locations.

Spatially interpolating pollution exposure is a particular type of prediction. It differs critically from other types of prediction like forecasting – extrapolating forward in time – or replication – predicting an entirely new realization of the data – in the relationship between the training data and the data to be predicted. These different relationships have a profound impact on the suitability of evaluation procedures. Selecting an appropriate evaluation procedure for any prediction problem begins with a clear articulation of the relationship between training data and prediction cases.

More formally, let $Y(s)$ denote a random field indexed by spatial location $s \in \mathcal{S}$, where $\mathcal{S}$ is a spatial domain, usually assumed to be a subset of two-dimensional Euclidean space, e.g. $\mathcal{S} \subset \mathcal{R}^2$. The goal of spatial interpolation is to predict the value of the field Y at new spatial locations given observations of the field at a set of n locations $\{s_1, ..., s_n\}$ in $\mathcal{S}$. Letting $\mathbf{y}(\mathbf{s}) = [y_1(s_1), ..., y_n(s_n)]'$ denote n observations of the field, and s_0 an unobserved location, the goal is to predict $Y(s_0)$ from $\mathbf{y}(\mathbf{s})$. This can be thought of as predicting $Y(s_0)|\mathbf{y}(\mathbf{s})$, the random variable $Y(s_0)$ conditional on the observed values, $\mathbf{y}(\mathbf{s})$. In practice, interpolation proceeds by selecting a function f of $\mathbf{y}(\mathbf{s})$ to define $\hat{Y}(s_0)$, a prediction for $Y(s_0)|\mathbf{y}(\mathbf{s})$, e.g.

$$\hat{Y}(s_0) = f[\mathbf{y}(\mathbf{s})]$$

A variety of spatial interpolation models have been employed, which amount to different choices of f. Theissen triangulation, an example of Voronoi tessellation, simply uses the nearest observed value (Thiessen 1911). Its simplicity is appealing, but the resulting discontinuities in the exposure surface are unrealistic and result in high variance predictions. Inverse distance weighting interpolates the field as a weighted average of its value at observed locations, where the weights are proportional to the inverse distance between locations raised to some exponent (Shepard 1968). This approach has intuitive appeal, but in practice the distance weighting function may be challenging to select and is often chosen in an ad hoc fashion.

Kriging historically has been the most commonly used method of spatial interpolation (Krige 1951). It is a special case of the best linear unbiased prediction (BLUP) for generalized least squares (Goldberger 1962) and the maximum a posteriori of a Bayesian Gaussian process (GP) with an uninformative prior. It relies upon a covariance function that defines the spatial dependence between observations. Kriging has many attractive properties, but selecting or estimating the covariance function may be difficult. Typically, some parametric form is assumed for this function and fit to the data, generally to the observed semivariogram (Cressie and Wikle 2015).

The availability of ancillary data potentially predictive of the environmental stressor across the geographic region of interest led to the development of interpolation models that could incorporate these data as features, which may also be referred to as covariates, predictors, independent variables, or explanatory variables depending on the discipline. The earliest such models were referred to as land use regression models, because they incorporated land use characteristics as features, e.g. urban versus vegetation. The label land use regression is still used in certain segments of the literature though features are no longer restricted to land use characteristics. As more spatially referenced data has become available, the number and variety of features used in interpolating exposure models have increased. In addition to land use characteristics, it is common to see meteorological data such as temperature, surface pressure, relative humidity, elevation, and boundary layer height, satellite retrievals such as those from NASA's Ozone Monitoring Instrument, and the output of atmospheric chemical transport models like Weather Research and Forecasting with Chemistry (WRF-Chem) (Powers et al. 2017).

Most spatial interpolation techniques that incorporate these or similar features require values for these features at both the observed locations $\mathbf{s} = (s_1, ..., s_n)$ and at the locations at which interpolation is desired. To formalize the addition of features to interpolation models, let $X_1(s), ..., X_p(s)$ denote the features at spatial location s, and let $\mathbf{X}(s) = [X_1(s), ..., X_p(s)]'$ denote a vector of length p containing those features. The features at $\mathbf{s}$, the locations at which the environmental stressor Y is observed, are denoted by $\mathbf{X}(\mathbf{s}) = [\mathbf{X}(s_1), ..., \mathbf{X}(s_n)]'$. Interpolating Y at a new location s_0 is then usually approached as a supervised learning regression problem in which $Y(s)$ given $\mathbf{X}(s)$ is assumed to be some function μ of $\mathbf{X}(s)$ plus a mean-zero error term, $\varepsilon(s)$, e.g.

$$Y(s)|\mathbf{X}(s) = \mu[\mathbf{X}(s)] + \varepsilon(s) \tag{10.1}$$

where $E\varepsilon(s) = 0$. Interpolating Y at a new location s_0 amounts to predicting $Y(s_0)$ given the feature vector at s_0, $\mathbf{X}(s_0)$, and the observed data, $\{\mathbf{y}(\mathbf{s}), \mathbf{X}(\mathbf{s})\}$, which can be written as

$$Y(s_0)|\mathbf{X}(s_0), \mathbf{y}(\mathbf{s}), \mathbf{X}(\mathbf{s}) = \mu[\mathbf{X}(s_0)] + \varepsilon[s_0|\mathbf{y}(\mathbf{s}), \mathbf{X}(\mathbf{s})] \tag{10.2}$$

The optimal prediction, e.g. the prediction that minimizes the mean square error (MSE), is the conditional expectation,

$$EY(s_0)|\mathbf{X}(s_0), \mathbf{y}(\mathbf{s}), \mathbf{X}(\mathbf{s}) = \mu[\mathbf{X}(s_0)] + E\varepsilon[s_0|\mathbf{y}(\mathbf{s}), \mathbf{X}(\mathbf{s})] \tag{10.3}$$

For independent data, the conditional expectation of the error term is 0, and the conditional expectation in Eq. (1.3) is simply $\mu[\mathbf{X}(s_0)]$. While spatially distributed exposure data are clearly not independent, this assumption is often made at least implicitly in the construction of exposure models. In particular, this is the case when models developed for independent data, including most machine learning models, are applied to estimate $\mu[\mathbf{X}(s)]$. In contrast, spatial statistics models, which are mostly based on universal kriging (kriging with features), or very closely related Bayesian Gaussian process models, explicitly model this dependence (Banerjee et al. 2003).

In the simple case of linear regression, $\mu[\mathbf{X}(s)]$ is assumed to be a linear combination of the features, e.g. $\mu[\mathbf{X}(s)] = \mathbf{X}(s)'\beta$, where β is a vector of p regression coefficients. An interpolating prediction for $Y(s_0)$ is given by $\hat{Y}(s_0) = \hat{\mu}[\mathbf{X}(s_0)] = \mathbf{X}(s_0)'\hat{\beta}$, where $\hat{\beta}$ is estimated from the training data via least squares. This mean function is very restrictive and certainly not a realistic model for the true relationship between $\mathbf{X}(s)$ and $Y(s)$. It is, however, easy to specify and trivial to fit and so is still sometimes used for exposure models (Saunders et al. 2014; Yao et al. 2018).

Geographically weighted regression uses spatially varying coefficients to more flexible model μ as $\mathbf{X}(s)'\beta(s)$, where $\beta(s)$ is usually estimated via weighted least squares with the weights depending upon the spatial dependence between observations (Fotheringham et al. 2003). The spatially heterogeneous effects of features allow for more realistic exposure models, and consequently geographically weighted regression has been used in a number of recent exposure assessment studies (Berrocal et al. 2010; He and Huang 2018; Shi et al. 2018; Yang et al. 2020; Zhao et al. 2020).

Focusing the exposition thus far on spatial exposure modeling mimics the historical development of these models, which initially emphasized the spatial case primarily for computational concerns. It is common, however, for environmental stressor data to have a temporal component, e.g. repeated measurements. The desire to learn from this data without collapsing it into spatial summaries has motivated the use of space–time, e.g. spatiotemporal, models. In this context, the outcome and covariates may be conceptualized as random fields indexed by space and time, e.g. $Y(s, t)$ and $X_j(s, t)$, where $s \in \mathcal{S}$ and $t \in \mathcal{T}$.

Exposure assessment in this context amounts to estimating pollution concentrations across the spatial domain repeatedly in the case of discrete time measurements or continuously in the case of arbitrarily spaced timepoints. This complicates exposure modeling in two ways. First, the addition of a time component to the data often greatly increases the amount of data, which can pose substantial computational challenges, especially to Gaussian process models (including kriging). Second, the time index adds an additional dimension to be accounted for in varying coefficient models and models that incorporate dependence between observations. It may be substantially more challenging to estimate space–time varying coefficients and space–time covariance functions.

10.3 Machine Learning Exposure Models

The formulation of exposure assessment as a supervised learning regression problem naturally led to the use of machine learning models. The well-known success of machine learning tools at a variety of prediction problems motivated their use as interpolation models for environmental exposure assessment. As might be expected, a wide variety of machine learning models have been used, including generalized additive models (GAMs) (Liu et al. 2009), multivariate adaptive regression splines (Reid et al. 2015; Watson et al. 2019), kernel-based regularized least squares (Weichenthal et al. 2016), support vector regression (SVR) (Weizhen et al. 2014; Hu et al. 2017a), deletion/substitution/addition (Beckerman et al. 2013), cubist regression (Xu et al. 2018), cluster-based bagging of mixed-effect models (Li et al. 2019), random forest (Hu et al. 2017b; Meng et al. 2018), gradient boosting (Zhan et al. 2017), treeging (Watson et al. 2021), neural networks (Di et al. 2016, 2017; Li et al. 2017), Bayesian regularized neural networks (Xu et al. 2018), generative adversarial networks (Gao et al. 2020), and approaches combining several of these (Xiao et al. 2018; Di et al. 2019).

The variety of machine learning models applied to this task indicates the ongoing interest of exposure assessment researchers in developing more accurate prediction models. These models differ substantially in form and complexity, corresponding to different procedures for estimating the mean function μ. In general, machine learning models provide more flexible mean structures than traditional alternatives, which largely accounts for their superior predictive accuracy in many contexts. Greater flexibility allows a model to more accurately approximate the true regression surface than a more restrictive model, thus accommodating more complicated relationships between features and the outcome.

It can be difficult to assess the relative superiority of machine learning models across studies due to differences in the informativeness of the training data. Several studies comparing the accuracy of machine learning models for interpolating air pollution have been published in recent years. Singh et al. (2013) found that bagged trees and boosted trees both outperformed SVR or a single tree for predicting air quality in Lucknow, India. Reid et al. (2015) compared 10 machine learning models at interpolating $PM_{2.5}$ during wildfires in Northern California in 2008. They found gradient boosting had the smallest estimated prediction error, followed by random forest. Watson et al. (2019) conducted a similar comparison of machine learning models for ozone exposure during the same wildfire events, also finding that gradient boosting and random forest were the best performing models. Hu et al. (2017a) compared seven machine learning models at modeling carbon monoxide exposure in Sydney, finding that random forest was the most accurate followed by decision trees and SVR. Kerckhoffs et al. (2019) compared a number of machine learning models against linear models augmented with feature selection procedures, using two different training sets and an external evaluation set. Random forest and bagged trees were the most accurate models in one set of comparisons, but the more rigid linear models outperformed them in the second comparison. This somewhat surprising result is likely explained by the fact that the external evaluation data was recorded by a different type of monitor and residential sites rather than roadways. Berrocal et al. (2020) compared random forest, SVR and a neural network against ordinary least squares, inverse distance weighting, kriging, and a downscaling spatial statistics model for interpolating $PM_{2.5}$ across the United States. They found that the spatial statistics models predicted more accurately, with SVR having the lowest cross-validation (CV) root mean square error (RMSE) of the three machine learning models. Random forest and gradient boosting again were the best performing models in a comparison of 13 models at interpolating ozone across the contiguous United States (Ren et al. 2020).

Random forest and boosting stand out across these comparisons consistently among the best performing models. This is not particularly surprising in light of their well-known performance on a number of other prediction problems. Both are heuristic ensembles of regression tree base learners, and their success as exposure interpolators may be attributed to their ability to automatically accommodate nonlinear effects, nonadditive effects, and perform feature selection. In the context of exposure assessment, the relationships between features and outcome may be quite complex, indeed nonlinear effects and interactions between features are to be expected, especially when the outcome depends upon atmospheric chemistry. The role of feature selection is also important in this context. It is quite likely that the data available as features exclude some of the quantities relevant to the true data generating process and include some features that are spurious. These spurious features can cause models without selection procedures to over-fit.

It is possible to nest models without built-in feature selection within a selection procedure as in Kerckhoffs et al. (2019) and Reid et al. (2015) or to perform selection as a preprocessing step, e.g. using principal component analysis as in Zhao et al. (2020). However, there are downsides to both approaches. The former is computationally demanding, and feasibility demands a heuristic selection procedure with no guarantee of optimality, while the latter approach typically ignores any uncertainty from the preprocessing step in the subsequent exposure model. The built-in capacity for feature selection in tree-based models is thus advantageous and contributes to their strong predictive performance.

The study by Berrocal et al. (2020) is notable for the relatively poor performance of the machine learning models, including random forest. They were bested not only by the downscaling model, which allows for effects that vary in space and time, but also by universal kriging, which has an additive, linear mean structure. The details of this comparison suggest two possibly interrelated reasons: (i) the covariates do not appear to be very informative, as evidenced by the addition of covariates beyond output of the Community Multiscale Air Quality (CMAQ) model *reducing* the accuracy of kriging predictions; (ii) the dense, gridded nature of the observed data is advantageous for models that exploit spatial (or space–time) dependence. This situation plays to the strengths of spatial statistics models and weaknesses of many machine learning models. In contrast, machine learning models tend to be at their best when the covariates are informative but dependence is weaker or the data are less dense in the spatial (or space–time) domain.

Artificial neural networks merit further discussion on account of their current popularity among machine learning algorithms. This popularity has resulted in a proliferation of deep learning neural networks being applied to exposure modeling problems (Di et al. 2016; Li et al. 2016; Zhang et al. 2016; Bui et al. 2018; Qi et al. 2018; Wen et al. 2019; Amato et al. 2020; Gao et al. 2020). A variety of architectures employed have been often aimed at incorporating spatial or space–time effects, including convolutional layers and long short-term memory recurrent networks. Deep learning certainly offers some attractive features that may be very useful for environmental exposure modeling, but most examples in the literature have been constructed in a somewhat ad hoc manner (Cabaneros et al. 2019). More thorough investigations are needed to clarify the conditions under which each architecture performs well and to establish how they fare against alternatives like tree-based ensembles. For example, the neural networks with inverse distance weighted convolutional layers that were used to predict particulate matter (Di et al. 2016) and ozone (Di et al. 2017) across the United States did not fare well against alternatives in Watson et al. (2019). This may be due at least in part to differences in the input data. More detailed comparisons and simulation studies are required to assess performance under different data scenarios.

The differing strengths of machine learning models with flexible mean structures and spatial statistical models that model dependence suggest a combined approach. A few models have attempted to marry regression tree or tree-based ensemble mean structure with a kriging-style (Gaussian process) covariance model. Most of these are limited to the spatial case, have not been applied to exposure assessment, and have substantial drawbacks. Watson et al. (2021) provide a detailed discussion of the situations under which each class of models performs well and propose treeging, an ensemble of regression trees enriched with kriging for spatial or space–time prediction that combine their strengths.

Most exposure models, including nearly all the approaches cited so far, construct an exposure surface using a supervised learning approach. However, since the features

$\mathbf{X}(s)$ must be available wherever predictions are desired, the problem may be framed as semi-supervised learning. In semi-supervised learning, models are trained using both labeled data, $\{\mathbf{y}(\mathbf{s}), \mathbf{X}(\mathbf{s})\}$, as well as a typically much larger set of unlabeled data, $\mathbf{X}(\mathbf{s}^*)$, where $\mathbf{s}^* \subset \mathcal{S}$ is the set of locations at which prediction is desired. This contrasts with supervised learning (which includes traditional regression models), which use only the labeled data $\{\mathbf{y}(\mathbf{s}), \mathbf{X}(\mathbf{s})\}$ in model training. A few examples of semi-supervised classification algorithms have been applied to discretized air pollution exposure (Bougoudis et al. 2016; Chen et al. 2016), but semi-supervised regression techniques may offer advantages for exposure assessment. Deep learning approaches may prove useful in this regard (Qi et al. 2018).

10.4 Model Evaluation

Model evaluation is a critical but often underappreciated step in the development of exposure models. Model comparison, tuning, averaging, and feature selection all require reliable evaluation procedures. Similarly, assessing the predictive accuracy of an exposure model requires a reliable estimation procedure. This is true of all models, but the spatial or space–time dependence of exposure models means that evaluation procedures developed for independent data may not be appropriate.

Cross-validation (CV) (Stone 1974) and the bootstrap (Efron and Tibshirani 1994) are two such procedures that are commonly used to evaluate exposure models. Both evaluate models by repeatedly training them on different subsets of the data and evaluating their performance on test subsets that are withheld from training. CV randomly divides the training data into nonoverlapping subsets called folds. In each repetition, onefold is used for testing a model trained on the remaining folds. The bootstrap, in contrast, constructs training sets by randomly sampling with replacement from the training data and testing on the unsampled data. The estimates are combined, typically by simple average, to produce a nonparametric estimate of model performance. In the case of exposure models, performance is typically assessed using prediction error (MSE or RMSE) and R^2. Prediction error estimation for independent data was described by Efron (1983). For spatially and temporally dependent data, however, the dependence between training and test subsets may render these procedures overly optimistic (Roberts et al. 2017; Meyer et al. 2018).

This is particularly true for space–time data, e.g. when repeated measurements are taken by air pollution monitors at fixed spatial locations. The goal of an exposure assessment model is to interpolate across the space–time field, e.g. to predict the time series of air pollution measurements at an unobserved location. A reasonable definition of prediction error in this context is the average error made by a model predicting across unobserved locations in the spatial domain of the data. However, naive application of CV or bootstrap will very likely include observations from the same monitors in both the training and test sets. This results in overly optimistic estimates of predictive performance, because the correlation between observations taken at the same location provides an unrealistic amount of information in the training data on the test fold. This tends to bias downward the resulting estimates of prediction error. This tendency is stronger for more flexible models, which are more prone to over-fitting, e.g. better able to exploit the additional information in the training data, than for more rigid models. The true interpolation error of predicting at new locations within the spatial domain would be greater than these estimates, because the data on which the model was trained would not include data from the new locations.

Structured resampling strategies can recover consistent estimates of prediction error if dependence is confined to observations within clusters or groups that are independent of each other. Spatial and space–time dependence are not so confined, however, leading to a variety of modified CV strategies for estimating prediction error in this context. These include partitioning the data into a spatial grid (Zhan et al. 2017), resampling monitor locations (Lee et al. 2016; Zhan et al. 2018), and spatially buffered CV (Le Rest et al. 2014; Pohjankukka et al. 2017). Some studies report multiple types of structured CV estimates without providing guidance as to their appropriateness for the type of prediction under evaluation (Hu et al. 2017b; Brokamp et al. 2018). For spatial interpolation, location-based CV is an appropriate estimator of prediction error with leave-one-location-out CV preferable to *k*-fold location-based CV when it is feasible (Watson et al. 2020). Naive estimators are not only inappropriate for prediction error estimation but also unreliable for model selection, tuning, or averaging. In addition to being overly optimistic, they may not rank model accuracy appropriately and misevaluate even ordinal comparisons of prediction models (Watson et al. 2019).

10.5 Case Study

Figure 10.1 depicts the locations of 98 ozone monitors spread throughout the state of California that each recorded daily eight-hour maximum average ground-level ozone for each day of June, 2008, during which a number of wildfires were ignited. The mean concentration was 40.4 ppb, and the standard deviation was 13.7 ppb. The log transformation of these observations serves as the outcome of interest in an exposure modeling analysis. In addition

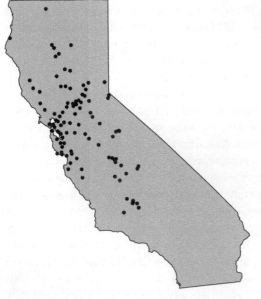

Figure 10.1 Ground-level ozone monitors in California during June 2008 wildfires.

to ozone, 18 features were recorded to be used in interpolating ozone across the region. These are listed in Table 10.1 along with their sources.

A generalized linear model (GLM), support vector machine with radial basis kernel, and random forest were trained on this data. Table 10.2 reports naive 10-fold and LOLO cross-validated estimates of their predictive accuracy. As discussed above in Section 10.4, naive CV is inappropriate for estimating interpolation error for space–time data, and LOLO CV is recommended. Random forest was the best performing model with a LOLO CV estimated $\hat{R}^2$ of 0.729, followed by SVM at 0.675 and GLM at 0.637. Model accuracy appears to improve with increased flexibility, with the tree-based ensemble performing especially well, as has been found in many similar studies (see Section 10.3).

Table 10.1   Features used to predict ozone and their data sources.

Feature	Data source
Monitor longitude	US Environmental Protection Agency
Monitor latitude	US Environmental Protection Agency
Date	US Environmental Protection Agency
Elevation (m)	National digital elevation model
Normalized difference vegetation index	Landsat data
Annual average traffic within 1 km	Dynamap 2000, TeleAtlas
Urban land use within 1 km (%)	2006 National Land Cover Database
Agricultural land use within 1 km (%)	2006 National Land Cover Database
Vegetation land use within 1 km (%)	2006 National Land Cover Database
Temperature at 2 m (°K)	Rapid update cycle
Boundary layer height (m)	Rapid update cycle
Nitrogen dioxide (log molecules cm^{-2})	Ozone monitoring instrument satellite
Surface pressure (Pa)	Rapid update cycle
Relative humidity (%)	Rapid update cycle
U-component of wind speed (m s^{-1})	Rapid update cycle
V-component of wind speed (m s^{-1})	Rapid update cycle
Inverse distance to nearest fire (m^{-1})	Fire inventory from NCAR v1.5
WRF-Chem ozone (log 8 Hour Maximum)	WRF-Chem

Table 10.2   Estimates of exposure model predictive accuracy on the California wildfire ozone data.

Model	10-Fold CV $\hat{R}^2$	LOLO CV $\hat{R}^2$
GLM	0.642	0.637
SVM	0.689	0.675
Random forest	0.706	0.729

10.6 Other Topics

10.6.1 Bias and Fairness

Bias and fairness are important concerns when training environmental exposure models. Literature on preferential sampling has pointed out that in many applications, the spatial sampling intensity of phenomena is not uniform (Diggle et al. 2010; Gelfand et al. 2012), leading to potentially biased training data. For example, air pollution monitors may be concentrated in highly populated areas, biologists may search for endangered species only in the most promising habitats, and mineral exploration may be focused where high-grade ore is expected. This literature emphasizes the bias that can result when uniform, e.g. non-preferential, sampling is mistakenly assumed. When sampling bias is associated with population demographics like race or income, exposure models naively trained on that data may be unfairly weighted for or against particular segments of the population. Correcting for training data bias and avoiding unfair outcomes is an important and historically under-emphasized aspect of environmental exposure assessment, as it is of machine learning in general.

10.6.2 Wearable Sensors

The development of wearable sensors capable of measuring air pollution concentrations provide more precise estimates of an individual's exposure (Dons et al. 2017). This approach allows detailed data to be collected on individual subjects that reflects differing exposure even for individuals living or working in proximity due to mobility and differential exposure indoors and outdoors. This is a promising avenue of future work, although there are still substantial challenges connecting exposure to health outcomes and generalizing to larger populations, since these collect exposure only for the individuals wearing them.

10.6.3 Interpretability

Interpretability is often a challenge when using machine learning algorithms, and exposure models are no exception. The ability of machine learning tools to fit very flexible regression surfaces and the opaqueness of the algorithms themselves can hinder inference. This may be somewhat less concerning when prediction is the primary goal, which is typically the case in the construction of exposure surfaces for use in health analyses. Even in a strictly predictive context, however, understanding the relationship between model inputs and outputs is essential for model robustness. Without this understanding, it is difficult to anticipate under what circumstances model accuracy may degrade. The drastic and unforeseen changes to patterns of human behavior across the world in 2020 caused by the COVID-19 pandemic have taught painful lessons in the negative consequences of domain shift to many machine learning practitioners.

10.6.4 Extreme Events

Extreme value prediction is an alternative to predicting pollution concentration, the usual target of exposure models. This may be a useful approach if extreme events, e.g. short-term exposure to high concentrations of an environmental stressor, are of primary concern.

There may also be utility in this approach related to governmental standards that are often defined in terms of maximum concentrations (Ministry of Ecology and Environment, People's Republic of China 2012; Environmental Protection Agency, USA 2015; World Health Organization 2021). Traditionally, this type of prediction relies on models with parametric forms, and there may be an opportunity for machine learning tools to make a valuable contribution in this area.

10.7 Conclusion

Environmental exposure assessment is critical for understanding the health effects of exposure to environment stressors. Machine learning algorithms have become a mainstay of the exposure modeling community, offering powerful tools for the spatial and space–time interpolation of exposure surfaces. The flexible mean structures they afford tend to provide better predictions than more rigid models. Tree-based ensembles, including random forest and gradient boosting, have performed particularly well in comparative studies on account of their ability to automatically incorporate nonlinear effects, nonadditive effects, and feature selection.

Deep learning neural networks are an attractive alternative, but the examples developed so far for exposure modeling differ substantially in architecture and often appear to have been constructed in an ad hoc manner. More thorough comparisons are required to understand the relative merits of deep learning architectures, especially against tree based ensembles.

While flexible mean structures empower machine learning algorithms to efficiently extract information from the training data, particularly when informative features are available, they may be outperformed when spatial (or space–time) dependence is strong or the features are uninformative. Recent work augmenting flexible machine learning models with covariance structures that explicitly model this dependence offers a combined approach.

Model evaluation is a critical aspect of exposure modeling, especially for machine learning tools. The dependence of the data renders some evaluation procedures developed for independent data inappropriate, and location-based CV (preferably leave-one-location-out CV) is the recommended approach for prediction error estimation and model selection, tuning, and averaging. The future is bright for machine learning exposure assessment models, and a number of directions for future work present themselves, especially related to bias, fairness, and interpretability.

References

Amato, F., Guignard, F., Robert, S. et al. (2020). A novel framework for spatio-temporal prediction of environmental data using deep learning. *Scientific Reports* 10 (1): 1–11.

Banerjee, S., Carlin, B.P., and Gelfand, A.E. (2003). *Hierarchical Modeling and Analysis for Spatial Data*. Chapman and Hall/CRC.

Beckerman, B.S., Jerrett, M., Serre, M. et al. (2013). A hybrid approach to estimating national scale spatiotemporal variability of $PM_{2.5}$ in the contiguous United States. *Environmental Science & Technology* 47 (13): 7233–7241.

Berrocal, V.J., Gelfand, A.E., and Holland, D.M. (2010). A spatio-temporal downscaler for output from numerical models. *Journal of Agricultural, Biological, and Environmental Statistics* 15 (2): 176–197.

Berrocal, V.J., Guan, Y., Muyskens, A. et al. (2020). A comparison of statistical and machine learning methods for creating national daily maps of ambient $PM_{2.5}$ concentration. *Atmospheric Environment* 222: 117130.

Bougoudis, I., Demertzis, K., Iliadis, L. et al. (2016). Semi-supervised hybrid modeling of atmospheric pollution in urban centers. In: *International Conference on Engineering Applications of Neural Networks* (ed. C. Jayne and L. Iliadis), 51–63. Cham: Springer.

Brokamp, C., Jandarov, R., Hossain, M. et al. (2018). Predicting daily urban fine particulate matter concentrations using a random forest model. *Environmental Science & Technology* 52 (7): 4173–4179.

Bui, T.C., Le, V.D., and Cha, S.K. (2018). A deep learning approach for forecasting air pollution in South Korea using LSTM. *arXiv preprint arXiv:1804.07891*.

Cabaneros, S.M., Calautit, J.K., and Hughes, B.R. (2019). A review of artificial neural network models for ambient air pollution prediction. *Environmental Modelling & Software* 119: 285–304.

Chen, L., Cai, Y., Ding, Y. et al. (2016). Spatially fine-grained urban air quality estimation using ensemble semi-supervised learning and pruning. In: *Proceedings of the 2016 ACM International Joint Conference on Pervasive and Ubiquitous Computing* (ed. K. Kise, J. Cheng, D. Roggen and Y. Enokibori), 1076–1087. New York, New York: The Association for Computing Machinery.

Cressie, N. and Wikle, C.K. (2015). *Statistics for Spatio-Temporal Data*. Wiley.

Di, Q., Kloog, I., Koutrakis, P. et al. (2016). Assessing $PM_{2.5}$ exposures with high spatiotemporal resolution across the continental United States. *Environmental Science & Technology* 50 (9): 4712–4721.

Di, Q., Rowland, S., Koutrakis, P. et al. (2017). A hybrid model for spatially and temporally resolved ozone exposures in the continental United States. *Journal of the Air & Waste Management Association* 67 (1): 39–52.

Di, Q., Amini, H., Shi, L. et al. (2019). An ensemble-based model of $PM_{2.5}$ concentration across the contiguous United States with high spatiotemporal resolution. *Environment International* 130: 104909.

Diggle, P.J., Menezes, R., and Su, T.L. (2010). Geostatistical inference under preferential sampling. *Journal of the Royal Statistical Society: Series C (Applied Statistics)* 59 (2): 191–232.

Dons, E., Laeremans, M., Orjuela, J.P. et al. (2017). Wearable sensors for personal monitoring and estimation of inhaled traffic-related air pollution: evaluation of methods. *Environmental Science & Technology* 51 (3): 1859–1867.

Efron, B. (1983). Estimating the error rate of a prediction rule: improvement on cross-validation. *Journal of the American Statistical Association* 78 (382): 316–331.

Efron, B. and Tibshirani, R.J. (1994). *An Introduction to the Bootstrap*. CRC Press.

Environmental Protection Agency, USA (2015). National ambient air quality standards for ozone, volume 80. Federal Register.

Fotheringham, A.S., Brunsdon, C., and Charlton, M. (2003). *Geographically Weighted Regression: The Analysis of Spatially Varying Relationships.* Wiley.

Gao, Y., Liu, L., Zhang, C. et al. (2020). SI-AGAN: spatial interpolation with attentional generative adversarial networks for environment monitoring. In: *Proceedings of the 24th European Conference on Artificial Intelligence* (ed. G. De Giacomo, A. Catala, B. Dilkina, et al.), 1786–1793. IOS Press.

Gelfand, A.E., Sahu, S.K., and Holland, D.M. (2012). On the effect of preferential sampling in spatial prediction. *Environmetrics* 23 (7): 565–578.

Goldberger, A.S. (1962). Best linear unbiased prediction in the generalized linear regression model. *Journal of the American Statistical Association* 57 (298): 369–375.

Gryparis, A., Paciorek, C.J., Zeka, A. et al. (2009). Measurement error caused by spatial misalignment in environmental epidemiology. *Biostatistics* 10 (2): 258–274.

He, Q. and Huang, B. (2018). Satellite-based high-resolution $PM_{2.5}$ estimation over the Beijing-Tianjin-Hebei region of China using an improved geographically and temporally weighted regression model. *Environmental Pollution* 236: 1027–1037.

Hu, K., Rahman, A., Bhrugubanda, H. et al. (2017a). HazeEst: machine learning based metropolitan air pollution estimation from fixed and mobile sensors. *IEEE Sensors Journal* 17 (11): 3517–3525.

Hu, X., Belle, J.H., Meng, X. et al. (2017b). Estimating $PM_{2.5}$ concentrations in the conterminous United States using the random forest approach. *Environmental Science & Technology* 51 (12): 6936–6944.

Kerckhoffs, J., Hoek, G., Portengen, L. et al. (2019). Performance of prediction algorithms for modeling outdoor air pollution spatial surfaces. *Environmental Science & Technology* 53 (3): 1413–1421.

Krige, D.G. (1951). A statistical approach to some basic mine valuation problems on the Witwatersrand. *Journal of the Southern African Institute of Mining and Metallurgy* 52 (6): 119–139.

Le Rest, K., Pinaud, D., Monestiez, P. et al. (2014). Spatial leave-one-out cross-validation for variable selection in the presence of spatial autocorrelation. *Global Ecology and Biogeography* 23 (7): 811–820.

Lee, M., Kloog, I., Chudnovsky, A. et al. (2016). Spatiotemporal prediction of fine particulate matter using high-resolution satellite images in the Southeastern US 2003–2011. *Journal of Exposure Science & Environmental Epidemiology* 26 (4): 377–384.

Li, X., Peng, L., Hu, Y. et al. (2016). Deep learning architecture for air quality predictions. *Environmental Science and Pollution Research* 23 (22): 22408–22417.

Li, T., Shen, H., Yuan, Q. et al. (2017). Estimating ground-level $PM_{2.5}$ by fusing satellite and station observations: a geo-intelligent deep learning approach. *Geophysical Research Letters* 44 (23): 11–985.

Li, L., Girguis, M., Lurmann, F. et al. (2019). Cluster-based bagging of constrained mixed-effects models for high spatiotemporal resolution nitrogen oxides prediction over large regions. *Environment International* 128: 310–323.

Liu, Y., Paciorek, C.J., and Koutrakis, P. (2009). Estimating regional spatial and temporal variability of $PM_{2.5}$ concentrations using satellite data, meteorology, and land use information. *Environmental Health Perspectives* 117 (6): 886–892.

Meng, X., Hand, J.L., Schichtel, B.A. et al. (2018). Space-time trends of $PM_{2.5}$ constituents in the conterminous United States estimated by a machine learning approach, 2005–2015. *Environment International* 121: 1137–1147.

Meyer, H., Reudenbach, C., Hengl, T. et al. (2018). Improving performance of spatio-temporal machine learning models using forward feature selection and target-oriented validation. *Environmental Modelling & Software* 101: 1–9.

Ministry of Ecology and Environment, People's Republic of China (2012). Ambient air quality standards. Document GB 3095.

Pohjankukka, J., Pahikkala, T., Nevalainen, P. et al. (2017). Estimating the prediction performance of spatial models via spatial k-fold cross validation. *International Journal of Geographical Information Science* 31 (10): 2001–2019.

Powers, J.G., Klemp, J.B., Skamarock, W.C. et al. (2017). The weather research and forecasting model: overview, system efforts, and future directions. *Bulletin of the American Meteorological Society* 98 (8): 1717–1737.

Qi, Z., Wang, T., Song, G. et al. (2018). Deep air learning: interpolation, prediction, and feature analysis of fine-grained air quality. *IEEE Transactions on Knowledge and Data Engineering* 30 (12): 2285–2297.

Reid, C.E., Jerrett, M., Petersen, M.L. et al. (2015). Spatiotemporal prediction of fine particulate matter during the 2008 northern California wildfires using machine learning. *Environmental Science & Technology* 49 (6): 3887–3896.

Ren, X., Mi, Z., and Georgopoulos, P.G. (2020). Comparison of machine learning and land use regression for fine scale spatiotemporal estimation of ambient air pollution: modeling ozone concentrations across the contiguous United States. *Environment International* 142: 105827.

Roberts, D.R., Bahn, V., Ciuti, S. et al. (2017). Cross-validation strategies for data with temporal, spatial, hierarchical, or phylogenetic structure. *Ecography* 40 (8): 913–929.

Saunders, R.O., Kahl, J.D., and Ghorai, J.K. (2014). Improved estimation of $PM_{2.5}$ using Lagrangian satellite-measured aerosol optical depth. *Atmospheric Environment* 91: 146–153.

Shepard, D. (1968). A two-dimensional interpolation function for irregularly-spaced data. In: *Proceedings of the 1968 23rd ACM National Conference* (ed. R. Blue and A. Rosenberg), 517–524. New York, New York: The Association for Computing Machinery.

Shi, Y., Ho, H.C., Xu, Y. et al. (2018). Improving satellite aerosol optical depth-$PM_{2.5}$ correlations using land use regression with microscale geographic predictors in a high-density urban context. *Atmospheric Environment* 190: 23–34.

Singh, K.P., Gupta, S., and Rai, P. (2013). Identifying pollution sources and predicting urban air quality using ensemble learning methods. *Atmospheric Environment* 80: 426–437.

Stone, M. (1974). Cross-validatory choice and assessment of statistical predictions. *Journal of the Royal Statistical Society: Series B (Methodological)* 36 (2): 111–133.

Szpiro, A.A. and Paciorek, C.J. (2013). Measurement error in two-stage analyses, with application to air pollution epidemiology. *Environmetrics* 24 (8): 501–517.

Thiessen, A.H. (1911). Precipitation averages for large areas. *Monthly Weather Review* 39 (7): 1082–1089.

Tobler, W.R. (1970). A computer movie simulating urban growth in the Detroit region. *Economic Geography* 46 (sup1): 234–240.

Watson, G.L., Telesca, D., Reid, C.E. et al. (2019). Machine learning models accurately predict ozone exposure during wildfire events. *Environmental Pollution* 254 (112): 792.

Watson, G.L., Reid, C.E., Jerrett, M. et al. (2020). Prediction and model evaluation for space-time data. *arXiv preprint arXiv:2012.13867*.

Watson, G.L., Jerrett, M., Reid, C.E. et al. (2021). Treeging. *arXiv preprint arXiv:2110.01053*.

Weichenthal, S., Van Ryswyk, K., Goldstein, A. et al. (2016). A land use regression model for ambient ultrafine particles in Montreal, Canada: a comparison of linear regression and a machine learning approach. *Environmental Research* 146: 65–72.

Weizhen, H., Zhengqiang, L., Yuhuan, Z. et al. (2014). Using support vector regression to predict PM_{10} and $PM_{2.5}$. *IOP Conference Series: Earth and Environmental Science*, 17(1), p. 012268. IOP Publishing.

Wen, C., Liu, S., Yao, X. et al. (2019). A novel spatiotemporal convolutional long short-term neural network for air pollution prediction. *Science of the Total Environment* 654: 1091–1099.

World Health Organization (2021). *WHO Global Air Quality Guidelines. Particulate Matter ($PM_{2.5}$ and PM_{10}), Ozone, Nitrogen Dioxide, Sulfur Dioxide and Carbon Monoxide. Geneva: World Health Organization.*

Xiao, Q., Chang, H.H., Geng, G. et al. (2018). An ensemble machine-learning model to predict historical $PM_{2.5}$ concentrations in China from satellite data. *Environmental Science & Technology* 52 (22): 13260–13269.

Xu, Y., Ho, H.C., Wong, M.S. et al. (2018). Evaluation of machine learning techniques with multiple remote sensing datasets in estimating monthly concentrations of ground-level $PM_{2.5}$. *Environmental Pollution* 242: 1417–1426.

Yang, Q., Yuan, Q., Yue, L. et al. (2020). Investigation of the spatially varying relationships of $PM_{2.5}$ with meteorology, topography, and emissions over China in 2015 by using modified geographically weighted regression. *Environmental Pollution* 262: 114257.

Yao, F., Si, M., Li, W. et al. (2018). A multidimensional comparison between MODIS and VIIRS AOD in estimating ground-level $PM_{2.5}$ concentrations over a heavily polluted region in China. *Science of the Total Environment* 618: 819–828.

Zhan, Y., Luo, Y., Deng, X. et al. (2017). Spatiotemporal prediction of continuous daily $PM_{2.5}$ concentrations across China using a spatially explicit machine learning algorithm. *Atmospheric Environment* 155: 129–139.

Zhan, Y., Luo, Y., Deng, X. et al. (2018). Satellite-based estimates of daily NO_2 exposure in China using hybrid random forest and spatiotemporal kriging model. *Environmental Science & Technology* 52 (7): 4180–4189.

Zhang, C., Yan, J., Li, C. et al. (2016). *On estimating air pollution from photos using convolutional neural network* (ed. A. Hanjalic, C. Snoek and M. Worring) *Proceedings of the 24th ACM International Conference on Multimedia*, 297–301. New York, New York: The Association for Computing Machinery.

Zhao, R., Zhan, L., Yao, M. et al. (2020). A geographically weighted regression model augmented by Geodetector analysis and principal component analysis for the spatial distribution of PM2.5. *Sustainable Cities and Society* 56: 102–106.

11

Air Quality Prediction Using Machine Learning

Lan Gao[1], Changjie Cai[2], and Xiao-Ming Hu[1,3]

[1] *School of Meteorology, The University of Oklahoma, Norman, OK, USA*
[2] *Department of Occupational and Environmental Health, Hudson College of Public Health, The University of Oklahoma, Oklahoma City, OK, USA*
[3] *Center for Analysis and Prediction of Storms, The University of Oklahoma, Norman, OK, USA*

11.1   Introduction

Air pollution is a global concern due to its significant effects on human health, agriculture and ecosystem, and even climate (Myhre et al. 2013; Fuhrer et al. 2016; Cohen et al. 2017). Over the past decades, observing the atmospheric compositions and pollutants using various instrumental platforms has become an increasingly powerful tool for understanding atmospheric processes and air quality. The platforms include Earth observation satellite instruments, ground-based *in situ*, remote-sensing stations, instrumented aircrafts, etc. (Laj et al. 2009). Many species, e.g. particulate matter (PM), nitrogen dioxide (NO_2), sulfur dioxide (SO_2), carbon monoxide (CO), ozone (O_3), lead, volatile organic compounds (VOCs), carbon, and other reactive gases observed from these platforms can be directly used in air quality studies, including near real-time monitoring, source apportionment, dispersion modeling, and air quality prediction (Maykut et al. 2003; Engel-Cox et al. 2004; Cai et al. 2010; Li et al. 2011; Iskandaryan et al. 2020).

Air quality prediction can be achieved by applying atmospheric chemical transport models (CTMs), statistical models, and machine learning algorithms (Carmichael et al. 2008; Singh et al. 2012; Xing et al. 2020; Mao et al. 2021). The CTMs simulate the concentrations of ambient air pollutants by considering meteorology, emissions, transports, dry and wet depositions, and other chemical and physical processes (Grell et al. 2005; Byun and Schere 2006). Therefore, the model performance largely depends on the biases within meteorological prediction, emission estimation, initial and boundary conditions, chemical and physical process parameterizations, and many other inputs. The CTMs serve as numerical laboratories and integrate our knowledge on how physical and chemical processes affect pollutant concentrations to investigate air pollution (Russell 1997); however, their limitations are also relatively "hard" constraints. For example, the inaccurate weather predictions could greatly

Machine Learning in Chemical Safety and Health: Fundamentals with Applications, First Edition.
Edited by Qingsheng Wang and Changjie Cai.
© 2023 John Wiley & Sons Ltd. Published 2023 by John Wiley & Sons Ltd.

affect the predicting performance (Hu 2015; Hu et al. 2019). The performance also largely depends on an updated parameterizations and emission inventory (Thomas et al. 2019), which are often compromised in developing countries. In addition, the CTM simulations often demand high-performance computational resources.

In the last decade, machine learning algorithms have been actively developed and applied in air quality prediction (Rybarczyk and Zalakeviciute 2018; Iskandaryan et al. 2020). Refined data is the key to train machine learning models. The instrumental platforms, including Earth observation satellite instruments and ground-based *in situ*, provide an invaluable database for training various machine learning models, such as shallow learning (or conventional machine learning) and deep learning (Zhu et al. 2018; Mao et al. 2021). Therefore, in this chapter, we mainly focus on the applications of shallow learning and deep learning. Figure 11.1 describes the basic flowchart of supervised machine learning applications in air quality studies.

In this chapter, we first introduce the air quality associated data acquisition such as the main instrumental platforms where researchers can obtain the data (mainly on Earth observation satellite instruments, and ground-based *in situ* observations), and then summarize the recent studies of implementing these datasets for air quality studies using shallow learning and deep learning. A practice example is provided at the end from our previous study. After this chapter, readers are expected to be able to (i) obtain data from a wide range of various instrumental platforms and (ii) practice implementing conventional machine learning and deep learning using the given example at the end of this chapter.

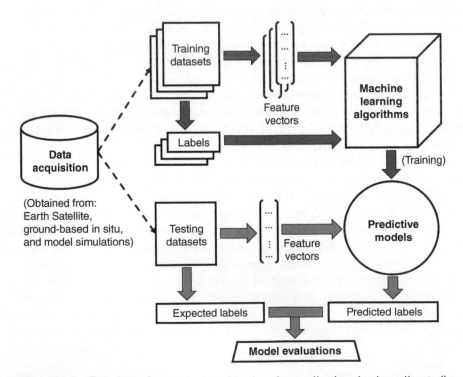

Figure 11.1 Flowchart of supervised machine learning applications in air quality studies.

11.2 Air Quality and Climate Data Acquisition

In this section, the commonly used datasets associated with air pollution and climate from both Earth satellite (Section 11.2.1) and ground-based *in situ* observations (Section 11.2.2) are introduced. Refined data is the key to train machine learning algorithms. After this section, readers are expected to have a general understanding of the characteristics of various datasets and where to acquire them, so that they could choose the best product(s) for their study area and specific applications.

11.2.1 Earth Satellite Observation Datasets

Existing satellite data for air quality study is primarily from the collaboration among several federal agencies and international space organizations, including the National Aeronautics and Space Administration (NASA), National Oceanic and Atmospheric Administration (NOAA), United States Geological Survey (USGS), and the European Space Agency (ESA). These air quality data include primary (directly emitted) and secondary pollutants (formed by chemical reactions), and most of them can be found from Earthdata and Copernicus Open Access Hub. In next few sections, some of the commonly used air quality relevant satellite products are introduced, including the data description and acquisition.

One of the most significant advantages of satellite measurements is their extensive spatial coverage, which can fill in air quality information in areas without any ground-based *in situ* observations and offers unique information on pollution transport and prediction, and links to climate change. The selection of satellite data to address specific air quality issues depends on data accuracy and spatial and temporal resolution. This requires the users to have some basic knowledge of satellite measurement and geolocation gridded datasets.

11.2.1.1 Basics of Earth Satellite Observations

In remote sensing theory, satellite measurements of air pollutants are based on species' spectral characteristics. Each of the atmospheric constituents has its own unique spectral feature for the emission and absorption of radiation. Passive remote sensors (air quality relevant) aboard satellites are designed to detect the scattering, absorption, and emission of radiation at a specific wavelength, which usually contains those spectral features. The geophysical variables of interest can be inferred from the measured radiances via a process called "retrieval." Current satellites providing air-quality-related measurements are in polar orbit, which can only overpass the same place twice a day. Therefore, such polar-orbiting satellites have a poor temporal resolution compared to geostationary satellites in higher orbit (far away from Earth), which can continuously monitor one place on the Earth. However, the spatial resolution, both horizontal and vertical, is much higher than geostationary satellites. The next generation of regional air quality monitoring focuses on developing high-resolution continuous air quality observation from the geostationary satellite, e.g. Tropospheric Emissions: Monitoring Pollution (TEMPO) mission (Zoogman et al. 2017), which will be the first geostationary satellite to monitor major air pollutants across the North American continent and aims to create a revolutionary new dataset of atmospheric chemistry measurements from space.

In addition to directly applying satellite data in air quality monitoring, these data may also be used to evaluate, initialize, and improve the air quality modeling. To evaluate the model simulations, the selection of time and spatial resolution and coverage of satellite observation is more important for comparing satellite data and model output. This usually requires some pre-processing of data, e.g. data aggregation or interpolation. Boundary conditions in a model simulation play a significant role in determining the simulation accuracy. Satellite data provides the spatial coverage needed for the boundary and initial conditions of regional air quality models. However, not every satellite data is good enough to improve models. Some chemical species are not well retrieved due to either instrument uncertainties or retrieval algorithm issues. This requires users to be familiar with different satellite products. The best way to obtain the information of specific data is from the product user manual and/or Algorithm Theoretical Basis Document (ATBD). Some of the available and validated datasets will be discussed in the next section.

11.2.1.2 Earth Satellite Products

Table 11.1 summarizes the commonly used earth satellite products for air quality studies, including the data of aerosols, gases, and climate. In this table, we provide the Variable Names, Satellite Platform and Instrument Sensor, Measurement Spatial Resolution, Temporal Resolution & Coverage, and Data Archive web links for readers. More details for each variable are provided in the following sub-sections.

11.2.1.2.1 Aerosol Optical Depth (AOD)

AOD indicates the level at which particles in the air prevent light from traveling through the atmosphere, and those particles are called "aerosol." Sources of aerosols include pollution from human activities (e.g. industry, traffic, cooking), smoke from wildfires, dust from dust storms, sea salt, and volcanic ash and smog. Aerosols compromise human health when inhaled by people, particularly those with asthma or other respiratory illnesses. Air quality studies usually define an AOD of less than 0.1 as "clean" atmosphere, associated with clear blue sky, bright Sun, and maximum visibility. As AOD increases, the amount of aerosol increases, the Sun becomes obscured, and visibility decreases.

In atmospheric science, AOD is used to indicate the amount of aerosol suspended in the air, and mathematically it is the integral of aerosol extinction coefficient through the whole atmosphere column. Before retrieving AOD from satellite-measured top-of-atmosphere reflected radiance, the non-aerosol contributions (e.g. surface, atmospheric gases) to the total radiance need to be separated from the aerosol signal to obtain AOD accurately. This is challenging because (i) the aerosol contribution to the total radiance is relatively small; (ii) the satellite instrument cannot penetrate cloud layer and highly reflective surface, producing misrepresentations of the data. Therefore, many algorithms have been developed for different types of scenes. For example, the Moderate Resolution Imaging Spectroradiometer (MODIS) has two distinct algorithms for aerosol retrievals over dark surface (e.g. ocean, vegetated land) and bright-reflecting surface (e.g. desert, urban area) (Sayer et al. 2014; Levy et al. 2007), i.e. Dark Target and Deep Blue. Although recent data collection has merged those two products providing global coverage, users need to pay attention to the quality of the data before applying it to the air quality study.

Besides MODIS, AOD is also provided by some other space-borne instruments, e.g. the Visible Infrared Imaging Radiometer Suite (VIIRS) aboard the joint NASA/NOAA Suomi

Table 11.1 Satellite datasets for air quality and climate variables.

Air quality and meteorological variables	Satellite platform and instrument sensor	Spatial resolution	Temporal resolution and coverage	Data archive
AOD, FRP	Terra and aqua: moderate resolution imaging spectroradiometer (MODIS)	250, 500 m, 1 km	1–2 days, 2000–present	Earthdata (https://cmr.earthdata.nasa.gov/search/concepts/C1000000505-LPDAAC_ECS.html)
AOD, FRP, $PM_{2.5}$[a]	NASA/NOAA suomi-NPP: visible infrared imaging radiometer suite (VIIRS)	375–750 m	1–2 days, 2011–present	Earthdata (https://earthdata.nasa.gov/earth-observation-data/near-real-time)
AOD, AI, SO_2	NASA/NOAA suomi national polar-orbiting partnership (suomi NPP): ozone mapping and profiler suite (OMPS)	50 × 50 km	101 minute, daily, 2017–present	Earthdata (https://earthdata.nasa.gov/eosdis/sips/ozone-sips)
AOD	Terra: multiangle imaging spectroradiometer (MISR)	17.6 × 17.6 km, 1 × 1°	Daily, monthly, 2000–present	NASA (https://asdc.larc.nasa.gov/project/MISR)
AOD (fine and coarse mode)	PARASOL: polarization and directionality of the earth's reflectances (POLDER)	18.5 × 18.5 km	Daily, 2004–present	GRASP open (https://www.grasp-open.com/products/polder-data-release)
AOD, AI, NO_2, SO_2, O_3	Aura: ozone monitoring instrument (OMI)	13 × 24 km	Daily, 2004–present	Earthdata (https://earthdata.nasa.gov)
AOD (profile), aerosol type	CALIPSO: cloud-aerosol Lidar with orthogonal polarization (CALIOP)	5 km H, 60 mV[b]	Daily, 2006–present	Earthdata (https://earthdata.nasa.gov) https://www-calipso.larc.nasa.gov
AI, NO_2, SO_2, CO, O_3,	ESA sentinel-5P: tropospheric monitoring instrument (TROPOMI)	7 × 3.5 km	Daily, 2017–present	Copernicus open access hub (https://scihub.copernicus.eu) http://www.tropomi.eu/data-products/level-2-products
$PM_{2.5}$	Derived product from a combination of MODIS, MISR, SeaWiFS data	0.01 × 0.01°	Yearly, 1998–2016	https://sedac.ciesin.columbia.edu/data/set/sdei-global-annual-gwr-pm2-5-modis-misr-seawifs-aod/data-download
NO_2, SO_2, O_3,	ERS-2: global ozone monitoring experiment (GOME)	40 × 320 km	Daily, 28 June 1995–02 July 2011	Earth online (https://earth.esa.int/eogateway/catalog/ers-2-gome-total-column-amount-of-trace-gases-product)
NO_2, SO_2, O_3,	MetOp-A: GOME-2	80 × 40 km	Daily, 2012–2020	CLASS (https://www.avl.class.noaa.gov/saa/products/welcome)

(Continued)

Table 11.1 (Continued)

Air quality and meteorological variables	Satellite platform and instrument sensor	Spatial resolution	Temporal resolution and coverage	Data archive
NO_2, SO_2, O_3,	Envisat: the scanning imaging absorption spectrometer for atmospheric chartography (SCIAMACHY)	30×30 km	March 2002–April 20124	https://www.sciamachy.org/products
CO	Terra: measurement of pollution in the troposphere (MOPITT)	17.6×17.6 km, $1 \times 1°$	Daily, monthly, 2000–present	UCAR (https://www2.acom.ucar.edu/mopitt)
CO, O_3, dust storms	Aqua: atmospheric infrared sounder (AIRS)	$1 \times 1°$	Daily, 8-days, monthly, 2002–present	Earthdata (https://earthdata.nasa.gov)
Precipitation	Polar operational environmental satellites (POES) spacecraft: ensemble tropical rainfall potential (eTRaP) forecast	4×4 km	3 hours, 2015–present	NOAA (https://www.ssd.noaa.gov/PS/TROP/etrap.html)
Precipitation	POES spacecraft: hydro-estimator rainfall	4×4 km	6 hours, 2002–present	https://www.ospo.noaa.gov/Products/atmosphere/ghe/index.html
Precipitation	Tropical rainfall measuring mission (TRMM) satellite: TRMM climate rainfall products	$0.25 \times 0.25°$	3 hours, December 1997–April 2015	http://rain.atmos.colostate.edu/CRDC/datasets/trmm.html
Precipitation	The global precipitation measurement mission (GPM): The integrated multi-satellite retrievals for GPM (IMERG)	10×10 km/ $0.1 \times 0.1°$	30 minutes, 3 hours, 1 day, 2000-06–present	https://gpm.nasa.gov/data/directory
Total atmospheric moisture	POES spacecraft: total precipitable water (SSM/I)	25×25 km	4 hours, 2002–present	https://www.ospo.noaa.gov/Products/land/spp/sharedprocessing.html#WV
Average solar radiation absorbed, average solar radiation absorbed, and outgoing longwave radiation	POES spacecraft	$1 \times 1°$	Daily, 1998–present	https://www.ospo.noaa.gov/Products/atmosphere/rad_budget.html
Wind speed and wind direction	POES spacecraft: high-density infrared cloud drift winds	30×30 km	2 hours, 1998–present	https://www.ospo.noaa.gov/Products/atmosphere/wind.html
Wind speed and wind direction	Geostationary operational environmental satellites (GOES) spacecraft: GOES spacecraft	60×60 km	1 hours, 2016–present	https://www.ospo.noaa.gov/Products/atmosphere/hdwinds/goes.html

[a] Need users to convert VIIRS AOD to surface $PM_{2.5}$.
[b] H: horizontal; V: vertical.

National Polar-orbiting Partnership (Suomi NPP) satellite, Multiangle Imaging SpectroRadiometer (MISR) aboard NASA Terra satellite, POLarization and Directionality of the Earth's Reflectances (POLDER) aboard the French Polarization & Anisotropy of Reflectances for Atmospheric Sciences coupled with Observations from a Lidar (PARASOL) satellite, Ozone Monitoring Instrument (OMI) from Aura Satellite. Aerosol profile data is available from Cloud-Aerosol Lidar with Orthogonal Polarization (CALIOP) aboard the Cloud-Aerosol Lidar and Infrared Pathfinder Satellite Observation (CALIPSO) satellite. The temporal and spatial resolutions of those AOD products are different (see Table 11.1). Users need to choose the best product for their study area and specific applications.

11.2.1.2.2 Aerosol Index (AI)

As mentioned above, aerosols absorb and scatter incoming sunlight, which reduces visibility and increases the optical depth. The AI, different from AOD, is the multiplication of AOD and the Angstrom Exponent, which is the slope of wavelength-dependent AOD in log scale and contains the information of particle size and type.

Satellite-retrieved AI product is useful for identifying and tracking the long-range transport of volcanic ash from volcanic eruptions, smoke from wildfires or biomass burning events, and dust from desert dust storms, even tracking over clouds and areas of snow and ice. The satellite products include OMI from Aura Satellite, TROPOspheric Monitoring Instrument (TROPOMI) from ESA Sentinel-5P Satellite, and Ozone Mapping and Profiler Suite (OMPS) from NASA/NOAA Suomi NPP.

11.2.1.2.3 Particulate Matter (PM)

PM is one of six criteria pollutants regulated by the US EPA (and in many other countries). $PM_{2.5}$, by its name, is defined as the particulate matter with an aerodynamic diameter less than 2.5 μm within a given volume of air near the surface. There are some differences between AOD and $PM_{2.5}$ from the satellite product: (i) AOD is an optical measurement, $PM_{2.5}$ is usually a mass concentration measurement; (ii) AOD is an integrated total column measurement, $PM_{2.5}$ is a ground measurement; (iii) AOD is an area-averaged measurement, $PM_{2.5}$ is a point measurement.

Despite the differences between the two parameters, they do correlate with each other, and there are several different techniques to convert AOD to $PM_{2.5}$. It is important to note that while there is a relationship between AOD and $PM_{2.5}$, there are other factors that can affect AOD, like humidity, the vertical distribution of aerosols, and the shape of the particles. For example, in a humid environment, aerosol particles can absorb water and swell up. This process, called the humidification effect, will increase the aerosol scattering, thus AOD, without increasing the concentration of the particles. Another example is in the vicinity of wildfire, and the smoke plume usually injects into the atmosphere at a higher altitude; therefore, the total column AOD is higher, but the surface $PM_{2.5}$ level is low because particles will be transported downwind instead of immediately settled down to the surface near the wildfire. Under such conditions, the relationship between AOD and $PM_{2.5}$ is rather complicated than a simple linear relationship. Hence, some regression models (Paciorek et al. 2008; Liu et al. 2005; Liu et al. 2004) are proposed to estimate the surface $PM_{2.5}$ using AOD together with other meteorological parameters.

One existing global $PM_{2.5}$ dataset is derived from the combination of MODIS, MISR, and SeaWiFS AOD products (Van Donkelaar et al. 2016), and this dataset can be accessed through the NASA Socioeconomic Data and Applications Center (SEDAC). Another way to obtain global $PM_{2.5}$ data is converting VIIRS AOD to surface $PM_{2.5}$, but this needs to be produced by users. The input data and conversion script can be accessed through Applied Remote Sensing Training Program (ARSET) GitHub site (https://github.com/NASAARSET).

11.2.1.2.4 Nitrogen Dioxide (NO₂)

NO_2 is also a criteria pollutant regulated by the US EPA. Most of the NO_2 pollution is at ground level, where ecosystems and people live and breathe. Once in the air, NO_2 can aggravate respiratory conditions in humans, especially those with asthma, leading to an increase of symptoms, hospital admissions, and emergency visits. Long-term exposure can lead to the development of asthma and potentially increase susceptibility to respiratory infections. The primary sources of NO_2 are lighting, burning of fossil fuels, automobiles, power plants, and industrial facilities. NO_2 reacts with other chemicals in the atmosphere, forming particulate matter and ozone, producing haze and even acid rain, and contributing to nitrogen pollution in coastal waters.

The total and tropospheric NO_2 columns can be retrieved from the Global Ozone Monitoring Experiment (GOME) aboard the European Remote-Sensing Satellite-2 (ERS-2) satellite, the GOME-2 aboard the MetOp-A satellite, and the SCanning Imaging Absorption spectroMeter for Atmospheric CHartographY (SCIAMACHY) aboard the Envisat satellite. In addition, there are two instruments in space that provide NO_2 data: OMI on NASA Aura satellite and the TROPOMI on the ESA Sentinel-5 satellite. Note that since 2008, there has been a 50% data loss from OMI due to its row anomaly issue. Users who are interested in using OMI data need to apply additional filters or quality flags contained in the data files. TROPOMI has a higher spatial resolution than OMI. However, the historical data record is shorter since Sentinel-5 was launched in 2017. NO_2 has a short lifetime near the surface (a few hours) and a longer lifetime in the upper troposphere. The satellite product only provides the total column of NO_2 concentration, users who are interested in surface NO_2 need to use information about pollution sources and atmospheric chemistry to infer information about the vertical distribution of a pollutant, including the amount near the surface. A nice description of inferring ground-level NO_2 from TROPOMI data can be found in Cooper et al. (2020).

11.2.1.2.5 Sulfur Dioxide (SO₂)

SO_2 is another US EPA regulated criteria pollutant. It is a short lifetime gas primarily produced by power plants, refineries, metal smelting, burning of fossil fuels, and volcanoes. SO_2 can significantly influence the atmosphere in many ways depending on its vertical location. Surface SO_2 is extremely toxic to humans by irritating the eyes, nose, and lungs (Chen et al. 2007). It is also a precursor to sulfuric acid, which is a major constituent of acid rain, causing severe environmental problems. When SO_2 is lofted or injected into the free troposphere, it forms aerosols that can alter cloud reflectivity and precipitation (Boucher and Lohmann 2017). In the stratosphere, volcanic SO_2 forms sulfate aerosols that can result in climate change (Rasch et al. 2008).

Recent satellite measurements of SO_2 are from some hyperspectral instruments, such as the GOME, GOME-2, OMI, and SCIAMACHY, which offer advantages for the detection of SO_2 due to their spectrum coverage and resolution (Krueger et al. 2009; Fioletov et al. 2013), and valuable information of SO_2 can be retrieved from using differential optical absorption spectroscopy (DOAS) algorithms (Lee et al. 2008).

11.2.1.2.6 *Carbon Monoxide (CO)*

CO, another US EPA regulated criteria pollutant, is a poisonous, odorless, and colorless gas and is usually produced by incomplete combustion of fossil fuels, biomass burning, or any carbon contained materials. It is one of the longest-lived, naturally occurring atmospheric carbon compounds, therefore, plays an important role in atmospheric chemistry (Holloway et al. 2000).

Satellite measurements of CO are useful for analyzing the distribution, transport, sources, and sinks of CO in the troposphere and can be used to observe how it interacts with land and ocean biospheres (Palve et al. 2018; Yurganov et al. 2008). Existing satellite measurements are from Measurements of Pollution in the Troposphere (MOPITT), Atmospheric InfraRed Sounder (AIRS), and TROPOMI, mainly based on the CO absorption bands around 4.6 μm.

11.2.1.2.7 *Ozone (O₃)*

O_3 is an US EPA regulated criteria pollutant as well. It can be either good or bad, depending on where it is found in the atmosphere. In the stratosphere, O_3 absorbs UV radiation, protecting humans, animals, and surface plants. In the troposphere or near ground level, however, O_3 can aggravate existing health problems in humans, destroy vegetation and materials like rubber. O_3 is not emitted directly into the atmosphere but instead forms from the chemical reaction between nitrogen oxides (NO_x) and VOCs (Finlayson-Pitts and Pitts 1993) in the presence of sunlight. NO_x and VOCs are emitted primarily from cars, power plants, and other industrial facilities.

O_3 column data can be retrieved from various instrument sensors, including OMI aboard the Aura satellite, TROPOMI aboard the ESA Sentinel-5P satellite, GOME aboard the ERS-2 satellite, GOME-2 aboard the MetOp-A satellite, and SCIAMACHY aboard the Envisat satellite. Again, they have different spatial resolution, temporal resolution and coverage. Users need to choose the best product for their study area and specific applications.

11.2.1.2.8 *Fire Radiative Power (FRP)*

The FRP is useful for studying the spatial and temporal distribution of fire, locating persistent hot spots and the source of air pollution from smoke, and even indicating the among of pollutants emitted from the fire (Kaiser et al. 2012). Fire is not one of the air pollutants. However, it is always associated with air quality problems. For example, wildfire caused by lightning or human activities usually emits a large amount of PM, CO, and NO_x, which can induce server regional air quality issues.

Satellite measurement also provides valuable data to detect wildfire from space. The VIIRS and MODIS Fire and Thermal Anomalies, usually expressed as FRP, can be used to detect active fire and thermal anomalies (Li et al. 2018).

11.2.1.2.9 Climate Data

Climate data, such as meteorological variables (such as precipitation, winds, boundary layer height, temperature, humidity, and radiation) and large-scale meteorological conditions (e.g. front), also play significant roles in air quality (Pearce et al. 2011; Hu et al. 2021a). For example, precipitation washout could contribute over half of the $PM_{2.5}$ removal locally (Hu et al. 2021b). Air pollutants can be transported easily from the upstream source regions to the downwind receptor areas by strong winds, and local air pollutant concentrations are largely governed by the boundary layer height (Hu et al. 2013b; Wang et al. 2017; Wu et al. 2017; Hu et al. 2019; Miao et al. 2019).

There are many valuable satellite measurement products (see Table 11.1). For example, the Tropical Rainfall Measuring Mission (TRMM), which is a joint mission from NASA and Japan Aerospace Exploration Agency (JAXA), is providing the tropical and subtropical precipitation process data over a long time period from 1998 to 2015 (Maggioni et al. 2016). After that, building upon the success of the TRMM, another international network of satellites called the Global Precipitation Measurement (GPM) mission centers on measuring precipitation from space (Hou et al. 2014). The Ensemble Tropical Rainfall Potential (eTRaP) is a rain forecast product from Polar Operational Environmental Satellites (POES) spacecraft (Ebert et al. 2011). Some other climate data, such as atmospheric moisture, radiations and winds, are available from NOAA satellites that are also included in Table 11.1.

11.2.2 Ground-Based *In Situ* Observation Datasets

Using United States as an example, the existing ground-based observations for air pollutants are primarily from the US EPA. In order to reduce and control air pollution nationwide in the United States, the Clean Air Act (CAA), which is the primary federal air quality law, was initially enacted in 1963 and amended many times since. The CAA requires that the air pollution control agencies from state, local, and tribal level to monitor the concentrations of certain air pollutants, including the six criteria air pollutants (PM, NO_2, SO_2, CO, O_3, and lead). There are also some superfund sites, which were established starting from the Comprehensive Environmental Response, Compensation and Liability Act (CERCLA) in 1980. For example, besides of the criteria air pollutants, the Fresno Supersite observables include *in situ*, continuous, short-duration measurements of SO_4^{-2}, NO_3^-, carbon, light absorption, light extinction, numbers of particles in discrete size bins (0.01 to ~10 μm), reactive gases (NO_y, HNO_3, peroxyacetyl nitrate [PAN], NH_3), and single particle characterization by time-of-flight mass spectrometry (Watson et al. 2000). This dataset is very useful for researches related to air quality, health, and control policy (Fann et al. 2016).

Advantages of the EPA air pollutant measurements are their extensive temporal (more than 40 years) and spatial coverage (see Figure 11.2 since the data is required to be reported by the law (available at https://aqs.epa.gov/aqsweb/airdata/download_files.html). For example, Figure 11.2 shows the O_3 ground-based monitoring sites in the US continent.

11.2.2.1 Basics of the Ground-Based *In Situ* Observations

Ambient ground-based *in situ* air monitoring is the systematic, long-term assessment of pollutant levels by measuring the concentrations of various types of pollutants in the ambient air. Again, using the EPA Air Quality System (AQS) Database as an example, it collects the

AQS site

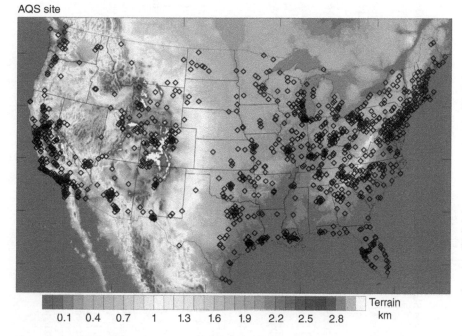

Figure 11.2 EPA Air Quality System (AQS) O_3 ground-based monitoring sites in the US continent.

measured data, along with metadata about the site and monitoring equipment and associated quality assurance data from the state, local, and tribal monitoring agencies. The purpose of the monitoring program is to (i) provide air pollution data to the public in a timely manner, (ii) support compliance with ambient air quality standards and emissions strategy development, and (iii) support for air pollution research studies (40 CFR Part 58 Appendix D.1).

The primary portal for public access to EPA air quality data is the Air Data website (https://www.epa.gov/outdoor-air-quality-data), where daily data can be downloaded directly. The real-time data is also available within AirData using the AirNow application programming interface (API) (https://docs.airnowapi.org). Data from this API is available to anyone but is targeted at researchers and application developers who are familiar with the data. New users need to sign up an account to obtain a key to use this API. Our previous study used the EPA API real-time data and an improved machine learning approach using model regularization and optimization (Zhu et al. 2018). More information can be found in this paper, and the data can be downloaded from the GitHub (https://github.com/OEH-AI-CAI/air-quality). In the next section, we summarize the ground-based *in situ* observations over the world, not only for air pollutants but also for meteorological variables.

11.2.2.2 Ground-Based *In Situ* Products

In this section, we summarized the commonly used ground-based *in situ* observations over the world, including United States, Canada, Europe, Japan, Taiwan, Hong Kong, and Australia (Table 11.2). All these countries/areas have been monitoring at least the

Table 11.2 Ground-based *in situ* air pollutants and meteorological observations over the world.

Air pollutants and meteorological variables	Observation products	Countries/areas	Temporal resolution	Data archive
PM_{10}, $PM_{2.5}$, PM_1, NO_2, SO_2, CO, O_3, and lead	US EPA air quality system (AQS) database	United States	Hourly	https://docs.airnowapi.org
PM_{10}, $PM_{2.5}$, PM_1, NO_2, SO_2, CO, O_3, lead, SO_4^{-2}, NO_3^-, carbon, particle size distributions, NO_y, HNO_3, PAN, NH_3, etc.	Superfund Sites in the US EPA air quality system (AQS) database	United States	Hourly	https://www.epa.gov/superfund/search-superfund-sites-where-you-live
PM_{10}, $PM_{2.5}$, NO_2, SO_2, CO, O_3	The National Air Pollution Surveillance (NAPS) program, Canada-Wide Air Quality Database (CWAQD)	Canada	Hourly	https://open.canada.ca/data/en/dataset/1b36a356-defd-4813-acea-47bc3abd859b
PM_{10}, $PM_{2.5}$, NO_2, SO_2, CO, O_3	The European Environment Agency (EEA) gathers air pollution data from a wide range of sources	Europe	Hourly	https://www.eea.europa.eu/data-and-maps/data/aqereporting-9 https://discomap.eea.europa.eu/map/fme/AirQualityUTDExport.htm
PM_{10}, $PM_{2.5}$, NO_2, SO_2, CO, O_3	National Institute for Environmental Studies (NIES)	Japan	Hourly	https://www.nies.go.jp/db/index-e.html
PM_{10}, $PM_{2.5}$, NO_2, SO_2, CO, O_3	Taiwan Air Quality Monitoring Network (AQMN)	Taiwan	Hourly	https://airtw.epa.gov.tw/ENG/EnvMonitoring/Central/CentralMonitoring,aspx
PM_{10}, $PM_{2.5}$, NO_2, SO_2, CO, O_3	Hong Kong Environmental Protection Department	Hong Kong	Hourly	www.epd.gov.hk/epd/english/environmentinhk/air/data/air_data.html
PM_{10}, $PM_{2.5}$, NO_2, SO_2, CO, O_3	ACT Health	Australia	Hourly	https://www.data.act.gov.au/Environment/Air-Quality-Monitoring-Data/94a5-zqnn
Temperature, pressure, humidity, wind speed, wind direction, precipitation, etc.	NOAA- NCDC	Worldwide (mainly in the US, South Pacific, and Europe)	Hourly	https://www.ncdc.noaa.gov/oa/climate
Temperature, pressure, humidity, wind speed, wind direction, precipitation, etc.	MesoWest	United States	Hourly	http://mesowest.utah.edu
Temperature, pressure, humidity, wind speed, wind direction, precipitation, etc.	Oklahoma Mesonet	Oklahoma State	5 minutes	https://www.mesonet.org
Temperature, pressure, humidity, wind speed, wind	New York Mesonet	New York State	5 minutes	http://nysmesonet.org

six criteria air pollutants. Some supersites collect much more atmospheric composition data.

The NOAA-National Climatic Data Center (NCDC) open-access database has been collecting climate data from more than 75 000 stations in 180 countries and territories, located (mainly) in the United States, South Pacific, and Europe (Arguez et al. 2012). Meteorological data can also be obtained from MesoWest (http://mesowest.utah.edu), a project within the Department of Meteorology at the University of Utah, which has been aggregating meteorological data since 2002 (Horel et al. 2002). The Mesonet is a regional network of automatic weather observing stations designed to diagnose weather features. For example, The Oklahoma Mesonet is a rural network of 120 meteorological stations with minimal influences from urban landscapes (Basara et al. 2008; McPherson et al. 2007). Each of the Mesonet stations is located within a fenced 100 m^2 plot of land and measures more than 20 environmental variables, including wind at 2 and 10 m, air temperature at 1.5 and 9 m AGL, and short-wave radiation. Data from Oklahoma Mesonet have been widely used in weather forecasting and atmospheric research (e.g. Hu et al. 2013c, d, 2016). The New York State Mesonet (NYSM; http://nysmesonet.org) consists of 181 state-of-the-art environmental monitoring stations, including 126 Standard sites that serve as the foundation of an Early Warning Severe Weather Detection network for the entire state of New York (Brotzge et al. 2020; Shrestha et al. 2021). The Oklahoma Mesonet and New York State Mesonet represent the best observing network in the United States with high quality data, and such high-quality network is not available in most other states.

11.3 Applications of Machine Learning in Air Quality Study

The application of satellite data and machine learning algorithms to study air quality only emerged recently. The main reasons for that are (i) improved satellite measurements and retrieval algorithms make more and more data become available in the past decade, especially the near real-time global coverage data; (ii) increased ability of machine learning algorithms to capture features of independent variables and solve complex nonlinear problems without pre-existing physical-based knowledge; (iii) high-performance computing becomes more powerful and cheaper for many applications, including satellite data processing and machine training. Most existing studies apply machine learning algorithms to train data from satellite retrievals, ground-based sensors, CTM simulations, and many other sources. After training, models are tested on new data to evaluate the accuracy and generalizability of model predictions to situations not encountered during the training process (Figure 11.1). Once a validated trained model is determined, then it can be used to forecast air quality. The trained models can be updated once more predictor parameters become available.

The most focused topic is predicting surface $PM_{2.5}$ concentration using satellite-retrieved AOD. Since satellite data provides broader spatial coverage and moderate spatial resolution, it can efficiently fill in data for areas lacking station observation. However, as mentioned in Section 11.2, AOD reflects the total extinction of atmospheric column due to aerosols, while $PM_{2.5}$ concentration reflects the localized near-surface particle mass concentration. Therefore, directly using AOD to represent ground-level $PM_{2.5}$ may be inaccurate (Lee et al. 2011).

Some studies have incorporated geographical and meteorological variables as additional predictors to improve model prediction performance (Hu et al. 2013a; Liu et al. 2009). These widely used statistical models include linear regression models (Arvani et al. 2016), multiple regression models (Gupta and Christopher 2009), land-use regression models (Kloog et al. 2011), and geographically weighted regression (Chu et al. 2019).

Statistical models use historical data to predict air quality without considering physical and chemical processes. The principle of those models is exploring the relationships between variables based on possibility and statistical average. Under some circumstances, a well-specified regression model may provide reasonable results. However, the relationships between air pollutants and influential factors are highly nonlinear. Therefore, more advanced algorithms, e.g. machine learning, are needed to account for proper nonlinear modeling of air quality prediction. Below is a summary of existing studies applying machine learning algorithms and satellite observations to predict air quality, primarily for predicting $PM_{2.5}$. The same methodology can also be applied to other pollutants based on specific research interests.

11.3.1 Shallow Learning

The biggest challenge of applying satellite data to forecast the surface $PM_{2.5}$ is to convert satellite measurement to surface relevant parameters. To improve the prediction performance, one typical way is to combine satellite-retrieved AOD together with CTM simulations as input to train models. Murray et al. (2018) combined AOD from MODIS retrieval and $PM_{2.5}$ simulated from Community Multiscale Air Quality (CMAQ) model coupled with Random Forests algorithm to predict $PM_{2.5}$ in the United States and argued their approach to be highly applicable for estimating other environmental risks from both satellite imagery and numerical model simulation. Xue et al. (2019) utilized MODIS AOD, Weather Research and Forecasting (WRF) model, and CMAQ model outputs to train data using elastic-net regression and predicted the spatial–temporal variation of surface $PM_{2.5}$, which is comparable to ground measurement. However, previous studies also reported decreased prediction accuracy when predicting $PM_{2.5}$ levels outside the model-training period. Hence, a set of ensemble learning methods, including random forests, generalized additive models, and extreme gradient boosting, have been performed to predict daily and monthly $PM_{2.5}$ concentration in China (Xiao et al. 2018). In addition to ensemble learning algorithms, Just et al. (2018) first trained a model to validate and correct AOD measurement from Aqua and Terra satellites using AERONET retrievals, then applied the corrected AOD to predict the surface $PM_{2.5}$ concentration with increased accuracy of 10%. Besides the $PM_{2.5}$, Zhan et al. (2018) developed a random-forest-spatiotemporal-kriging model to estimate the daily ambient NO_2 concentrations from OMI data and geographic covariates from multiple sources. Their models show good prediction performance for daily and spatial predictions.

11.3.2 Deep Learning

Deep learning, a subfield of machine learning, usually trains very large neural networks of multiple hidden layers to extract the features in the dataset and learns the representations linking inputs and outputs automatically based on nonlinear transformations of data (Lecun et al. 2015; Alzubaidi et al. 2021). It is performed over successive layers of the network. During the training procedure, the computer is shown repeated examples of inputs and expected outputs and learns weights for each layer that define the input–output function of the machine (Lecun et al. 2015). These weights are adjusted iteratively to minimize the error between model predictions and the known output labels in the training dataset (Lecun et al. 2015). After training, models are evaluated using the new dataset to predict the situations that are not encountered during training. Deep learning performs better in extracting the spatiotemporal features than shallow learning and can be ideal for satellite imagery learning since satellite data is highly spatiotemporal correlated. The spatiotemporal deep learning (STDL) architectures can be constructed by connections or coupling of temporal modules (e.g. recurrent neural network, long short-term memory network, gated recurrent unit network) and spatial modules (e.g. convolutional neural network [CNN], stacked autoencoder, deep belief network). The STDL architectures can be ideal for air quality forecasts due to their ability to capture complex spatiotemporal features across scales of pollutant evolutions. A more detailed summary of different deep learning modules and air quality applications can be found in the review paper by Liao et al. (2020).

The application of deep learning in air quality prediction usually employs spatiotemporal data from satellites using one or some of the deep learning modules mentioned above. Li et al. (2017) developed a geo-intelligent approach to estimate surface $PM_{2.5}$ from MODIS AOD and NDVI data, considering the geographical distance and spatiotemporal correlation of $PM_{2.5}$ in a deep belief network. This approach performs significantly better than the traditional neural network. Using the same deep belief networks, Shen et al. (2018) further predicted the surface $PM_{2.5}$ with high accuracy by only using satellite-measured TOA reflectance without using the retrieved AOD. This study provides an alternative technique to estimate $PM_{2.5}$ avoiding the potential errors in AOD retrievals. Another example of applying the deep learning approach to predict the surface $PM_{2.5}$ is from Sun et al. (2019). They proposed a deep neural network-based $PM_{2.5}$ prediction model to capture the spatiotemporal variability of ground $PM_{2.5}$ using AOD retrieved from the Himawari-8 satellite along with the meteorological variables. Their study indicates that the deep learning model achieves the best performance among many other models and demonstrates the power of deep learning in environmental research.

The spatial features in satellite data products can be well extracted by the deep CNN module. Hong et al. (2019) predicted the global variations in $PM_{2.5}$ concentrations using satellite true color images and ground-level $PM_{2.5}$ measurements. Their results indicate that the exception architecture performs the best among some well-known convolutional-based architectures in predicting $PM_{2.5}$. The methodologies described in the above literature can also be adapted to predict other surface pollutants with inputs from satellite measurement. For example, Eslami et al. (2019) used deep CNN architecture to successfully predict hourly ozone concentrations over Seoul, South Korea. Readers need to choose the best machine learning modules based on their research interests and the features of different modules.

11.4 An Application Practice Example

In this section, we provide an application practice example of implementing satellite data and machine learning for dust detection from our previous study (Lee et al. 2021). The processed data and machine learning codes can be downloaded from Github (https://github.com/OEH-AI-CAI/dust-detection).

11.4.1 Satellite Data Acquisition and Variable Selections

In this study, our goal is to predict dust aerosol using machine learning from the VIIRS measurements on a global scale. The dust pixel is labeled using the state-of-the-art aerosol product from the CALIOP. Since the two satellites have different orbit tracks, we first collocated daytime VIIRS and CALIOP data. The collocated data contain merged aerosol/cloud layer retrievals from CALIOP version 3 (V3) operational products and the level-1b radiance observations from the 16 moderate-resolution bands of VIIRS along with the viewing/illumination geometries. In the V3 operational algorithm, the CALIP pixels are classified into multiple categories. We include both "pure dust" and "dust mixtures" pixels labeled as "dust," and others are labeled as "non-dust."

The predictor variables include the 16 moderate-resolution bands of VIIRS, four geometric variables, pixel time, and locations (latitude and longitude). The geometric variables are included in the training because the reflection of sunlight by dust aerosols is dependent on solar and viewing geometries. In addition, the transmittance and emission of a dust layer in the thermal infrared region are also dependent on viewing geometry. The time and geo-location of the pixel are also included because dust events are known to be dependent on season and geo-location.

11.4.2 Machine Learning and Deep Learning Algorithms

For algorithm selection, according to the published research, both conventional machine learning and deep-learning-based algorithms are selected, including the logistical regression (LR), K nearest neighbors (KNN), random forests (RFs), feedforward neural networks (FFNNs), and convolutional neural networks (CNNs). LR is a classification model that uses a logistic function, which converts the multivariate predictor variables into the output between 0 (non-dust) and 1 (dust). The main advantage of using LR is that the model is relatively easy to interpret in a physical sense due to its simple nature. As a nonparametric method, KNN methods classify the input features into two or more classes by assigning the input features to the class that is most common among its K nearest neighbors. The main advantage of using KNN is that, since it considers the K nearest testing data, the classes do not have to be linearly separable. RFs are an ensemble learning technique that performs classification or regression by building a structure of multiple decision trees. The performance of RFs is considered to be comparable with the best supervised learning algorithms. Artificial neural networks (ANNs) have been widely used in remote-sensing data. Here, we implement one of the most basic structures of ANN, the FFNNs. The main advantage of using CNNs is that CNNs are able to capture the spatial dependencies in an image by applying filters. Furthermore, CNNs extract and reduces images without removing critical features.

The processed data and machine learning codes are available from Github. Readers can practice of using various machine learning and deep learning algorithms by manipulating the data and sample code.

References

Alzubaidi, L., Zhang, J., Humaidi, A.J. et al. (2021). Review of deep learning: concepts, CNN architectures, challenges, applications, future directions. *Journal of Big Data* 8: 53 https://doi.org/10.1186/s40537-021-00444-8.

Arguez, A., Durre, I., Applequist, S. et al. (2012). NOAA's 1981–2010 US climate normals: an overview. *Bulletin of the American Meteorological Society* 93 (11): 1687–1697.

Arvani, B., Pierce, R.B., Lyapustin, A.I. et al. (2016). Seasonal monitoring and estimation of regional aerosol distribution over Po valley, northern Italy, using a high-resolution MAIAC product. *Atmospheric Environment* 141: 106–121, https://doi.org/10.1016/j.atmosenv.2016.06.037.

Basara, J.B., Hall, P.K., Schroeder, A.J. et al. (2008). Diurnal cycle of the Oklahoma city urban heat island. *Journal of Geophysical Research-Atmospheres* 113 (D20): https://doi.org/10.1029/2008jd010311.

Boucher, O. and Lohmann, U. (2017). The sulfate-CCN-cloud albedo effect. *Tellus Series B: Chemical and Physical Meteorology* 47: 281–300, https://doi.org/10.3402/tellusb.v47i3.16048.

Brotzge, J.A., Wang, J., Thorncroft, C.D. et al. (2020). A technical overview of the New York state Mesonet standard network. *Journal of Atmospheric and Oceanic Technology* 37 (10): 1827–1845. https://doi.org/10.1175/jtech-d-19-0220.1.

Byun, D. and Schere, K.L. (2006). Review of the governing equations, computational algorithms, and other components of the models-3 community multiscale air quality (CMAQ) modeling system. *Applied Mechanics Reviews* 59 (2): 51–77.

Cai, C., Geng, F., Tie, X. et al. (2010). Characteristics and source apportionment of VOCs measured in Shanghai, China. *Atmospheric Environment* 44 (38): 5005–5014.

Carmichael, G.R., Sandu, A., Chai, T. et al. (2008). Predicting air quality: improvements through advanced methods to integrate models and measurements. *Journal of Computational Physics* 227 (7): 3540–3571.

Chen, T.-M., Kuschner, W.G., Gokhale, J., and Shofer, S. (2007). Outdoor air pollution: nitrogen dioxide, sulfur dioxide, and carbon monoxide health effects. *The American Journal of the Medical Sciences* 333: 249–256.

Chu, H.-J., Yang, C.-H., and Chou, C. (2019). Adaptive non-negative geographically weighted regression for population density estimation based on nighttime light. *ISPRS International Journal of Geo-Information* 8, https://doi.org/10.3390/ijgi8010026.

Cohen, A.J., Brauer, M., Burnett, R. et al. (2017). Estimates and 25-year trends of the global burden of disease attributable to ambient air pollution: an analysis of data from the global burden of diseases study 2015. *The Lancet* 389 (10082): 1907–1918.

Cooper, M.J., Martin, R.V., McLinden, C.A., and Brook, J.R. (2020). Inferring ground-level nitrogen dioxide concentrations at fine spatial resolution applied to the TROPOMI satellite instrument. *Environmental Research Letters* 15, https://doi.org/10.1088/1748-9326/aba3a5.

van Donkelaar, A., Martin, R.V., Brauer, M. et al. (2016). Global estimates of fine particulate matter using a combined geophysical-statistical method with information from satellites. *Environmental Science and Technology* 50: 3762–3772, https://doi.org/10.1021/acs.est.5b05833.

Ebert, E.E., Turk, M., Kusselson, S.J. et al. (2011). Ensemble tropical rainfall potential (eTRaP) forecasts. *Weather and Forecasting* 26 (2): 213–224.

Engel-Cox, J.A., Hoff, R.M., and Haymet, A.D. (2004). Recommendations on the use of satellite remote-sensing data for urban air quality. *Journal of the Air & Waste Management Association (1995)* 54: 1360–1371, https://doi.org/10.1080/10473289.2004.10471005.

Eslami, E., Choi, Y., Lops, Y., and Sayeed, A. (2019). A real-time hourly ozone prediction system using deep convolutional neural network. *Neural Computing and Applications* 32: 8783–8797, https://doi.org/10.1007/s00521-019-04282-x.

Fann, N., Wesson, K., and Hubbell, B. (2016). Characterizing the confluence of air pollution risks in the United States. *Air Quality, Atmosphere and Health* 9 (3): 293–301.

Finlayson-Pitts, B.J. and Pitts, J.N. (1993). Atmospheric chemistry of tropospheric ozone formation: scientific and regulatory implications. *Journal of Air & Waste Management Association* 43: 1091–1100, https://doi.org/10.1080/1073161x.1993.10467187.

Fioletov, V.E., McLinden, C.A., Krotkov, N. et al. (2013). Application of OMI, SCIAMACHY, and GOME-2 satellite SO_2 retrievals for detection of large emission sources. *Journal of Geophysical Research-Atmospheres* 118 (19): 11–399.

Fuhrer, J., Val Martin, M., Mills, G. et al. (2016). Current and future ozone risks to global terrestrial biodiversity and ecosystem processes. *Ecology and Evolution* 6 (24): 8785–8799.

Grell, G.A., Peckham, S.E., Schmitz, R. et al. (2005). Fully coupled "online" chemistry within the WRF model. *Atmospheric Environment* 39 (37): 6957–6975.

Gupta, P. and Christopher, S.A. (2009). Particulate matter air quality assessment using integrated surface, satellite, and meteorological products: 2. A neural network approach. *Journal of Geophysical Research* 114, https://doi.org/10.1029/2008jd011497.

Holloway, T., Levy, H., and Kasibhatla, P. (2000). Global distribution of carbon monoxide. *Journal of Geophysical Research-Atmospheres* 105: 12123–12147, https://doi.org/10.1029/1999jd901173.

Hong, K.Y., Pinheiro, P.O., and Weichenthal, S. (2019). Predicting global variations in outdoor $PM_{2.5}$concentrations using satellite images and deep convolutional neural networks. *arXiv* preprint arXiv:1906.03975.

Horel, J., Splitt, M., Dunn, L. et al. (2002). Mesowest: cooperative mesonets in the western United States. *Bulletin of the American Meteorological Society* 83 (2): 211–226.

Hou, A.Y., Kakar, R.K., Neeck, S. et al. (2014). The global precipitation measurement mission. *Bulletin of the American Meteorological Society* 95 (5): 701–722.

Hu, X.M. (2015). Boundary layer (atmospheric) and air pollution | air pollution meteorology. In: *Encyclopedia of Atmospheric Sciences*, 2e (ed. G.R. North, J. Pyle and F. Zhang), 227–236. Oxford: Academic Press https://doi.org/10.1016/B978-0-12-382225-3.00499-0.

Hu, X., Waller, L.A., Al-Hamdan, M.Z. et al. (2013a). Estimating ground-level PM(2.5) concentrations in the southeastern U.S. using geographically weighted regression. *Environmental Research* 121: 1–10, https://doi.org/10.1016/j.envres.2012.11.003.

Hu, X.M., Klein, P.M., and Xue, M. (2013b). Evaluation of the updated YSU planetary boundary layer scheme within WRF for wind resource and air quality assessments. *Journal of Geophysical Research-Atmospheres* 118 (18): 10–490.

Hu, X.-M., Klein, P.M., Xue, M. et al. (2013c). Impact of low-level jets on the nocturnal urban heat island intensity in Oklahoma City. *Journal of Applied Meteorology and Climatology* 52 (8): 1779–1802. https://doi.org/10.1175/jamc-d-12-0256.1.

Hu, X.-M., Klein, P.M., Xue, M. et al. (2013d). Enhanced vertical mixing associated with a nocturnal cold front passage and its impact on near-surface temperature and ozone concentration. *Journal of Geophysical Research-Atmospheres* 118 (7): 2714–2728. https://doi.org/10.1002/jgrd.50309.

Hu, X.M., Xue, M., Klein, P.M. et al. (2016). Analysis of urban effects in Oklahoma City using a dense surface observing network. *Journal of Applied Meteorology and Climatology* 55 (3): 723–741.

Hu, X.M., Xue, M., Kong, F., and Zhang, H. (2019). Meteorological conditions during an ozone episode in Dallas-Fort Worth, Texas, and impact of their modeling uncertainties on air quality prediction. *Journal of Geophysical Research-Atmospheres* 124 (4): 1941–1961.

Hu, X.M., Hu, J., Gao, L. et al. (2021a). Multisensor and multimodel monitoring and investigation of a wintertime air pollution event ahead of a cold front over eastern China. *Journal of Geophysical Research-Atmospheres* 126 (10): https://doi.org/10.1029/2020jd033538.

Hu, W., Zhao, T., Bai, Y. et al. (2021b). Importance of regional $PM_{2.5}$ transport and precipitation washout in heavy air pollution in the Twain-Hu Basin over Central China: observational analysis and WRF-Chem simulation. *Science of the Total Environment* 758: 143710.

Iskandaryan, D., Ramos, F., and Trilles, S. (2020). Air quality prediction in smart cities using machine learning technologies based on sensor data: a review. *Applied Sciences* 10 (7): 2401.

Just, A.C., De Carli, M.M., Shtein, A. et al. (2018). Correcting measurement error in satellite aerosol optical depth with machine learning for modeling $PM_{2.5}$ in the northeastern USA. *Remote Sensing* 10, https://doi.org/10.3390/rs10050805.

Kaiser, J.W., Heil, A., Andreae, M.O. et al. (2012). Biomass burning emissions estimated with a global fire assimilation system based on observed fire radiative power. *Biogeosciences* 9: 527–554, https://doi.org/10.5194/bg-9-527-2012.

Kloog, I., Koutrakis, P., Coull, B.A. et al. (2011). Assessing temporally and spatially resolved $PM_{2.5}$ exposures for epidemiological studies using satellite aerosol optical depth measurements. *Atmospheric Environment* 45: 6267–6275, https://doi.org/10.1016/j.atmosenv.2011.08.066.

Krueger, A.J., Krotkov, N.A., Yang, K. et al. (2009). Applications of satellite-based sulfur dioxide monitoring. *IEEE Journal of Selected Topics in Applied Earth Observations and Remote Sensing* 2: 293–298, https://doi.org/10.1109/jstars.2009.2037334.

Laj, P., Klausen, J., Bilde, M. et al. (2009). Measuring atmospheric composition change. *Atmospheric Environment* 43: 5351–5414, https://doi.org/10.1016/j.atmosenv.2009.08.020.

LeCun, Y., Bengio, Y., and Hinton, G. (2015). Deep learning. *Nature* 521: 436–444, https://doi.org/10.1038/nature14539.

Lee, C., Richter, A., Weber, M., and Burrows, J.P. (2008). SO_2 retrieval from SCIAMACHY using the weighting function DOAS (WFDOAS) technique: comparison with standard DOAS retrieval. *Atmospheric Chemistry and Physics* 8: 6137–6145.

Lee, H.J., Liu, Y., Coull, B.A. et al. (2011). A novel calibration approach of MODIS AOD data to predict $PM_{2.5}$ concentrations. *Atmospheric Chemistry and Physics* 11: 7991–8002, https://doi.org/10.5194/acp-11-7991-2011.

Lee, J., Shi, Y.R., Cai, C. et al. (2021). Machine learning based algorithms for global dust aerosol detection from satellite images: inter-comparisons and evaluation. *Remote Sensing* 13 (3): 456.

Levy, R. C., Remer, L. A., Mattoo, S., Vermote, E. F., and Kaufman, Y. J. (2007). Second-generation operational algorithm: Retrieval of aerosol properties over land from inversion of Moderate Resolution Imaging Spectroradiometer spectral reflectance. Journal of Geophysical Research-Atmospheres 112(D13211): https://doi.org/10.1029/2006JD007811.

Li, C., Hsu, N.C., and Tsay, S.-C. (2011). A study on the potential applications of satellite data in air quality monitoring and forecasting. *Atmospheric Environment* 45: 3663–3675, https://doi.org/10.1016/j.atmosenv.2011.04.032.

Li, T., Shen, H., Yuan, Q. et al. (2017). Estimating ground-level $PM_{2.5}$ by fusing satellite and station observations: a geo-intelligent deep learning approach. *Geophysical Research Letters* 44: 11985–11993, https://doi.org/10.1002/2017gl075710.

Li, F., Zhang, X., Kondragunta, S., and Csiszar, I. (2018). Comparison of fire radiative power estimates from VIIRS and MODIS observations. *Journal of Geophysical Research-Atmospheres* 123: 4545–4563, https://doi.org/10.1029/2017jd027823.

Liao, Q., Zhu, M., Wu, L. et al. (2020). Deep learning for air quality forecasts: a review. *Current Pollution Reports* 6: 399–409, https://doi.org/10.1007/s40726-020-00159-z.

Liu, Y., Park, R.J., Jacob, D.J. et al. (2004). Mapping annual mean ground-level $PM_{2.5}$ concentrations using multiangle imaging Spectroradiometer aerosol optical thickness over the contiguous United States. *Journal of Geophysical Research-Atmospheres* 109: n/a–n/a, https://doi.org/10.1029/2004jd005025.

Liu, Y., Sarnat, J.A., Kilaru, V. et al. (2005). Estimating ground-level $PM_{2.5}$ in the eastern United States using satellite remote sensing. *Environmental Science & Technology* 39: 3269–3278.

Liu, Y., Paciorek, C.J., and Koutrakis, P. (2009). Estimating regional spatial and temporal variability of PM(2.5) concentrations using satellite data, meteorology, and land use information. *Environmental Health Perspectives* 117: 886–892, https://doi.org/10.1289/ehp.0800123.

Maggioni, V., Meyers, P.C., and Robinson, M.D. (2016). A review of merged high-resolution satellite precipitation product accuracy during the tropical rainfall measuring Mission (TRMM) era. *Journal of Hydrometeorology* 17 (4): 1101–1117.

Mao, W., Wang, W., Jiao, L. et al. (2021). Modeling air quality prediction using a deep learning approach: method optimization and evaluation. *Sustainable Cities and Society* 65: 102567.

Maykut, N.N., Lewtas, J., Kim, E., and Larson, T.V. (2003). Source apportionment of $PM_{2.5}$ at an urban IMPROVE site in Seattle, Washington. *Environmental Science & Technology* 37 (22): 5135–5142.

McPherson, R.A., Fiebrich, C.A., Crawford, K.C. et al. (2007). Statewide monitoring of the mesoscale environment: a technical update on the Oklahoma Mesonet. *Journal of Atmospheric and Oceanic Technology* 24 (3): 301–321. https://doi.org/10.1175/Jtech1976.1.

Miao, Y., Liu, S., Sheng, L. et al. (2019). Influence of boundary layer structure and low-level jet on $PM_{2.5}$ pollution in Beijing: a case study. *International Journal of Environmental Research and Public Health* 16 (4): 616.

Murray, N., Chang, H.H., Holmes, H., and Liu, Y. (2018). Combining satellite imagery and numerical model simulation to estimate ambient air pollution: an ensemble averaging approach. *arXiv preprint arXiv:1802.03077*.

Myhre, G., Samset, B.H., Schulz, M. et al. (2013). Radiative forcing of the direct aerosol effect from AeroCom Phase II simulations. *Atmospheric Chemistry and Physics* 13 (4): 1853–1877.

Paciorek, C.J., Liu, Y., Moreno-Macias, H., and Kondragunta, S. (2008). Spatiotemporal associations between GOES aerosol optical depth retrievals and ground-level $PM_{2.5}$. *Environmental Science & Technology* 42: 5800–5806.

Palve, S.N., Nemade, P.D., and Ghude, S.D. (2018). MOPITT carbon monoxide its source distributions, interannual variability and transport pathways over India during 2005–2015.

International Journal of Remote Sensing 39: 5952–5964, https://doi.org/10.1080/01431161.2018.1452076.

Pearce, J.L., Beringer, J., Nicholls, N. et al. (2011). Quantifying the influence of local meteorology on air quality using generalized additive models. *Atmospheric Environment* 45 (6): 1328–1336.

Rasch, P.J., Crutzen, P.J., and Coleman, D.B. (2008). Exploring the geoengineering of climate using stratospheric sulfate aerosols: the role of particle size. *Geophysical Research Letters* 35, https://doi.org/10.1029/2007gl032179.

Russell, A. (1997). Regional photochemical air quality modeling: model formulations, history, and state of the science. *Annual Review of Energy and the Environment* 22: 537–588. https://doi.org/10.1146/annurev.energy.22.1.537.

Rybarczyk, Y. and Zalakeviciute, R. (2018). Machine learning approaches for outdoor air quality modelling: a systematic review. *Applied Sciences* 8 (12): 2570.

Sayer, A.M., Munchak, L.A., Hsu, N.C. et al. (2014). MODIS collection 6 aerosol products: comparison between Aqua's e-deep blue, dark target, and "merged" data sets, and usage recommendations. *Journal of Geophysical Research-Atmospheres* 119: 13965–13989, https://doi.org/10.1002/2014jd022453.

Shen, H., Li, T., Yuan, Q., and Zhang, L. (2018). Estimating regional ground-level PM$_{2.5}$ directly from satellite top-of-atmosphere reflectance using deep belief networks. *Journal of Geophysical Research-Atmospheres* 123, https://doi.org/10.1029/2018jd028759.

Shrestha, B., Brotzge, J.A., Wang, J. et al. (2021). Overview and applications of the New York state Mesonet profiler network. *Journal of Applied Meteorology and Climatology* https://doi.org/10.1175/jamc-d-21-0104.1.

Singh, K.P., Gupta, S., Kumar, A., and Shukla, S.P. (2012). Linear and nonlinear modeling approaches for urban air quality prediction. *Science of the Total Environment* 426: 244–255.

Sun, Y., Zeng, Q., Geng, B. et al. (2019). Deep learning architecture for estimating hourly ground-level PM$_{2.5}$ using satellite remote sensing. *IEEE Geoscience and Remote Sensing Letters* 16: 1343–1347, https://doi.org/10.1109/lgrs.2019.2900270.

Thomas, A.M., Huff, A.K., Hu, X.-M., and Zhang, F. (2019). Quantifying uncertainties of ground-level ozone within WRF-Chem simulations in the mid-Atlantic region of the United States as a response to variability. *Journal of Advances in Modeling Earth Systems* 11: 1100–1116. https://doi.org/10.1029/2018MS001457.

Wang, Y., Bao, S., Wang, S. et al. (2017). Local and regional contributions to fine particulate matter in Beijing during heavy haze episodes. *Science of the Total Environment* 580: 283–296.

Watson, J.G., Chow, J.C., Bowen, J.L. et al. (2000). Air quality measurements from the Fresno supersite. *Journal of the Air & Waste Management Association* 50 (8): 1321–1334.

Wu, J., Li, G., Cao, J. et al. (2017). Contributions of trans-boundary transport to summertime air quality in Beijing, China. *Atmospheric Chemistry and Physics* 17 (3): 2035–2051.

Xiao, Q., Chang, H.H., Geng, G., and Liu, Y. (2018). An ensemble machine-learning model to predict historical PM$_{2.5}$ concentrations in China from satellite data. *Environmental Science & Technology* 52: 13260–13269, https://doi.org/10.1021/acs.est.8b02917.

Xing, J., Zheng, S., Ding, D. et al. (2020). Deep learning for prediction of the air quality response to emission changes. *Environmental Science & Technology* 54 (14): 8589–8600.

Xue, T., Zheng, Y., Tong, D. et al. (2019). Spatiotemporal continuous estimates of PM$_{2.5}$ concentrations in China, 2000–2016: a machine learning method with inputs from satellites, chemical transport model, and ground observations. *Environment International* 123 (345–357) https://doi.org/10.1016/j.envint.2018.11.075.

Yurganov, L.N., McMillan, W.W., Dzhola, A.V. et al. (2008). Global AIRS and MOPITT CO measurements: validation, comparison, and links to biomass burning variations and carbon cycle. *Journal of Geophysical Research* 113, https://doi.org/10.1029/2007jd009229.

Zhan, Y., Luo, Y., Deng, X. et al. (2018). Satellite-based estimates of daily NO_2 exposure in China using hybrid random forest and spatiotemporal kriging model. *Environmental Science & Technology* 52: 4180–4189, https://doi.org/10.1021/acs.est.7b05669.

Zhu, D., Cai, C., Yang, T., and Zhou, X. (2018). A machine learning approach for air quality prediction: model regularization and optimization. *Big Data and Cognitive Computing* 2 (1): 5.

Zoogman, P., Liu, X., Suleiman, R.M. et al. (2017). Tropospheric emissions: monitoring of pollution (TEMPO). *Journal of Quantitative Spectroscopy and Radiation Transfer* 186: 17–39, https://doi.org/10.1016/j.jqsrt.2016.05.008.

12

Current Challenges and Perspectives

Changjie Cai[1] and Qingsheng Wang[2]

[1] Department of Occupational and Environmental Health, Hudson College of Public Health, The University of Oklahoma, Oklahoma City, OK, USA
[2] Artie McFerrin Department of Chemical Engineering, Texas A&M University, College Station, TX, USA

This book identifies the machine learning fundamentals and various applications in chemical safety and health areas. It is the first to (i) comprehensively summarize their application history, (ii) organize relevant databases, and (iii) analyze their up-to-date implementations in a variety of fields in chemical safety and health. This book can provide guidance for students, professionals, engineers, and scientists who are interested in studying and applying various machine learning (e.g. shallow learning and deep learning) methods in their studies, work, and research. This chapter discusses some main challenges and provides some perspectives based on previous chapters.

12.1   Current Challenges

12.1.1   Data Development and Cleaning

A large amount of high-quality data with proper data structure is the key to train accurate machine learning models (Baier et al. 2019). The development of comprehensive and high-quality databases remains a priority. Failure to do so can result in inaccurate analytics and unreliable decisions. Regarding the database development, some commonly used meaningful data are often overlooked in chemical safety and health. For example, surround sounds (audio data) and photos (image data) encode important information related to exposures and risks in occupational safety and health and can be easily collected using sensors, cameras, and other methods (Weichenthal et al. 2019; Dang and Le 2020). In addition, audio data may provide behavioral information, which is important for estimating the risks of exposure by tracking time-activity patterns determined by "acoustic scenes" (Weichenthal et al. 2019). The lack of this type of data hinders the training and development of the models. It should be noted here that simulation data could be obtained from computational fluid dynamics (CFD) modeling (see Chapter 8) or process modeling when experimental data are not available (Hu et al. 2022). However, CFD methods require significant computing time and knowledge in fluid mechanics (Shen et al. 2020).

During data deployment, finding existing data with high frequency and large quantity can be challenging (Polyzotis et al. 2017; Baier et al. 2019). The validity of predictions made from models depends significantly on the quality of data entered. However, the efforts toward cleaning data are relatively less devoted than the efforts toward improving the quality of models (Jain et al. 2020). For instance, real-time monitoring equipment and low-cost sensors can easily generate a large amount of data. However, the signals of various sensors might shift differently over time (Rajendran et al. 2018). In order to enter the data into the models, datasets from various sources (such as sensors, images, audios, and simulations) should be paired. Large databases of paired data must be developed to support this initiative. It is challenging to pair the data, especially when they have different temporal/spatial resolutions/dimensions, and imbalanced or biased data pose another problem (Kocheturov et al. 2019; Lee et al. 2021).

Another example is for chemical process fault detections using real-time process operation data. Such data include process historical data, process design data, process hazard analysis (PHA) data, and process maintenance data. Since these data may also come from different chemical processes, the integration process entails cleaning to avoid the junk, resolving redundancy, and checking for integrity to ensure the data quality. It is extremely challenge to clean, merge, or integrate data, of any configuration, into a single format that users can access for the development of machine learning models. It is important to transfer these cleaned real-time data to operational data store systems for model building purpose.

12.1.2 Hardware Issues

Aside from the accuracy, the energy consumption, throughput/latency, and cost are key metrics of machine learning as well, especially for embedded machine learning with big datasets (Sze et al. 2017; Kocheturov et al. 2019), which is invaluable in the field of chemical safety and health. Sufficiently large and high-quality datasets determine the accuracy. Meanwhile, the high dimensionality of data and the need to store/process the data might significantly increase the computations, which poses a challenge for energy efficiency (Horowitz 2014). Training the data often happens in the cloud using graphics processing unit (GPUs), central processing units (CPUs), field programmable gate array (FPGAs), and application-specific integrated circuits (ASICs). Increased computational capacities have been catalyzing the growth of machine learning applications. However, due to the concerns of privacy, latency, and security, the need to perform the analysis locally rather than sending the raw data to the cloud has been increasing (Sze et al. 2017). Therefore, it is important to balance the accuracy, energy consumption, throughput/latency, and cost.

12.1.3 Data Confidentiality

It is well known how challenge it is to preserve the safety and privacy of sensitive datasets and to ensure legal compliance with new regulations. Currently, it is difficult to access process operational historic data because chemical process industries keep these data as confidential. For example, assuming a profit organization is working on a project using machine learning, the data development and deployment may be severely affected by the privacy policy of this and other organizations and the legal framework (Amodei et al. 2016; Flaounas 2017; Amershi et al. 2019; Baier et al. 2019). Paullada et al. (2021) advocate that, during the creation and usage of datasets, we could use both qualitative and quantitative approaches to more carefully document and analyze datasets. They advocate a turn in the culture toward more cautious practices of dataset development, maintenance, distribution, as well as respect of the intellectual property and privacy of data creators and subjects.

12.1.4 Other Challenges

One limitation of the applications of machine learning in chemical safety and health might be the coding and programming barrier. Traditionally, in addition to specialized knowledge, machine learning applications require a considerable amount of programming and coding skills to clean data and develop models. In recent decades, the development of easy-to-use machine learning toolkits (see Chapter 1) have been greatly propelling the applications of machine learning as introduced in previous chapters of this book. However, a substantial barrier still exists for researchers and practitioners who do not possess the programming/coding skills to develop machine learning models (Ozan 2021). In other words, deployment of machine learning algorithms on large datasets or on various infrastructures still require dedicated knowledge (Dyck 2018). Nowadays, the deployment is usually performed by dedicated machine learning technical team(s) in collaboration with the corresponding department(s), which is (are) requesting solutions for its problems. Easily applicable tools need to be developed in order to enable nontechnical researchers or practitioners to apply machine learning models (Kocheturov et al. 2019).

We have discussed in Chapter 3 on current research in numerous machine-learning-based quantitative structure–property relationships (QSPR) models. There are various approaches for the replacement of first-principle methods by machine learning. It is noted that most chemicals involved in these QSPR models are hydrocarbon vapors/liquids or small molecules (Jiao et al. 2019). When it comes to solid-state materials, the computational chemistry becomes more complicated, which will increase the computing time significantly. Two major questions are always the interpretability of and the physical understanding gained from machine-learning-based QSPR models. The lack of interpretability is one of the main challenges for a wider adoption of neural networks in industry and experimental sciences (Schmidt et al. 2019). Visualization of the response and understanding the inner workings of neural networks should be developed along with the machine learning models.

12.2 Perspectives

12.2.1 Real-Time Monitoring and Forecast of Chemical Hazards

Machine learning has rarely been used to real-time monitor and forecast chemical safety and health hazards on occupational sites (Dang and Le 2020), although the chemical hazards involved with workplace activities are well known (Hathaway and Proctor 2014).

Workers are routinely exposed to a variety of chemical safety and health hazards (ACGIH 1995; Wolkoff et al. 1998; Glanz et al. 2007). The Occupational Safety and Health Administration (OSHA) has indicated that businesses spend $170 billion a year on costs associated with occupational fatalities, injuries, and illness. In 2017, 5147 US workers died from occupational injuries (Statistics 2018). An on-site industrial hygienist/safety professional can readily identify unsafe actions and conditions. However, this professional is not often on-site throughout the workday or will not be devoted full time to the task of observing the work for unsafe actions or conditions. Therefore, it is critical to develop cost-effective automated ways to recognize chemical safety and health hazards on occupational sites and alert workers of their presence in a timely manner. Weichenthal et al. (2019) indicated that just a picture could identify a thousand exposures using deep learning image analyses. Low-cost sensors have been successfully distributed on various occupational sites for mapping occupational hazards and monitoring worker exposures (Sousan et al. 2016; Thomas et al. 2018; Zuidema et al. 2019; Zuidema et al. 2020), and they can generate a large amount of datasets with high spatial and temporal coverages. Therefore, we believe that machine learning techniques have the potential to shift the paradigm of chemical safety and health hazard recognition and prevention on occupational sites.

In the future, we could develop a cloud-based or embedded machine learning system that can check and forecast the real-time chemical safety and health hazards. If we use the cloud for training, and implement the prediction model on the edge computing or a small datacenter, then a workplace supervisor can quickly monitor, predict, and take appropriate decisions for hazard exposure.

12.2.2 Toolkits for Dummies

As previously mentioned, coding and programming can be a big barrier preventing the application of machine learning methods in this field. Most practitioners in chemical safety and health areas would expect some simple tools that can be installed in phones or laptops. To address this issue, no-code machine learning application tools have been developed recently. For example, Teachable Machine, which is developed by Google, is a web-based graphical user interface (GUI) tool for creating custom classification models without specialized technical expertise (Carney et al. 2020). A novel browser-based no-code application development tool was introduced by Ozan (2021) that is geared toward the needs of researchers who have limited knowledge in programming. Similar tools could enable users to create and test their machine learning models without downloading any software modules, which can significantly boost the applications of machine learning in the chemical safety and health field, allowing the users to address their specific research and practical questions. In the same manner for data cleaning, developing dummies' tools for this purpose might be beneficial as well.

12.2.3 Physics-Informed Machine Learning

Although the recent development of machine learning increases the opportunities in chemical safety and health, sometimes the results from such machine learning models are difficult to explain, especially for deep learning applications (Jiao et al. 2020). Both machine learning modeling and theoretical analysis are needed. Physics-informed machine learning model integrates data, partial differential equations, and mathematical models to solve supervised learning tasks while respecting any given laws of physics described by general nonlinear equations. Such physics-informed learning integrates data and mathematical models and implements them through neural networks or other kernel-based regression networks. Moreover, it may be possible to design specialized network architectures that automatically satisfy some of the physical invariants for better accuracy, faster training, and improved generalization (Karniadakis et al. 2021).

Overall, physics-informed machine learning will play an important role in fluid dynamics, quantum mechanics, computational resources, and data storage. Future studies on chemical safety and health should use interpretable and explainable machine learning models to extract important scientific insights from comprehensive datasets. One step toward achieving this goal is to integrate model consistency in addition to model prediction accuracy to evaluate the overall performance (Wadoux et al. 2020). We must be able to interpret and generalize associations discovered using machine learning, especially the deep learning. The models used in future studies must be not only accurate but also valid in light of the current knowledge and theories, so that such models can serve as a source to generate hypotheses as well (Wadoux et al. 2020).

References

American Conference of Governmental Industrial Hygienists (1995). Threshold limit values for chemical substances and physical agents and biological exposure indices. American Conference of Governmental Industrial Hygienists.

Amershi, S., Begel, A., Bird, C., et al. (2019). Software engineering for machine learning: a case study. *2019 IEEE/ACM 41st International Conference on Software Engineering: Software Engineering in Practice (ICSE-SEIP)*, Montreal, QC, Canada (25–31 May 2019), pp. 291–300. IEEE.

Amodei, D., Olah, C., Steinhardt, J. et al. (2016). Concrete problems in AI safety. *arXiv* preprint arXiv:1606.06565.

Baier, L., Jöhren, F., and Seebacher, S. (2019). Challenges in the deployment and operation of machine learning in practice. *Twenty-Seventh European Conference on Information Systems (ECIS2019)*, Stockholm-Uppsala, Sweden (8–14 June).

Carney, M., Webster, B., Alvarado, I., et al. (2020). Teachable machine: approachable web-based tool for exploring machine learning classification. *Extended Abstracts of the 2020 CHI Conference on Human Factors in Computing Systems*, Honolulu, Hawaii (25–30 April 2020), 1–8. ACM.

Dang, K. and Le, T. (2020). A novel audio-based machine learning model for automated detection of collision hazards at construction sites. *ISARC. Proceedings of the International Symposium on Automation and Robotics in Construction*, Kitakyushu, Japan (27–28 October 2020), Vol. 37, 829–835. IAARC Publications.

Dyck, J. (2018). Machine learning for engineering. *2018 23rd Asia and South Pacific Design Automation Conference (ASP-DAC)*, Jeju Republic of Korea (22–25 January 2018), 422–427. IEEE.

Flaounas, I. (2017). Beyond the technical challenges for deploying machine learning solutions in a software company. *arXiv preprint arXiv:1708.02363*.

Glanz, K., Buller, D.B., and Saraiya, M. (2007). Reducing ultraviolet radiation exposure among outdoor workers: state of the evidence and recommendations. *Environmental Health* 6 (1): 22.

Hathaway, G.J. and Proctor, N.H. (2014). *Proctor and Hughes' Chemical Hazards of the Workplace*. Wiley.

Horowitz, M. (2014). 1.1 Computing's energy problem (and what we can do about it). *2014 IEEE International Solid-State Circuits Conference Digest of Technical Papers (ISSCC)*, San Francisco, CA, USA (9–13 February 2014), 10–14. IEEE.

Hu, P., Cai, C., Yi, H. et al. (2022). Aiding airway obstruction diagnosis with computational fluid dynamics and convolutional neural network: a new perspective and numerical case study. *Journal of Fluids Engineering.* 144 (8): 081206. https://doi.org/10.1115/1.4053651.

Jain, A., Patel, H., Nagalapatti, L., et al. (2020). Overview and importance of data quality for machine learning tasks. *Proceedings of the 26th ACM SIGKDD International Conference on Knowledge Discovery & Data Mining*, CA USA (6–10 July 2020), 3561–3562. New York, NY: Association for Computing Machinery.

Jiao, Z., Escobar-Hernandez, H.U., Parker, T., and Wang, Q. (2019). Review of recent developments of quantitative structure-property relationship models on fire and explosion-related properties. *Process Safety and Environmental Protection* 129: 280–290.

Jiao, Z., Hu, P., Xu, H., and Wang, Q. (2020). Machine learning and deep learning in chemical health and safety: a systematic review of techniques and applications. *ACS Chemical Health & Safety* 27 (6): 316–334.

Karniadakis, G.E., Kevrekidis, I.G., Lu, L. et al. (2021). Physics-informed machine learning. *Nature Reviews Physics* 3: 422–440.

Kocheturov, A., Pardalos, P.M., and Karakitsiou, A. (2019). Massive datasets and machine learning for computational biomedicine: trends and challenges. *Annals of Operations Research* 276 (1): 5–34.

Lee, J., Shi, Y.R., Cai, C. et al. (2021). Machine learning based algorithms for global dust aerosol detection from satellite images: inter-comparisons and evaluation. *Remote Sensing* 13 (3): 456.

Ozan, E. (2021). A novel browser-based no-code machine learning application development tool. *2021 IEEE World AI IoT Congress (AIIoT)*, Seattle, USA (10–13 May 2021), 0282–0284. IEEE.

Paullada, A., Raji, I.D., Bender, E.M. et al. (2021). Data and its (dis) contents: a survey of dataset development and use in machine learning research. *Patterns* 2 (11): 100336.

Polyzotis, N., Roy, S., Whang, S.E., and Zinkevich, M. (2017). Data management challenges in production machine learning. *Proceedings of the 2017 ACM International Conference on Management of Data*, Chicago IL (14–19 May 2017), 1723–1726. New York, NY: Association for Computing.

Rajendran, S., Meert, W., Giustiniano, D. et al. (2018). Deep learning models for wireless signal classification with distributed low-cost spectrum sensors. *IEEE Transactions on Cognitive Communications and Networking* 4 (3): 433–445.

Schmidt, J., Marques, M.R.G., Botti, S., and Marques, M.A.L. (2019). Recent advances and applications of machine learning in solid-state materials science. *npj Computational Materials* 5: 83.

Shen, R., Jiao, Z., Parker, T. et al. (2020). Recent application of computational fluid dynamics (CFD) in process safety and loss prevention: a review. *Journal of Loss Prevention in the Process Industries* 67: 104252.

Sousan, S., Koehler, K., Thomas, G. et al. (2016). Inter-comparison of low-cost sensors for measuring the mass concentration of occupational aerosols. *Aerosol Science and Technology* 50 (5): 462–473.

Statistics, B. O. L. (2018). *Census of Fatal Occupational Injuries*. Washington, DC: Bureau of Labor Statistics.

Sze, V., Chen, Y.H., Emer, J. et al. (2017). Hardware for machine learning: challenges and opportunities. In: *2017 IEEE Custom Integrated Circuits Conference (CICC)*, 1–8. IEEE.

Thomas, G.W., Sousan, S., Tatum, M. et al. (2018). Low-cost, distributed environmental monitors for factory worker health. *Sensors* 18 (5): 1411.

Wadoux, A.M.C., Minasny, B., and McBratney, A.B. (2020). Machine learning for digital soil mapping: applications, challenges and suggested solutions. *Earth-Science Reviews* 210: 103359.

Weichenthal, S., Hatzopoulou, M., and Brauer, M. (2019). A picture tells a thousand ... exposures: opportunities and challenges of deep learning image analyses in exposure science and environmental epidemiology. *Environment International* 122: 3–10.

Wolkoff, P., Schneider, T., Kildesø, J. et al. (1998). Risk in cleaning: chemical and physical exposure. *Science of the Total Environment* 215 (1–2): 135–156.

Zuidema, C., Sousan, S., Stebounova, L.V. et al. (2019). Mapping occupational hazards with a multi-sensor network in a heavy-vehicle manufacturing facility. *Annals of Work Exposures and Health* 63 (3): 280–293.

Zuidema, C., Stebounova, L.V., Sousan, S. et al. (2020). Estimating personal exposures from a multi-hazard sensor network. *Journal of Exposure Science & Environmental Epidemiology* 30 (6): 1013–1022.

Index

Machine Learning in Chemical Safety and Health: Fundamentals with Applications, First Edition.
Edited by Qingsheng Wang and Changjie Cai.
© 2023 John Wiley & Sons Ltd. Published 2023 by John Wiley & Sons Ltd.